Completion Problems on Operator Matrices

Mathematical Surveys and Monographs

Volume 267

Completion Problems on Operator Matrices

Dragana S. Cvetković Ilić

AMERICAN MATHEMATICAL SOCIETY
Providence, Rhode Island

2020 *Mathematics Subject Classification.* Primary 47A08, 47B01, 47B02, 47A53, 15A83.

For additional information and updates on this book, visit
www.ams.org/bookpages/surv-267

Library of Congress Cataloging-in-Publication Data

Cataloging-in-Publication Data has been applied for by the AMS.
See http://www.loc.gov/publish/cip/.
DOI: https://doi.org/10.1090/surv/267

∞ The paper used in this book is acid-free and falls within the guidelines
established to ensure permanence and durability.
Visit the AMS home page at `https://www.ams.org/`

10 9 8 7 6 5 4 3 2 1 27 26 25 24 23 22

Contents

Preface

Completion problems for operator matrices are concerned with the question of whether a partially specified operator matrix can be completed to form an operator of a desired type. The research devoted to this topic constitutes an excellent means to investigate the structure of operators, which is one of the most important problems in operator theory.

This book provides an overview of completion problems dealing with completions to invertible, injective, Fredholm, Kato, Browder, and Fredholm consistent operators and can be considered as a sequel to the results presented in the monographs (see [**82, 83**]), which are mostly concerned with matrix completions. The study of operator matrices arises naturally from a well-known fact that if $\mathcal{Z}$ is a Banach space, decomposed as a direct sum of two closed and complementary subspaces $\mathcal{X}$ and $\mathcal{Y}$, then each bounded linear operator $T : \mathcal{Z} \to \mathcal{Z}$ can be expressed in the operator matrix form

$$T = \begin{bmatrix} A & C \\ D & B \end{bmatrix},$$

with respect to that particular space decomposition, where $A : \mathcal{X} \to \mathcal{X},\ B : \mathcal{Y} \to \mathcal{Y}$, $C : \mathcal{Y} \to \mathcal{X}$, and $D : \mathcal{X} \to \mathcal{Y}$ are bounded linear operators. Studying an operator through these component operators can simplify its analysis.

A particular version of the completion problem, in which a matrix is to be completed so that it becomes positive definite, has received considerable attention due, in part, to the role it plays in a number of applications in probability and statistics, image enhancement, systems engineering, geophysics, etc., but also due to the fact that it is closely linked to certain other completion problems, related to spectral norm contractions and Euclidean distance matrices, both of which are important in the molecular conformation problem in chemistry (see [**6, 7, 15, 16**]). Another way to view the positive definite completion problem is as a mechanism for addressing the following fundamental problem in Euclidean geometry: Which potential geometric configurations of vectors (i.e., configurations with angles between some vectors specified) are realizable in Euclidean space (see [**8, 126, 158**])? It is also worth mentioning that unitary completions of complex symmetric and skew symmetric matrices are exploited in atomic physics, in so-called phase-integral halfway-house variational continuum distorted wave theory (PIVCDW) (see [**159**]).

A good theoretical background for this book can be found in the monographs (see [**1, 2, 99**]). Besides applications of completion techniques in theoretical research (see [**150**], [**168**]), completion problems occur also in a variety of research areas of applied origin, such as the following: computer science, in particular in the problem of data compression, recommender systems (see [**109, 153**]), recovery of missing data in electricity distribution systems (see [**80**]), in urban systems (see [**81**]) and others (see [**33, 85, 138, 166**]), in statistics for missing data analysis and entropy methods (see [**25**–**27**, **33**, **37**, **38**, **131**, **154**]), for different prediction methods (see [**110**, **112**, **139**, **157**]), in chemistry (the molecular conformation problem), systems theory, discrete optimization (relaxation methods), etc. For an extensive reference list of matrix completion problems see [**129**].

The completions problems have long been a subject of research for many authors, and several relevant problems have already been resolved. However, there still remain problems that have been either only partially solved or remain open to this day. The present book aims at reviewing and summarizing these results (scattered in various literature sources) and at pointing out to the reader possible directions for further research in this area of mathematics as well as indicating their potential applications in different types of problems.

The problem of completion of operator matrices might be considered for the following types of matrices:

$$M_C = \begin{bmatrix} A & C \\ 0 & B \end{bmatrix} : \begin{bmatrix} \mathcal{H} \\ \mathcal{K} \end{bmatrix} \to \begin{bmatrix} \mathcal{H} \\ \mathcal{K} \end{bmatrix},$$

$$M_X = \begin{bmatrix} A & C \\ X & B \end{bmatrix} : \begin{bmatrix} \mathcal{H} \\ \mathcal{K} \end{bmatrix} \to \begin{bmatrix} \mathcal{H} \\ \mathcal{K} \end{bmatrix},$$

$$M_{(X\ Y)} = \begin{bmatrix} A & B \\ X & Y \end{bmatrix} : \begin{bmatrix} \mathcal{H} \\ \mathcal{K} \end{bmatrix} \to \begin{bmatrix} \mathcal{H} \\ \mathcal{K} \end{bmatrix}.$$

With respect to the first type of matrices, the following questions can be posed:

Question 1: Is there an operator $C \in \mathcal{B}(\mathcal{K}, \mathcal{H})$ such that M_C is of a certain type (invertible, right invertible, left invertible, regular, ...)?

Question 2: For given operators $A \in \mathcal{B}(\mathcal{H})$ and $B \in \mathcal{B}(\mathcal{K})$, can we compute $\bigcap_{C \in \mathcal{B}(\mathcal{K},\mathcal{H})} \sigma_*(M_C)$, where $\sigma_*(M_C)$ is any type of spectra of M_C, such as the following: essential, left, point, residual, right, Weyl, Browder, essential approximation spectrum, etc.?

Question 3: For given operators $A \in \mathcal{B}(\mathcal{H})$ and $B \in \mathcal{B}(\mathcal{K})$, is there an operator $C' \in \mathcal{B}(\mathcal{K}, \mathcal{H})$ such that

$$\sigma_*(M_{C'}) = \bigcap_{C \in \mathcal{B}(\mathcal{K},\mathcal{H})} \sigma_*(M_C),$$

where again $\sigma_*(.)$ is any type of spectra of an argument operator?

Clearly, analogous questions can be posed with respect to the other two types of matrices listed above, i.e., M_X and $M_{(X\ Y)}$. These questions play a focal role in the present book and will be discussed thoughtfully on the next pages.

The book consists of five chapters, among which three concern completion problems for three different types of operator matrices. The additional two chapters are concerned with the problem of completing the operator $A + CX$ to different types of operators and with applications of completion results, respectively.

The first chapter gives an exposition of completion problems for upper triangular operator matrices. The study of upper triangular operator matrices arises from the fact that if $A \in \mathcal{B}(\mathcal{H})$ and S is a closed, complemented A-invariant subspace of $\mathcal{H}$, then A has a representation in the form of an upper triangular operator matrix. Special emphasis is put on injectivity and boundedness below, invertibility, regularity, Drazin and generalized Drazin invertibility, Fredholmness, generalized left and right Weylness, the Browder property, Kato nonsingularity, and Fredholm consistency. Also, some related results for the spectra of upper triangular operator matrices are presented.

In the second chapter, we shift our attention from the problem of the completion of operator matrices to that of the completion by X of the operator $A + CX$, for given operators A, C, to different types of operators. The first to ever address the question of the existence of $X \in \mathcal{B}(\mathcal{H})$ such that $A + CX$ is of a desired type, for given $A \in \mathcal{B}(\mathcal{H})$ and $C \in \mathcal{B}(\mathcal{H})$, was K. Takahashi, who considered the invertibility of the sum. More precisely, he established a complete characterization of situations in which, for given $A \in \mathcal{B}(\mathcal{H})$ and $C \in \mathcal{B}(\mathcal{H})$, there exists $X \in \mathcal{B}(\mathcal{H})$ such that $A + CX$ is invertible. Based on this result, Takahashi derived a solution to a different, though related, problem, concerned with the completion of the operator matrix M_X, by an operator $X \in \mathcal{B}(\mathcal{H}, \mathcal{K})$ (see [**160**]). Therefore, Takahashi established a link between the invertibility completion problem and the existence of X such that $A + CX$ is invertible. For given operators $A \in \mathcal{B}(\mathcal{H}, \mathcal{K})$ and $C \in \mathcal{B}(\mathcal{L}, \mathcal{K})$, we consider the existence of $X \in \mathcal{B}(\mathcal{H}, \mathcal{L})$ such that $A + CX$ is injective, surjective, with dense range, with closed range, or Fredholm.

The next chapter deals with the problems concerned with completion of operator matrices of the form M_X to invertibility, right and left invertibility, injectivity, Fredholmness as well as to an operator with nondense or closed ranges and right and left semi-Fredholm operators.

The fourth chapter concerns the third type of completions, namely of operator matrices $M_{(X,Y)}$. The chapter covers such issues as completions of $M_{(X,Y)}$ to invertibility, right and left invertibility. Furthermore, the existence of operators X, Y such that a given operator $M_{(X,Y)}$ is consistent in invertibility and Fredholmness is considered as well.

In the fifth chapter necessary and sufficient conditions for the sum $A + B$ of two operators $A, B \in \mathcal{B}(\mathcal{H})$ to belong to a certain class of operators are identified. The task is achieved by exploiting the results on completion problems of the upper triangular operator matrices M_C.

The present book is the outcome of long-lasting research on this subject by the author. Nevertheless, an extensive impact on its content originated from works of several of my colleagues and coauthors. I am thankful for all of the inspirational contributions and partnerships, which affected the realization of this project.

Dragana S. Cvetković Ilić

Notation

For a subset S of a topological space, $\operatorname{int}(S)$, $\overline{S}$, r(S), $\operatorname{acc}(S)$, and $\operatorname{iso}(S)$ stand for interior, closure, boundary, set of all accumulation points, and set of all isolated points of S, respectively. By Y° we denote the annihilator of a subspace Y of a normed linear space X that is defined as $\{f \in X' : f(a) = 0, \text{ for all } a \in Y\}$.

Let $\mathcal{B}(\mathcal{H}, \mathcal{K})$ denote the set of all bounded linear operators from the space $\mathcal{H}$ to the space $\mathcal{K}$. For simplicity, we write $\mathcal{B}(\mathcal{H}, \mathcal{H})$ as $\mathcal{B}(\mathcal{H})$. For a given $A \in \mathcal{B}(\mathcal{H}, \mathcal{K})$, the symbols $\mathcal{N}(A)$ and $\mathcal{R}(A)$ stand for the null space and the range of A, respectively. Let $\alpha(A) = \dim(\mathcal{N}(A))$, $\beta(A) = \operatorname{codim}(\mathcal{R}(A))$. For subspaces L and M of $\mathcal{H}$ with $L \subseteq M$, we set $\operatorname{codim}_L(M) = \dim(L/M)$, where L/M is quotient space. If $\mathcal{H}$ is a Hilbert space and $\mathcal{M}$ a closed subspace of $\mathcal{H}$, the symbol $P_{\mathcal{M}}$ means the orthogonal projector onto $\mathcal{M}$. Also, we use the symbol $P_{\mathcal{M}}^{\mathtt{rst}}$ to denote an operator from $\mathcal{B}(\mathcal{H}, \mathcal{M})$ defined by $P_{\mathcal{M}}^{\mathtt{rst}} x = P_{\mathcal{M}} x$, for all $x \in \mathcal{H}$. Also, if $\mathcal{H}, \mathcal{K}$ are Hilbert spaces and $A \in \mathcal{B}(\mathcal{H}, \mathcal{K})$, we denote $d(A) = \dim \mathcal{R}(A)^\perp$. For subspaces $K, L, M \subseteq \mathcal{H}$, by $K \oplus L = M$ we understand that $K + L = M$ and $K \cap L = \{0\}$, i.e., that the sum is direct. If $A \in \mathcal{B}(\mathcal{H}, \mathcal{K})$ and $\mathcal{M}$ is a subspace of $\mathcal{K}$, then $A|_{\mathcal{M}}$ and $A[\mathcal{M}]$ stand for the restriction of the operator A to the subspace $\mathcal{M}$ and the image of $\mathcal{M}$ under A, respectively.

An operator $A \in \mathcal{B}(\mathcal{H}, \mathcal{K})$ is relatively regular (or simply, regular), if there exists an operator $X \in \mathcal{B}(\mathcal{K}, \mathcal{H})$ satisfying $AXA = A$. In that case any such X is an inner inverse of A which we denote by $A^{(1)}$ or A^-. The symbol $A\{1\}$ stands for the set of all inner inverses of A. If $X \in A\{1\}$ satisfies $XAX = X$, then X is a reflexive or $\{1,2\}$-inverse of A, denoted by $A^{(1,2)}$. The set of all reflexive inverses of A is denoted by $A\{1,2\}$. It is well known that A is regular if and only if $\mathcal{R}(A)$ and $\mathcal{N}(A)$ are closed and complemented subspaces of $\mathcal{K}$ and $\mathcal{H}$, respectively. Suppose that $A \in \mathcal{B}(\mathcal{H}, \mathcal{K})$ is relatively regular and let $\mathcal{T} = \mathcal{R}(XA)$ and $\mathcal{S} = \mathcal{N}(AX)$, for some $X \in A\{1\}$. Then we have the following decomposition of spaces $\mathcal{H} = \mathcal{T} \oplus \mathcal{N}(A)$ and $\mathcal{K} = \mathcal{R}(A) \oplus \mathcal{S}$, and, consequently, the matrix form of A reads:

$$A = \begin{bmatrix} A_1 & 0 \\ 0 & 0 \end{bmatrix} : \begin{bmatrix} \mathcal{T} \\ \mathcal{N}(A) \end{bmatrix} \to \begin{bmatrix} \mathcal{R}(A) \\ \mathcal{S} \end{bmatrix},$$

where $A_1 \in \mathcal{B}(\mathcal{T}, \mathcal{R}(A))$ is invertible. Evidently, X is of the form

$$X = \begin{bmatrix} A_1^{-1} & 0 \\ 0 & N \end{bmatrix} : \begin{bmatrix} \mathcal{R}(A) \\ \mathcal{S} \end{bmatrix} \to \begin{bmatrix} \mathcal{T} \\ \mathcal{N}(A) \end{bmatrix},$$

for some $N \in \mathcal{B}(\mathcal{S}, \mathcal{N}(A))$.

In the case when $\mathcal{H}, \mathcal{K}$ are Hilbert spaces and $A \in \mathcal{B}(\mathcal{H}, \mathcal{K})$, the Moore-Penrose inverse of A, which we denote by $A^\dagger$, is the unique generalized inverse of A satisfying the four Penrose conditions: $AA^\dagger A = A$, $A^\dagger AA^\dagger = A^\dagger$, $(AA^\dagger)^* = AA^\dagger$, and

$(A^\dagger A)^* = A^\dagger A$. If $\mathcal{R}(A)$ is closed, then $\mathcal{H} = \mathcal{N}(A) \oplus \mathcal{N}(A)^\perp$ and $\mathcal{K} = \mathcal{R}(A) \oplus \mathcal{R}(A)^\perp$, so

$$A = \begin{bmatrix} A_1 & 0 \\ 0 & 0 \end{bmatrix} : \begin{bmatrix} \mathcal{N}(A)^\perp \\ \mathcal{N}(A) \end{bmatrix} \to \begin{bmatrix} \mathcal{R}(A) \\ \mathcal{R}(A)^\perp \end{bmatrix}$$

and

$$A^\dagger = \begin{bmatrix} A_1^{-1} & 0 \\ 0 & 0 \end{bmatrix} : \begin{bmatrix} \mathcal{R}(A) \\ \mathcal{R}(A)^\perp \end{bmatrix} \to \begin{bmatrix} \mathcal{N}(A)^\perp \\ \mathcal{N}(A) \end{bmatrix}.$$

Visibly, in this case $A^\dagger$ satisfies

$$A^\dagger A = P_{\mathcal{R}(A^*)} \quad \text{and} \quad AA^\dagger = P_{\mathcal{R}(A)}.$$

If $\mathcal{R}(A)$ is not closed, then $A^\dagger$ is not a bounded operator, for which the identities

$$A^\dagger A = P_{\overline{\mathcal{R}(A^*)}} \quad \text{and} \quad AA^\dagger = P_{\overline{\mathcal{R}(A)}}|_{\mathcal{R}(A)\oplus\mathcal{R}(A)^\perp}$$

hold true.

An operator $A \in \mathcal{B}(\mathcal{H})$ is with finite ascent (finite descent) if there is a nonnegative integer m such that $\mathcal{N}(A^m) = \mathcal{N}(A^{m+1})$ ($\mathcal{R}(A^m) = \mathcal{R}(A^{m+1})$), in which case it is customary to use $\operatorname{asc}(A)$ ($\operatorname{dsc}(A)$) to denote the least such m; $\operatorname{asc}(A) = \infty$ ($\operatorname{dsc}(A) = \infty$) if A is not with finite ascent (descent).

We reserve the notation $\mathcal{H}'$ for the space of all linear bounded functionals on any space $\mathcal{H}$. The symbol A' stands for the dual operator of $A \in \mathcal{B}(\mathcal{H}, \mathcal{K})$. If $\mathcal{H}$ and $\mathcal{K}$ are separable Hilbert spaces and $A \in \mathcal{B}(\mathcal{H}, \mathcal{K})$ is such that $\mathcal{R}(A)$ is closed and $\alpha(A) < \infty$, then A is an upper semi-Fredholm operator (or equivalently left semi-Fredholm). If $\beta(A) < \infty$, then A is a lower semi-Fredholm operator (or equivalently right semi-Fredholm). An operator $A \in \mathcal{B}(\mathcal{H}, \mathcal{K})$ is called Fredholm if it is both upper semi-Fredholm and lower semi-Fredholm. The symbol $\mathcal{F}(\mathcal{H}, \mathcal{K})$ stands for the set of all Fredholm operators from the space $\mathcal{B}(\mathcal{H}, \mathcal{K})$, whereas $\mathcal{F}_+(\mathcal{H}, \mathcal{K})$ or $\mathcal{F}_l(\mathcal{H}, \mathcal{K})$ ($\mathcal{F}_-(\mathcal{H}, \mathcal{K})$ or $\mathcal{F}_r(\mathcal{H}, \mathcal{K})$) means the set of all upper (lower) semi-Fredholm operators from $\mathcal{B}(\mathcal{H}, \mathcal{K})$. For a Fredholm operator A, its Fredholm index is defined by $\operatorname{ind}(A) = \alpha(A) - \beta(A)$. By $\mathcal{F}_+^-(\mathcal{H}.\mathcal{K})$ we denote the class of all left semi-Fredholm operators from $\mathcal{B}(\mathcal{H}, \mathcal{K})$ with nonpositive index.

If an operator $A \in \mathcal{B}(\mathcal{H}, \mathcal{K})$ is such that $\mathcal{R}(A)$ is closed and $\alpha(A) \leq d(A)$, then A is a generalized left Weyl operator. If $A \in \mathcal{B}(\mathcal{H}, \mathcal{K})$ is such that $\mathcal{R}(A)$ is closed and $d(A) \leq \alpha(A)$, then A is a generalized right Weyl operator. Notice that in the cases of generalized left (right) Weyl operators, $\alpha(A)$ and $d(A)$ are allowed to be infinity.

An operator $A \in \mathcal{B}(\mathcal{H}, \mathcal{K})$ is a generalized Weyl operator if it is both generalized right Weyl and generalized left Weyl. The set of all generalized Weyl operators from $\mathcal{B}(\mathcal{H}, \mathcal{K})$ is denoted by $W^g(\mathcal{H}, \mathcal{K})$. The symbol $W_{gl}(\mathcal{H}, \mathcal{K})$ ($W_{gr}(\mathcal{H}, \mathcal{K})$) means the subset of $\mathcal{B}(\mathcal{H}, \mathcal{K})$ consisting of all generalized left (right) Weyl operators. An operator A is called Browder if it is Fredholm of finite ascent and descent.

From among various types of spectra of an operator $A \in \mathcal{B}(\mathcal{H})$ that can be spotted in the literature, we refer to the following:

- the spectrum: $\sigma(A) = \{\lambda \in \mathbb{C}\colon\ A - \lambda I \text{ is not invertible}\}$,
- the point spectrum: $\sigma_p(A) = \{\lambda \in \mathbb{C}\colon\ A - \lambda I \text{ is not injective}\}$,
- the left spectrum: $\sigma_l(A) = \{\lambda \in \mathbb{C}\colon\ A - \lambda I \text{ is not left invertible}\}$,
- the right spectrum: $\sigma_r(A) = \{\lambda \in \mathbb{C}\colon\ A - \lambda I \text{ is not right invertible}\}$,

- the compression spectrum: $\sigma_{cp}(A) = \{\lambda \in \mathbb{C}\colon \overline{\mathcal{R}(\lambda - A)} \neq \mathcal{H}\}$,
- the essential spectrum: $\sigma_e(A) = \{\lambda \in \mathbb{C}\colon\ A - \lambda I$ is not Fredholm$\}$,
- the left (right) essential spectrum: $\sigma_{le}(A)(\sigma_{re}(A)) = \{\lambda \in \mathbb{C}\colon\ A - \lambda I$ is not left (right) Fredholm$\}$,
- the generalized left (right) Weyl spectrum: $\sigma^g_{lw}(A)(\sigma^g_{rw}(A)) = \{\lambda \in \mathbb{C}\colon A - \lambda I$ is not generalized left (right) Weyl$\}$,
- the generalized Weyl spectrum: $\sigma^g_w(A) = \{\lambda \in \mathbb{C}\colon A - \lambda I$ is not generalized Weyl$\}$,
- the Browder spectrum: $\sigma_b(A) = \{\lambda \in \mathbb{C}\colon A - \lambda I$ is not Browder$\}$,
- the essential approximation spectrum: $\sigma_{ea}(A) = \{\lambda \in \mathbb{C}\colon A - \lambda I \notin \mathcal{F}^-_+(\mathcal{H})\}$,
- the Browder essential approximate point spectrum: $\sigma_{ab}(A) = \{\lambda \in \mathbb{C}\colon A - \lambda I \notin \mathcal{F}^-_+(\mathcal{H})$ or $\operatorname{asc}(A - \lambda I) = \infty\}$.

If $\sigma_*(A)$ is any of the spectra defined above, by $\rho_*(A)$ we denote the set $\mathbb{C} \setminus \sigma_*(A)$. In the case when A is right (left) invertible, $(A)_r^{-1}$ $((A)_l^{-1})$ stands for an arbitrary fixed right (left) inverse of A.

If for Banach spaces $\mathcal{X}$, $\mathcal{Y}$ there exists a left invertible operator $T \in \mathcal{B}(\mathcal{X}, \mathcal{Y})$, then we say that $\mathcal{X}$ is embedded in $\mathcal{Y}$, which is denoted by $\mathcal{X} \preceq \mathcal{Y}$.

CHAPTER 1

Completion problems for upper triangular operator matrices

In recent years an extensive study of the 2×2 upper triangular operator matrices has been carried out. The research was motivated primarily by the following fact: if $A \in \mathcal{B}(\mathcal{H})$ and $\mathcal{S}$ is a closed, complemented A-invariant subspace of $\mathcal{H}$, then A may be expressed as

$$A = \begin{bmatrix} * & * \\ 0 & * \end{bmatrix} : \begin{bmatrix} \mathcal{S} \\ \mathcal{P} \end{bmatrix} \to \begin{bmatrix} \mathcal{S} \\ \mathcal{P} \end{bmatrix},$$

where $\mathcal{H} = \mathcal{S} \oplus \mathcal{P}$, with $\mathcal{P}$ closed.

Many recently published papers are focussed on the issues of the invertibility and regularity for an upper triangular operator matrix

$$M_C = \begin{bmatrix} A & C \\ 0 & B \end{bmatrix} : \begin{bmatrix} \mathcal{H} \\ \mathcal{K} \end{bmatrix} \to \begin{bmatrix} \mathcal{H} \\ \mathcal{K} \end{bmatrix}, \tag{1.1}$$

as well as on various types of spectra of M_C. A particular problem originating from these investigations is the one dealing with completions of the partially specified operator matrix

$$\begin{bmatrix} A & ? \\ 0 & B \end{bmatrix} : \begin{bmatrix} \mathcal{H} \\ \mathcal{K} \end{bmatrix} \to \begin{bmatrix} \mathcal{H} \\ \mathcal{K} \end{bmatrix}$$

in order to obtain an operator with some requested property. More precisely, for given $A \in \mathcal{B}(\mathcal{H})$ and $B \in \mathcal{B}(\mathcal{K})$, the following three questions repeatedly arise:

Question 1: Is there an operator $C \in \mathcal{B}(\mathcal{K}, \mathcal{H})$ such that M_C is of a certain type (invertible, right invertible, left invertible, regular, $\dots$) ?

Question 2: For given operators $A \in \mathcal{B}(\mathcal{H})$ and $B \in \mathcal{B}(\mathcal{K})$, can we compute $\bigcap_{C\in\mathcal{B}(\mathcal{K},\mathcal{H})} \sigma_*(M_C)$, where $\sigma_*(M_C)$ is any type of spectra of M_C, such as the following: essential, left, point, residual, right, Weyl, Browder, essential approximation spectrum, etc.?

Question 3: For given operators $A \in \mathcal{B}(\mathcal{H})$ and $B \in \mathcal{B}(\mathcal{K})$, is there an operator $C' \in \mathcal{B}(\mathcal{K}, \mathcal{H})$ such that

$$\sigma_*(M_{C'}) = \bigcap_{C\in\mathcal{B}(\mathcal{K},\mathcal{H})} \sigma_*(M_C),$$

where again $\sigma_*(.)$ is any type of spectra of an argument operator?

In this chapter, for given operators $A \in \mathcal{B}(\mathcal{H})$ and $B \in \mathcal{B}(\mathcal{K})$, we discuss the existence of an operator $C \in \mathcal{B}(\mathcal{K}, \mathcal{H})$ such that an operator matrix M_C given by (1.1) belongs to any of the classes of right (left) invertible, injective, Drazin and generalized Drazin invertible, semi-Fredholm, Fredholm, Weyl, Browder, Kato, or closed range operators. In general, the methodology presented in the following sections consists of two steps: in the first one, for given operators A and B we are looking for the necessary and sufficient conditions for the existence of an operator C such that M_C belongs to a certain set of operators and in the second step we describe the set of all such operators C. Both steps jointly lead us to the complete answer to the question of when an upper triangular operator matrix is of a requested type.

1.1. Completions of upper triangular operator matrices to injective and left invertible operator

In this section we consider injectivity and left invertibility of upper triangular operator matrix M_C given by (1.1). Let us point out that the point spectrum of M_C is considered in [**179**] in the case of separable Hilbert spaces, where the set $\bigcap_{C\in\mathcal{B}(\mathcal{K},\mathcal{H})}\sigma_p(M_C)$ is described by

$$\bigcap_{C\in\mathcal{B}(\mathcal{K},\mathcal{H})}\sigma_p(M_C)=\sigma_p(A)\cup\{\lambda\in\mathbb{C}\colon \alpha(B-\lambda I)>\beta(A-\lambda I)\}.$$

However, in [**179**] neither the question of the existence of an operator $C'\in\mathcal{B}(\mathcal{K},\mathcal{H})$ such that $\sigma_p(M_{C'})=\bigcap_{C\in\mathcal{B}(\mathcal{K},\mathcal{H})}\sigma_p(M_C)$ nor the description of all C for which M_C is injective was considered. This is precisely what we discuss in the present section. In order to describe the set of all C such that M_C is injective, we first give different necessary and sufficient conditions than those given in [**179**] for their existence, which allows us later to describe the set of all such C. Let us point out that for a given $A\in\mathcal{B}(\mathcal{H})$ and $B\in\mathcal{B}(\mathcal{X})$, the set of all $C\in\mathcal{B}(\mathcal{K},\mathcal{H})$ such that M_C is injective (invertible) will be denoted by $S_p(A,B)(S_I(A,B))$. For given $A\in\mathcal{B}(\mathcal{H})$ and $B\in\mathcal{B}(\mathcal{K})$, we suppose that arbitrary $C\in\mathcal{B}(\mathcal{K},\mathcal{H})$ is given by

$$C=\begin{bmatrix} C_1 & C_2\\ C_3 & C_4\end{bmatrix}:\begin{bmatrix}\mathcal{N}(B)\\ \mathcal{N}(B)^\perp\end{bmatrix}\to\begin{bmatrix}\overline{\mathcal{R}(A)}\\ \mathcal{R}(A)^\perp\end{bmatrix}. \tag{1.2}$$

In general by $\mathcal{H}$ and $\mathcal{K}$ we denote separable Hilbert spaces and by $\mathcal{X}$ and $\mathcal{Y}$ we denote Banach spaces.

THEOREM 1.1. *Let $A\in\mathcal{B}(\mathcal{X})$ and $B\in\mathcal{B}(\mathcal{Y})$ be given operators. The operator matrix M_C given by* (1.1) *is injective for $C\in\mathcal{B}(\mathcal{Y},\mathcal{X})$ if and only if A is injective and $C|_{\mathcal{N}(B)}$ is injective with range disjoint with $\mathcal{R}(A)\setminus\{0\}$.*

PROOF. ($\Rightarrow$:) Suppose that M_C is injective for some $C\in\mathcal{B}(\mathcal{Y},\mathcal{X})$. Then for $x\in\mathcal{N}(A)$, we get that $\begin{bmatrix} x\\ 0\end{bmatrix}\in\mathcal{N}(M_C)$ which implies that $x=0$. Hence, $\mathcal{N}(A)=\{0\}$. Taking $y\in\mathcal{N}(B)$, we get that $M_C\begin{bmatrix} 0\\ y\end{bmatrix}=\begin{bmatrix} Cy\\ 0\end{bmatrix}$, so $\mathcal{N}(C)\cap\mathcal{N}(B)=\{0\}$. Now, under the assumptions $\mathcal{N}(A)=\{0\}$ and $\mathcal{N}(C)\cap\mathcal{N}(B)=\{0\}$, it is evident that the following holds:

$$\mathcal{N}(M_C)=\{0\}\Leftrightarrow C(\mathcal{N}(B))\cap\mathcal{R}(A)=\{0\}. \tag{1.3}$$

($\Leftarrow$:) If $\mathcal{N}(M_C)\neq\{0\}$, then there are three possibilities:

1. $\begin{bmatrix} x\\ 0\end{bmatrix}\in\mathcal{N}(M_C)$, $x\neq 0$.

2. $\begin{bmatrix} 0 \\ y \end{bmatrix} \in \mathcal{N}(M_C),\ y \neq 0.$

3. $\begin{bmatrix} x \\ y \end{bmatrix} \in \mathcal{N}(M_C), x \neq 0, y \neq 0.$

The first case holds if and only if $\mathcal{N}(A) \neq \{0\}$. The second holds if and only if $\mathcal{N}(C) \cap \mathcal{N}(B) \neq \{0\}$. If neither of the assumptions 1 and 2 is satisfied, by (1.3), we get that the third one is satisfied if and only if $C(\mathcal{N}(B)) \cap \mathcal{R}(A) \neq \{0\}$. □

As noticed before, in the next theorem which was presented in [**46**], certain necessary and sufficient conditions for the existence of an operator C for which M_C is injective are given as well as a description of the set of all such C.

THEOREM 1.2 ([**46**]). *Let $A \in \mathcal{B}(\mathcal{H})$ and $B \in \mathcal{B}(\mathcal{K})$ be given operators. There exists $C \in \mathcal{B}(\mathcal{K}, \mathcal{H})$ such that the operator matrix M_C is injective if and only if A is injective and one of the following conditions is satisfied:*

(i) *$\mathcal{R}(A)$ is closed, $\alpha(B) \leq \beta(A)$.*
(ii) *$\mathcal{R}(A)$ is not closed.*

Furthermore, if (i) *is satisfied, then the set of all $C \in \mathcal{B}(\mathcal{K}, \mathcal{H})$ such that M_C is injective is given by*

$$S_I(A, B) = \{C \in \mathcal{B}(\mathcal{K}, \mathcal{H}) \colon C \text{ is given by (1.2) and } C_3 \text{ is an injection}\},$$

while if (ii) *holds, it is given by*

$$S_I(A, B) = \{C \in \mathcal{B}(\mathcal{K}, \mathcal{H}) \colon C \text{ is given by (1.2) and}$$
$$\mathcal{N}(C_1) \cap \mathcal{N}(C_3) = \{0\}\}, C_1(\mathcal{N}(C_3)) \cap \mathcal{R}(A) = \{0\}\}.$$

PROOF. By Theorem 1.1, we have that M_C is an injection for some $C \in \mathcal{B}(\mathcal{K}, \mathcal{H})$, given by (1.2), if and only if A is an injection and $\begin{bmatrix} C_2 \\ C_4 \end{bmatrix} : \mathcal{N}(B) \to \begin{bmatrix} \overline{\mathcal{R}(A)} \\ \mathcal{R}(A)^\perp \end{bmatrix}$ is an injection with the range disjoint with $\mathcal{R}(A) \setminus \{0\}$. $\begin{bmatrix} C_2 \\ C_4 \end{bmatrix}$ is an injection if and only if $\mathcal{N}(C_2) \cap \mathcal{N}(C_4) = \{0\}$. The second condition, $C(\mathcal{N}(B)) \cap \mathcal{R}(A) = \{0\}$, holds if and only if $C_2(\mathcal{N}(C_4)) \cap \mathcal{R}(A) = \{0\}$. Hence, for $C \in \mathcal{B}(\mathcal{K}, \mathcal{H})$ given M_C is an injection if and only if A is an injection, $\mathcal{N}(C_2) \cap \mathcal{N}(C_4) = \{0\}$, and $C_2(\mathcal{N}(C_4)) \cap \mathcal{R}(A) = \{0\}$.

Now, under the assumption that A is injective, we are looking for conditions which guarantee the existence of such an operator C. We consider two cases: when $\mathcal{R}(A)$ is closed and when it is not.

First, if $\mathcal{R}(A)$ is closed, then $C_2(\mathcal{N}(C_4)) \cap \mathcal{R}(A) = \{0\}$ is equivalent with $\mathcal{N}(C_4) \subseteq \mathcal{N}(C_2)$ which together with $\mathcal{N}(C_2) \cap \mathcal{N}(C_4) = \{0\}$ gives $\mathcal{N}(C_4) = \{0\}$. Evidently, in this case $C_2 \in \mathcal{B}(\mathcal{N}(B), \mathcal{R}(A))$ can be arbitrary, while $C_4 \in \mathcal{B}(\mathcal{N}(B), \mathcal{R}(A)^\perp)$ must be an injection. The existence of such C_4 is equivalent with $\alpha(B) \leq \beta(A)$. □

Evidently, in the case of Banach spaces, we have the following:

$$\bigcap_{C \in \mathcal{B}(\mathcal{Y}, \mathcal{X})} \sigma_p(M_C) = \sigma_p(A) \cup \{\lambda \in \mathbb{C} \colon \nexists C \in \mathcal{B}(\mathcal{Y}, \mathcal{X}),\ C \colon \mathcal{N}(B - \lambda I) \overset{1-1}{\to} \mathcal{X},$$
$$C(\mathcal{N}(B - \lambda I)) \cap \mathcal{R}(A - \lambda I) = \{0\}\}.$$

As a corollary, we get the result from [**179**] in the case of separable Hilbert spaces:

COROLLARY 1.3 ([**179**]). For given operators $A \in \mathcal{B}(\mathcal{H})$ and $B \in \mathcal{B}(\mathcal{K})$, we have

$$\bigcap_{C \in \mathcal{B}(\mathcal{K},\mathcal{H})} \sigma_p(M_C) = \sigma_p(A) \cup \{\lambda \in \mathbb{C}\colon \alpha(B-\lambda I) > \beta(A-\lambda I)\}.$$

Theorem 1.2 allows us to partially answer the question of existence of an operator $C' \in \mathcal{B}(\mathcal{Y},\mathcal{X})$ such that

$$\sigma_*(M_{C'}) = \bigcap_{C \in \mathcal{B}(\mathcal{Y},\mathcal{X})} \sigma_*(M_C), \tag{1.4}$$

in the case of the point spectrum.

REMARK 1.4. Denote by P_p the complement of the set $\bigcap_{C \in \mathcal{B}(\mathcal{Y},\mathcal{X})} \sigma_p(M_C)$ in $\mathbb{C}$. Thus P_p consists of all those $\lambda \in \rho_p(A)$ for which there is a $C \in \mathcal{B}(\mathcal{Y},\mathcal{X})$ such that $\mathcal{N}(C) \cap \mathcal{N}(B-\lambda I) = \{0\}$ and $C(\mathcal{N}(B-\lambda I)) \cap \mathcal{R}(A-\lambda I) = \{0\}$.

In (1.4) in the case of point spectrum, we actually seek a $C' \in \mathcal{B}(\mathcal{Y},\mathcal{X})$ such that $\mathcal{N}(C') \cap \mathcal{N}(B-\lambda I) = \{0\}$ and $C'(\mathcal{N}(B-\lambda I)) \cap \mathcal{R}(A-\lambda I) = \{0\}$ for each $\lambda \in P_p$. If for each $\lambda \in R_0 = P_p \cap \sigma_p(B)$ we fix a $C_\lambda \in \mathcal{B}(\mathcal{Y},\mathcal{X})$ with $\mathcal{N}(C_\lambda) \cap \mathcal{N}(B-\lambda I) = \{0\}$ and $C_\lambda(\mathcal{N}(B-\lambda I)) \cap \mathcal{R}(A-\lambda I) = \{0\}$, then any common extension of the restrictions $C_\lambda | \mathcal{N}(B-\lambda I)$ for $\lambda \in P_p$ would be as required. Now, we point out in the sequel two particular cases in which such extensions exist.

(1) The first case is when the set R_0 is finite. Obviously, if $R_0 = \emptyset$, then any $C' \in \mathcal{B}(\mathcal{Y},\mathcal{X})$ is a common extension. If $R_0 = \{\lambda_1, \dots, \lambda_k\}$, then we can take $C' = \sum_{i=1}^{k} C_{\lambda_i}\pi_i\pi$, where $\pi_i : \bigoplus_{j=1}^{k} \mathcal{N}(B-\lambda_j I) \to \mathcal{N}(B-\lambda_i I)$, for $1 \le i \le k$, are projectors corresponding to the given decomposition of the space $\mathcal{P} = \bigoplus_{j=1}^{k} \mathcal{N}(B-\lambda_j I)$ and $\pi : \mathcal{Y} \to \mathcal{P}$ is an arbitrary (bounded) projector onto $\mathcal{P}$. The reason why there is such a projector is because $\mathcal{P}$ can easily be seen to be closed by induction on k. Note that R_0 is finite in particular if A is a compact operator (R_0 is finite also, when A is a Riesz or polynomial Riesz operator). Indeed, then $\sigma(A) \subseteq \sigma_p(A) \cup \{0\}$ as is well known. So if $\lambda \in P_p \setminus \{0\}$, then $A - \lambda I$ is an invertible operator; hence $\mathcal{R}(A-\lambda I) = \mathcal{X}$ and consequently $\mathcal{N}(B-\lambda I) = \{0\}$. This means that $R_0 \subseteq \{0\}$.

(2) When $\mathcal{Y}$ is a separable Hilbert space and B is a normal operator, as is well known, all the eigenvalues of B can be arranged in a sequence $(\lambda_n : n \in \mathbb{N})$ and the eigenspaces $\mathcal{N}(B-\lambda_i I)$ and $\mathcal{N}(B-\lambda_j I)$ are mutually orthogonal for $i \neq j$. Let $\pi_n : \mathcal{Y} \to \mathcal{N}(B-\lambda_n I)$ be the orthogonal projector onto the corresponding eigenspace for $n \in \mathbb{N}$. Set $D_n = \frac{1}{\|C_{\lambda_n}\| n^2} C_{\lambda_n}$ for $n \in \mathbb{N}$. Clearly $\mathcal{N}(D_n) \cap \mathcal{N}(B-\lambda_n I) = \{0\}$ and $D_n(\mathcal{N}(B-\lambda_n I)) \cap \mathcal{R}(A-\lambda_n I) = \{0\}$. From $\|D_n\pi_n\| \le \|D_n\| \le \frac{1}{n^2}$ it follows that the series $\sum_{n=1}^{\infty} D_n\pi_n$ converges in the Banach space $\mathcal{B}(\mathcal{Y},\mathcal{X})$ to its sum which

we denote by C'. Then C' is a common extension of the restrictions $D_n|\mathcal{N}(B-\lambda_n I)$. For if we take $x \in \mathcal{N}(B-\lambda_n I)$, then $\pi_m(x)=0$ for all $m \neq n$ so $C'(x) = D_n\pi_n(x) = D_n(x)$.

The left invertibility of an upper triangular operator matrix M_C was considered in [**107**] where the following characterization was given.

THEOREM 1.5 ([**107**]). *Let $A \in \mathcal{B}(\mathcal{H})$ and $B \in \mathcal{B}(\mathcal{K})$. An operator matrix M_C is left invertible for some $C \in \mathcal{B}(\mathcal{K},\mathcal{H})$ if and only if A is left invertible and one of the following conditions holds:*

(i) *$\mathcal{R}(B)$ is closed and $\alpha(B) \leq \beta(A)$.*
(ii) *$\mathcal{R}(B)$ is not closed and $\beta(A)=\infty$.*

In the following theorem, using the conditions given in Theorem 1.5, we completely describe the set of all $C \in \mathcal{B}(\mathcal{K},\mathcal{H})$ for which M_C is left invertible, for given operators $A \in \mathcal{B}(\mathcal{H})$ and $B \in \mathcal{B}(\mathcal{K})$ which we denote by $S_{BB}(A,B)$. Our method allows us to give a much simpler proof than the one given in [**107**] from which we directly obtain a characterization of the set $S_{BB}(A,B)$.

THEOREM 1.6. *Let $A \in \mathcal{B}(\mathcal{H})$ and $B \in \mathcal{B}(\mathcal{K})$. If $\mathcal{R}(B)$ is closed and $\alpha(B) \leq \beta(A)$, then the set of all $C \in \mathcal{B}(\mathcal{K},\mathcal{H})$ for which M_C is left invertible is given by*

$$S_{BB}(A,B) = \{C \in \mathcal{B}(\mathcal{K},\mathcal{H})\colon C \text{ is given by (1.2) and } C_3 \text{ is left invertible}\},$$

while if $\mathcal{R}(B)$ is not closed and $\beta(A)=\infty$, it is given by

$$S_{BB}(A,B) = \left\{C \in \mathcal{B}(\mathcal{K},\mathcal{H})\colon \begin{bmatrix} P_{\mathcal{R}(A)^\perp}C \\ B \end{bmatrix} : \mathcal{K} \to \mathcal{R}(A)^\perp \oplus \mathcal{K} \text{ is left invertible}\right\}.$$

PROOF. Since M_C is left invertible if and only if $\mathcal{R}(M_C)$ is closed and $\mathcal{N}(M_C) = \{0\}$, then, using Theorem 1.2, we conclude that M_C can be left invertible for some C only in one of the following two cases: Case 1. A is injective, $\mathcal{R}(A)$ is closed, $\alpha(B) \leq \beta(A)$; Case 2. A is injective, and $\mathcal{R}(A)$ is not closed.

Case 1. We consider two possibilities: when $\mathcal{R}(B)$ is closed and when it is not. First, suppose that $\mathcal{R}(B)$ is closed. Then M_C as an operator from $\mathcal{H}\oplus\mathcal{K} = \mathcal{N}(A)^\perp \oplus \mathcal{N}(B) \oplus \mathcal{N}(B)^\perp$ into $\mathcal{H}\oplus\mathcal{K} = (\mathcal{R}(A)\oplus\mathcal{R}(A)^\perp) \oplus (\mathcal{R}(B)\oplus\mathcal{R}(B)^\perp)$ has the following operator matrix form:

$$M_C = \begin{bmatrix} A_1 & C_1 & C_2 \\ 0 & C_3 & C_4 \\ 0 & 0 & B_1 \\ 0 & 0 & 0 \end{bmatrix},$$

where A_1 and B_1 are invertible. By Theorem 1.2, we have that $\mathcal{N}(M_C) = \{0\}$ if and only if C_3 is an injection. It is easy to check that $\mathcal{R}(M_C)$ is closed if and only if $\mathcal{R}(M_C')$ is closed, where

$$M_C' = \begin{bmatrix} A_1 & 0 & 0 \\ 0 & C_3 & 0 \\ 0 & 0 & 0 \\ 0 & 0 & B_1 \end{bmatrix}.$$

Since $\mathcal{R}(M'_C)$ is closed if and only if $\mathcal{R}((M'_C)^* M'_C)$ is closed and since

$$(M'_C)^* M'_C = \begin{bmatrix} A_1^* A_1 & 0 & 0 \\ 0 & 0 & 0 \\ 0 & C_3^* C_3 & 0 \\ 0 & 0 & B_1^* B_1 \end{bmatrix},$$

we get that $\mathcal{R}(M_C)$ is closed if and only if $\mathcal{R}(C_3)$ is closed. Hence, M_C is left invertible for any $C \in \mathcal{B}(\mathcal{K}, \mathcal{H})$ such that C_3 is left invertible (evidently since $\alpha(B) \le \beta(A)$, such C exists).

If we suppose that $\mathcal{R}(B)$ is not closed, M_C has an operator matrix representation

$$M_C = \begin{bmatrix} A_1 & C_1 \\ 0 & C_2 \\ 0 & B \end{bmatrix} : \mathcal{H} \oplus \mathcal{K} \to \mathcal{R}(A) \oplus \mathcal{R}(A)^\perp \oplus \mathcal{K},$$

where $A_1 : \mathcal{H} \to \mathcal{R}(A)$ is invertible. There is an invertible operator

$$V = \begin{bmatrix} I & -A_1^{-1} C_1 \\ 0 & I \end{bmatrix} : \mathcal{H} \oplus \mathcal{K} \to \mathcal{H} \oplus \mathcal{K}$$

such that

$$\begin{bmatrix} A_1 & C_1 \\ 0 & C_2 \\ 0 & B \end{bmatrix} V = \begin{bmatrix} A_1 & 0 \\ 0 & C_2 \\ 0 & B \end{bmatrix}. \tag{1.5}$$

Hence, M_C is left invertible if and only if $\begin{bmatrix} C_2 \\ B \end{bmatrix} = \begin{bmatrix} P_{\mathcal{R}(A)^\perp} C \\ B \end{bmatrix} : \mathcal{K} \to \mathcal{R}(A)^\perp \oplus \mathcal{K}$ is left invertible. Now, we prove that in the case when $\beta(A) < \infty$, such C does not exist: Suppose to the contrary that there exists C such that M_C is left invertible and that $\beta(A) < \infty$. Then, we have that C_2 is a finite rank operator, so $C_2^* C_2$ is a compact. Since from the fact that $\begin{bmatrix} C_2 \\ B \end{bmatrix}$ is left invertible, we have that $C_2^* C_2 + B^* B$ is Fredholm, we get that $B^* B$ is Fredholm as well, which contradicts the fact that $\mathcal{R}(B)$ is not closed.

Let us prove that in the case when $\beta(A) = \infty$ such C exists: If $\beta(A) = \infty$, there exists an invertible operator $S : \mathcal{K} \to \mathcal{R}(A)^\perp$. Let

$$C = \begin{bmatrix} 0 \\ S \end{bmatrix} : \mathcal{K} \to \begin{bmatrix} \mathcal{R}(A) \\ \mathcal{R}(A)^\perp \end{bmatrix}.$$

Then

$$M_C = \begin{bmatrix} A_1 & 0 \\ 0 & S \\ 0 & B \end{bmatrix} : \begin{bmatrix} \mathcal{H} \\ \mathcal{K} \end{bmatrix} \to \begin{bmatrix} \mathcal{R}(A) \\ \mathcal{R}(A)^\perp \\ \mathcal{K} \end{bmatrix}$$

is left invertible.

Case 2. Since A is not left invertible, using the equality

$$M_C = \begin{bmatrix} I & 0 \\ 0 & B \end{bmatrix} \begin{bmatrix} I & C \\ 0 & I \end{bmatrix} \begin{bmatrix} A & 0 \\ 0 & I \end{bmatrix},$$

we have that M_C can never be left invertible for any $C \in \mathcal{B}(\mathcal{K}, \mathcal{H})$. $\square$

REMARK 1.7. In the case when A is left invertible, $\alpha(B) \leq \beta(A)$, and $\mathcal{R}(B)$ is not closed, using the polar decomposition of $B = UP$, where $P = \int_0^{\|B\|} \lambda dE_\lambda$, we have that M_C as an operator from $\mathcal{H} \oplus \mathcal{K} = \mathcal{N}(A) \oplus E([\delta, \|B\|])\mathcal{K} \oplus E([0,\delta))\mathcal{K}$ into $\mathcal{H} \oplus \mathcal{K} = \mathcal{R}(A) \oplus \mathcal{R}(A)^\perp \oplus UE([\delta, \|B\|])\mathcal{K} \oplus (\mathcal{N}(B) \oplus E([0,\delta))\mathcal{K})$ has the following operator matrix representation:

$$M_C = \begin{bmatrix} A_1 & C_1 & C_2 \\ 0 & C_3 & C_4 \\ 0 & B_1 & 0 \\ 0 & 0 & B_2 \end{bmatrix}, \tag{1.6}$$

where A_1 and B_1 are invertible for $\delta > 0$ small enough. Remark that $\dim E([0,\delta))\mathcal{K}$ is infinite for small enough δ. Using the decomposition (1.6), we get that $\mathcal{N}(M_C) = \{0\}$ if and only if $\mathcal{N}(C_4) \cap \mathcal{N}(B_2) = \{0\}$. On the other hand, $\mathcal{R}(M_C)$ is closed if and only if $\mathcal{R}(M_C'')$ is closed, where

$$M_C'' = \begin{bmatrix} A_1 & 0 & 0 \\ 0 & 0 & C_4 \\ 0 & B_1 & 0 \\ 0 & 0 & B_2 \end{bmatrix},$$

which is equivalent with the fact that $\mathcal{R}(C_4^* C_4 + B_1^* B_1)$ is closed. Hence, in this case, we get that M_C is left invertible if and only if

$$\begin{bmatrix} C_4 \\ B_2 \end{bmatrix} : E([0,\delta))\mathcal{K} \to \begin{bmatrix} \mathcal{R}(A)^\perp \\ \mathcal{N}(B) \oplus E([0,\delta))\mathcal{K} \end{bmatrix}$$

is left invertible which is equivalent with the fact that $\mathcal{N}(C_4) \cap \mathcal{N}(B_2) = \{0\}$ and $\mathcal{R}(C_4^* C_4 + B_1^* B_1)$ is closed.

1.2. Completions of upper triangular operator matrices to invertible and right invertible operator

One of the first results that appeared concerning the completions of upper triangular operator matrices to invertibility was published by H. K. Du et al. [**72**]. They considered the invertibility of the operator matrix M_C for given operators $A \in \mathcal{B}(\mathcal{H})$ and $B \in \mathcal{B}(\mathcal{K})$ in the case when $\mathcal{H}$ and $\mathcal{K}$ are separable Hilbert spaces. Their results were generalized to the case of Banach spaces in [**98**], where necessary and sufficient conditions were given for the existence of an operator $C \in \mathcal{B}(\mathcal{K}, \mathcal{H})$ such that M_C is invertible. Necessary and sufficient conditions presented in either of these papers for the existence of an operator $C \in \mathcal{B}(\mathcal{K}, \mathcal{H})$ such that M_C is invertible are the same:

(i) A is left invertible.
(ii) B is right invertible.
(iii) $\mathcal{N}(B) \cong X/R(A)$.

The set $\bigcap_{C \in \mathcal{B}(\mathcal{Y},\mathcal{X})} \sigma(M_C)$ is completely described in [**72**] but question of the existence of an operator $C' \in \mathcal{B}(\mathcal{Y}, \mathcal{X})$ such that

$$\sigma(M_{C'}) = \bigcap_{C \in \mathcal{B}(\mathcal{Y},\mathcal{X})} \sigma(M_C)$$

is still not answered. Also, in [**72**] and [**98**], neither is the set of all such operators $C \in \mathcal{B}(\mathcal{Y}, \mathcal{X})$ described nor does the proof offer a clue as to what such operators could look like. In the main result of this section, we give necessary and sufficient conditions for the invertibility of M_C, using a method which is much simpler than the one used in [**72**] and completely different than the one used in [**98**]. Our method allows us to completely describe the set of all $C \in \mathcal{B}(\mathcal{Y}, \mathcal{X})$ for which M_C is invertible for given operators $A \in \mathcal{B}(\mathcal{X})$ and $B \in \mathcal{B}(\mathcal{Y})$, in the case when $\mathcal{X}$ and $\mathcal{Y}$ are Banach spaces. Also, in some cases we can find C' such that (1.4) holds in the case of ordinary spectrum.

For a given $A \in \mathcal{B}(\mathcal{X})$ and $B \in \mathcal{B}(\mathcal{Y})$, the set of all $C \in \mathcal{B}(\mathcal{Y}, \mathcal{X})$ such that M_C is invertible and right invertible will be denoted by $S(A, B)$ and $S_r(A, B)$, respectively.

THEOREM 1.8. *Let $A \in \mathcal{B}(\mathcal{X})$ and $B \in \mathcal{B}(\mathcal{Y})$ be given operators. The operator matrix M_C is invertible for some $C \in \mathcal{B}(\mathcal{Y}, \mathcal{X})$ if and only if:*

(i) *A is left invertible.*
(ii) *B is right invertible.*
(iii) *$\mathcal{N}(B) \cong \mathcal{X}/R(A)$.*

If conditions (i)–(iii) *are satisfied, the set of all $C \in \mathcal{B}(\mathcal{Y}, \mathcal{X})$ such that M_C is invertible is given by*

(1.7)
$$S(A,B) = \Big\{ C \in \mathcal{B}(\mathcal{Y},\mathcal{X}) \colon C = \begin{bmatrix} C_1 & 0 \\ 0 & C_4 \end{bmatrix} : \begin{bmatrix} \mathcal{P} \\ \mathcal{N}(B) \end{bmatrix} \to \begin{bmatrix} \mathcal{R}(A) \\ \mathcal{S} \end{bmatrix}, \quad C_4 \text{ is invertible}, \mathcal{X} = \mathcal{R}(A) \oplus \mathcal{S}, \text{ and } \mathcal{Y} = \mathcal{P} \oplus \mathcal{N}(B) \Big\}.$$

PROOF. Suppose that M_C is invertible for some $C \in \mathcal{B}(\mathcal{Y}, \mathcal{X})$. From

$$M_C = \begin{bmatrix} I & 0 \\ 0 & B \end{bmatrix} \begin{bmatrix} I & C \\ 0 & I \end{bmatrix} \begin{bmatrix} A & 0 \\ 0 & I \end{bmatrix}, \tag{1.8}$$

we have that B is right invertible and A is left invertible. Hence, $\mathcal{R}(A)$ and $\mathcal{N}(B)$ are complemented subspaces of $\mathcal{X}$ and $\mathcal{Y}$, respectively.

From invertibility of M_C, we get that there exists an operator

$$\begin{bmatrix} X & Y \\ Z & W \end{bmatrix} : \begin{bmatrix} \mathcal{X} \\ \mathcal{Y} \end{bmatrix} \to \begin{bmatrix} \mathcal{X} \\ \mathcal{Y} \end{bmatrix}$$

such that

$$\begin{bmatrix} A & C \\ 0 & B \end{bmatrix} \begin{bmatrix} X & Y \\ Z & W \end{bmatrix} = \begin{bmatrix} I & 0 \\ 0 & I \end{bmatrix} \tag{1.9}$$

and

$$\begin{bmatrix} X & Y \\ Z & W \end{bmatrix} \begin{bmatrix} A & C \\ 0 & B \end{bmatrix} = \begin{bmatrix} I & 0 \\ 0 & I \end{bmatrix}. \tag{1.10}$$

By (1.10), it follows that $X = (A)_l^{-1}$ and $XC + YB = 0$, so we get that

$$(A)_l^{-1} C (B)_r^{-1} B = (A)_l^{-1} C, \tag{1.11}$$

for any right inverse $(B)_r^{-1}$ of B. Analogously, by (1.9), it follows that $W = (B)_r^{-1}$ and $AY + CW = 0$, so we get that

$$A(A)_l^{-1} C (B)_r^{-1} = C(B)_r^{-1}, \tag{1.12}$$

for any left inverse $(A)_l^{-1}$ of A. Now, by (1.11) and (1.12), we get that

$$C(B)_r^{-1} B = A(A)_l^{-1} C, \tag{1.13}$$

for some $(A)_l^{-1}$ and $(B)_r^{-1}$. Now, we can take corresponding decompositions of $\mathcal{X}$ and $\mathcal{Y}$, $\mathcal{X} = \mathcal{R}(A) \oplus \mathcal{S}$ and $\mathcal{Y} = \mathcal{P} \oplus \mathcal{N}(B)$, such that

$$A(A)_l^{-1} = \begin{bmatrix} I & 0 \\ 0 & 0 \end{bmatrix} : \begin{bmatrix} \mathcal{R}(A) \\ \mathcal{S} \end{bmatrix} \to \begin{bmatrix} \mathcal{R}(A) \\ \mathcal{S} \end{bmatrix}$$

and

$$(B)_r^{-1}B = \begin{bmatrix} I & 0 \\ 0 & 0 \end{bmatrix} : \begin{bmatrix} \mathcal{P} \\ \mathcal{N}(B) \end{bmatrix} \to \begin{bmatrix} \mathcal{P} \\ \mathcal{N}(B) \end{bmatrix}.$$

Let C be given by

$$C = \begin{bmatrix} C_1 & C_2 \\ C_3 & C_4 \end{bmatrix} : \begin{bmatrix} \mathcal{P} \\ \mathcal{N}(B) \end{bmatrix} \to \begin{bmatrix} \mathcal{R}(A) \\ \mathcal{S} \end{bmatrix}.$$

Using (1.13), we can check that $C_2 = 0$ and $C_3 = 0$. Now, by (1.9), it follows that $A(A)_l^{-1} + CZ = I$, which using the above decomposition of C gives that C_4 is right invertible. Analogously, by (1.10) and the equation $ZC + (B)_r^{-1}B = I$, we get that C_4 is left invertible. Now, from the invertibility of C_4, we have that $\mathcal{N}(B) \cong X/R(A)$. Furthermore, we proved that if M_C is invertible, then there exist some decompositions of $\mathcal{X}$ and $\mathcal{Y}$ such that C is given by

$$C = \begin{bmatrix} C_1 & 0 \\ 0 & C_4 \end{bmatrix} : \begin{bmatrix} \mathcal{P} \\ \mathcal{N}(B) \end{bmatrix} \to \begin{bmatrix} \mathcal{R}(A) \\ \mathcal{S} \end{bmatrix},$$

where C_4 is invertible.

Now, let us assume that (i)–(iii) are satisfied. From (i) and (ii), it follows that there exist closed subspaces $\mathcal{S}$ and $\mathcal{P}$ such that $\mathcal{X} = \mathcal{R}(A) \oplus \mathcal{S}$ and $\mathcal{Y} = \mathcal{P} \oplus \mathcal{N}(B)$. Using these decompositions, we have that $A = \begin{bmatrix} A_1 \\ 0 \end{bmatrix} : \mathcal{X} \to \begin{bmatrix} \mathcal{R}(A) \\ \mathcal{S} \end{bmatrix}$ and $B = \begin{bmatrix} B_1 & 0 \end{bmatrix} : \begin{bmatrix} \mathcal{P} \\ \mathcal{N}(B) \end{bmatrix} \to \mathcal{Y}$, where A_1 and B_1 are invertible. By the condition (iii), there exists an invertible operator $C_4 : \mathcal{N}(B) \to \mathcal{S}$. Let $C_1 : \mathcal{P} \to \mathcal{R}(A)$ be arbitrary. It is easy to check that

$$M_C = \begin{bmatrix} A_1 & C_1 & 0 \\ 0 & 0 & C_4 \\ 0 & B_1 & 0 \end{bmatrix} : \begin{bmatrix} \mathcal{X} \\ \mathcal{P} \\ \mathcal{N}(B) \end{bmatrix} \to \begin{bmatrix} \mathcal{R}(A) \\ \mathcal{S} \\ \mathcal{Y} \end{bmatrix}$$

is invertible with the inverse given by

$$\begin{bmatrix} A_1^{-1} & 0 & -A_1^{-1}C_1B_1^{-1} \\ 0 & 0 & B_1^{-1} \\ 0 & C_4^{-1} & 0 \end{bmatrix}.$$

Now, it is clear that (1.7) holds. □

REMARK 1.9. We can check that if the conditions (i)–(iii) of Theorem 1.8 hold and if we take arbitrary but fixed decompositions of $\mathcal{X}$ and $\mathcal{Y}$, $\mathcal{X} = \mathcal{R}(A) \oplus \mathcal{S}$ and $\mathcal{Y} = \mathcal{P} \oplus \mathcal{N}(B)$, then

(1.14)

$$S(A,B) = \left\{ C \in \mathcal{B}(\mathcal{Y},\mathcal{X}) : C = \begin{bmatrix} C_1 & C_2 \\ C_3 & C_4 \end{bmatrix} : \begin{bmatrix} \mathcal{P} \\ \mathcal{N}(B) \end{bmatrix} \to \begin{bmatrix} \mathcal{R}(A) \\ \mathcal{S} \end{bmatrix}, \right.$$
$$\left. C_4 \text{ is invertible} \right\}.$$

To prove that, let us denote by $S_1(A,B)$ the right side of (1.14). For $C \in S_1(A,B)$, it is easy to check that M_C is invertible and in that case the inverse is given by

$$(M_C)^{-1} = \begin{bmatrix} A_1^{-1} & -A_1^{-1}C_2C_4^{-1} & -A_1^{-1}(C_1 - C_2C_4^{-1}C_3)B_1^{-1} \\ 0 & 0 & 0 \\ 0 & 0 & B_1^{-1} \\ 0 & (C_4)_r^{-1} & -C_4^{-1}C_3B_1^{-1} \end{bmatrix}.$$

Hence, $S_1(A,B) \subseteq S(A,B)$. If $C \in S(A,B)$, then using the given decompositions of $\mathcal{X}$ and $\mathcal{Y}$, we get that $A = \begin{bmatrix} A_1 \\ 0 \end{bmatrix} : \mathcal{X} \to \begin{bmatrix} \mathcal{R}(A) \\ \mathcal{S} \end{bmatrix}$ and $B = \begin{bmatrix} B_1 & 0 \end{bmatrix} : \begin{bmatrix} \mathcal{P} \\ \mathcal{N}(B) \end{bmatrix} \to \mathcal{Y}$, where A_1 and B_1 are invertible. In that case, any right inverse of B and any left inverse of A are given by $(B)_r^{-1} = \begin{bmatrix} B_1^{-1} \\ T \end{bmatrix} : \mathcal{Y} \to \begin{bmatrix} \mathcal{P} \\ \mathcal{N}(B) \end{bmatrix}$ and $(A)_l^{-1} = \begin{bmatrix} A_1^{-1} & R \end{bmatrix} : \begin{bmatrix} \mathcal{R}(A) \\ \mathcal{S} \end{bmatrix} \to \mathcal{X}$, for some operators T and R. Using (1.9) and (1.10), we get that

$$\begin{aligned} Z &= (I - (B)_r^{-1}B)Z(I - A(A)_l^{-1}), \\ A(A)_l^{-1} + CZ &= I, \\ ZC + (B)_r^{-1}B &= I. \end{aligned} \tag{1.15}$$

Now if we suppose that

$$C = \begin{bmatrix} C_1 & C_2 \\ C_3 & C_4 \end{bmatrix} : \begin{bmatrix} P \\ \mathcal{N}(B) \end{bmatrix} \to \begin{bmatrix} \mathcal{R}(A) \\ \mathcal{S} \end{bmatrix},$$

by (1.15), we get that C_4 is invertible, which means that $C \in S_1(A,B)$. Hence, (1.14) holds.

REMARK 1.10. As in **[98]**, we get that

$$\bigcap_{C \in \mathcal{B}(\mathcal{Y},\mathcal{X})} \sigma(M_C) = \sigma_l(A) \cup \sigma_r(B) \cup \{\lambda \in \mathbb{C} : \mathcal{N}(B - \lambda I) \cong \mathcal{X}/\mathcal{R}(A - \lambda I)\}$$

but using (1.7), we can find C' such that (1.4) holds in the case of ordinary spectrum. For arbitrary $C' \in \mathcal{B}(\mathcal{Y},\mathcal{X})$, it is evident that $\bigcap_{C \in \mathcal{B}(\mathcal{Y},\mathcal{X})} \sigma(M_C) \subseteq \sigma(M_{C'})$ while the reverse inclusion holds if and only if for every $\lambda \in P = \rho_l(A) \cap \rho_r(B) \cap \{\lambda \in \mathbb{C} \colon \mathcal{N}(B - \lambda I) \cong \mathcal{Y}/\mathcal{R}(A - \lambda I)\}$ the operator C' belongs to $S(A - \lambda I, B - \lambda I)$. In other words, there exists $C' \in \mathcal{B}(\mathcal{Y},\mathcal{X})$ such that

$$\sigma(M_{C'}) = \bigcap_{C \in \mathcal{B}(\mathcal{Y},\mathcal{X})} \sigma(M_C) \tag{1.16}$$

if and only if

$$\bigcap_{\lambda \in P} S(A - \lambda I, B - \lambda I) \neq \emptyset.$$

In the case of separable Hilbert spaces, $P = \rho_l(A) \cap \rho_r(B) \cap \{\lambda \in \mathbb{C} \colon \alpha(B - \lambda I) = \beta(A - \lambda I)\}$. In particular if A is a compact (or Riesz and polynomial Riesz) operator on a Banach space, then $P = \rho(A) \cap \rho(B)$, so it is evident that C' can be an arbitrary operator from $\mathcal{B}(\mathcal{Y},\mathcal{X})$. Analogously, as in Remark 1.10, we can find an operator C' which satisfies (1.16), in the case when B is normal and $\mathcal{Y}$ is separable Hilbert space.

Now, we point our attention to right invertible completions of M_C.

THEOREM 1.11. *Let $A \in \mathcal{B}(\mathcal{X})$ and $B \in \mathcal{B}(\mathcal{Y})$ be given operators. The operator matrix M_C is right invertible for $C \in \mathcal{B}(\mathcal{Y},\mathcal{X})$ if and only if:*

(i) *B is right invertible.*
(ii) *$\begin{bmatrix} A & CP_{\mathcal{N}(B)} \end{bmatrix}$ is right invertible.*

PROOF. There is $C \in \mathcal{B}(\mathcal{Y},\mathcal{X})$ such that the operator M_C is right invertible if and only if there exist appropriate operators X, Y, Z, W, T such that

$$\begin{bmatrix} A & C \\ 0 & B \end{bmatrix}\begin{bmatrix} X & Y \\ Z & W \end{bmatrix} = \begin{bmatrix} I & T \\ 0 & I \end{bmatrix},$$

i.e., if and only if the following system is solvable:

$$\begin{aligned} AX + CZ &= I, \\ BZ &= 0, \\ BW &= I. \end{aligned} \tag{1.17}$$

Evidently, (1.17) is solvable if and only if B is right invertible and the equation $AX + CP_{\mathcal{N}(B)}Q = I$ is solvable for some $Q \in \mathcal{B}(\mathcal{Y})$ (the last condition being equivalent to the right invertibility of $\begin{bmatrix} A & CP_{\mathcal{N}(B)} \end{bmatrix}$). □

REMARK 1.12. From Theorem 1.11, we can conclude that the right invertibility of M_C is equivalent to the right invertibility of B and the solvability of the equation

$$AX + CP_{\mathcal{N}(B)}Q = I\ , \tag{1.18}$$

for some $Q \in \mathcal{B}(\mathcal{Y})$.

If we impose the additional condition of the regularity of $A \in \mathcal{B}(\mathcal{X})$, we have the following modification of Theorem 1.11:

THEOREM 1.13. *Let $A \in \mathcal{B}(\mathcal{X})$ and $B \in \mathcal{B}(\mathcal{Y})$ be given operators. If A is regular, then the operator matrix M_C is right invertible for some $C \in \mathcal{B}(\mathcal{Y},\mathcal{X})$ if and only if:*

(i) *B is right invertible.*
(ii) *$\mathcal{X}/\mathcal{R}(A) \preceq \mathcal{N}(B)$.*

If conditions (i)–(ii) *are satisfied, the set of all $C \in \mathcal{B}(\mathcal{Y},\mathcal{X})$ such that M_C is right invertible is described by*

$$S_r(A,B) = \left\{ C \in \mathcal{B}(\mathcal{Y},\mathcal{X}) \colon\ C = \begin{bmatrix} C_1 & C_2 \\ C_3 & C_4 \end{bmatrix} : \begin{bmatrix} \mathcal{P} \\ \mathcal{N}(B) \end{bmatrix} \to \begin{bmatrix} \mathcal{R}(A) \\ \mathcal{S} \end{bmatrix}, \quad C_4 \text{ is right invertible} \right\}, \tag{1.19}$$

where $\mathcal{X} = \mathcal{R}(A) \oplus \mathcal{S}$ and $\mathcal{Y} = \mathcal{P} \oplus \mathcal{N}(B)$.

PROOF. Suppose that M_C is right invertible for some $C \in \mathcal{B}(\mathcal{Y},\mathcal{X})$. From Remark 1.12, we get that the right invertibility of M_C is equivalent to the right invertibility of $B \in \mathcal{B}(\mathcal{Y})$ and the solvability of the equation (1.18). Let X be any solutions of (1.18) for some Q. Since A is regular, we have that

$$A = \begin{bmatrix} A_1 & 0 \\ 0 & 0 \end{bmatrix} : \begin{bmatrix} \mathcal{T} \\ \mathcal{N}(A) \end{bmatrix} \to \begin{bmatrix} \mathcal{R}(A) \\ \mathcal{S} \end{bmatrix}, \tag{1.20}$$

where $A_1 \in \mathcal{B}(\mathcal{T}, \mathcal{R}(A))$ is invertible and $\mathcal{X} = \mathcal{T} \oplus \mathcal{N}(A) = \mathcal{R}(A) \oplus \mathcal{S}$. Also,

$$B = \begin{bmatrix} B_1 & 0 \end{bmatrix} : \begin{bmatrix} \mathcal{M} \\ \mathcal{N}(B) \end{bmatrix} \to \mathcal{Y}, \tag{1.21}$$

where $B_1 \in \mathcal{B}(\mathcal{M}, \mathcal{Y})$ is invertible and $\mathcal{Y} = \mathcal{M} \oplus \mathcal{N}(B)$. Suppose that

$$X = \begin{bmatrix} X_1 & X_2 \\ X_3 & X_4 \end{bmatrix} : \begin{bmatrix} \mathcal{R}(A) \\ \mathcal{S} \end{bmatrix} \to \begin{bmatrix} \mathcal{T} \\ \mathcal{N}(A) \end{bmatrix},$$

$$Q = \begin{bmatrix} Q_1 & Q_2 \\ Q_3 & Q_4 \end{bmatrix} : \begin{bmatrix} \mathcal{R}(A) \\ \mathcal{S} \end{bmatrix} \to \begin{bmatrix} \mathcal{M} \\ \mathcal{N}(B) \end{bmatrix},$$

and

$$C = \begin{bmatrix} C_1 & C_2 \\ C_3 & C_4 \end{bmatrix} : \begin{bmatrix} \mathcal{M} \\ \mathcal{N}(B) \end{bmatrix} \to \begin{bmatrix} \mathcal{R}(A) \\ \mathcal{S} \end{bmatrix}.$$

We have that (1.18) is equivalent to

$$\begin{bmatrix} A_1 & 0 \\ 0 & 0 \end{bmatrix} \begin{bmatrix} X_1 & X_2 \\ X_3 & X_4 \end{bmatrix} + \begin{bmatrix} C_1 & C_2 \\ C_3 & C_4 \end{bmatrix} \begin{bmatrix} 0 & 0 \\ 0 & I \end{bmatrix} \begin{bmatrix} Q_1 & Q_2 \\ Q_3 & Q_4 \end{bmatrix} = \begin{bmatrix} I & 0 \\ 0 & I \end{bmatrix};$$

i.e.,

$$A_1X_1 + C_2Q_3 = I, \quad A_1X_2 + C_2Q_4 = 0, \quad C_4Q_3 = 0, \quad C_4Q_4 = I. \tag{1.22}$$

Evidently, (1.18) is solvable if and only if C_4 is right invertible. Hence, $\mathcal{X}/\mathcal{R}(A) \preceq \mathcal{N}(B)$.

Let (i)–(iii) be satisfied. By (i) and (ii) we have the existence of

$$C = \begin{bmatrix} 0 & 0 \\ 0 & C_4 \end{bmatrix} : \begin{bmatrix} \mathcal{M} \\ \mathcal{N}(B) \end{bmatrix} \to \begin{bmatrix} \mathcal{R}(A) \\ \mathcal{S} \end{bmatrix},$$

where C_4 is right invertible. It is evident that the right inverse of M_C is given by

$$\begin{bmatrix} A_1^{-1} & 0 & 0 \\ 0 & 0 & 0 \\ 0 & 0 & B_1^{-1} \\ 0 & (C_4)_r^{-1} & 0 \end{bmatrix} : \begin{bmatrix} \mathcal{R}(A) \\ \mathcal{S} \\ \mathcal{Y} \end{bmatrix} \to \begin{bmatrix} \mathcal{T} \\ \mathcal{N}(A) \\ \mathcal{M} \\ \mathcal{N}(B) \end{bmatrix}.$$

To prove that the set of all $C \in \mathcal{B}(\mathcal{Y}, \mathcal{X})$ such that M_C is right invertible is described by (1.19), it is enough to note that if $C = \begin{bmatrix} C_1 & C_2 \\ C_3 & C_4 \end{bmatrix} : \begin{bmatrix} \mathcal{M} \\ \mathcal{N}(B) \end{bmatrix} \to \begin{bmatrix} \mathcal{R}(A) \\ \mathcal{S} \end{bmatrix}$ for decompositions $\mathcal{X} = \mathcal{R}(A) \oplus \mathcal{S}$ and $\mathcal{Y} = \mathcal{M} \oplus \mathcal{N}(B)$, then we have that A and B are given by (1.20) and (1.21), respectively, and in that case a right inverse of M_C is given by

$$(M_C)_r^{-1} = \begin{bmatrix} A_1^{-1} & -A_1^{-1}C_2(C_4)_r^{-1} & -A_1^{-1}(C_1B_1^{-1} - C_2(C_4)_r^{-1}C_3B_1^{-1}) \\ 0 & 0 & 0 \\ 0 & 0 & B_1^{-1} \\ 0 & (C_4)_r^{-1} & -(C_4)_r^{-1}C_3B_1^{-1} \end{bmatrix}.$$

□

REMARK 1.14. The condition (ii) of Theorem 1.13, in the case when $\mathcal{X}$ and $\mathcal{Y}$ are Hilbert spaces, is equivalent to the condition $\mathcal{R}(A) + \mathcal{R}(CP_{\mathcal{N}(B)}) = \mathcal{X}$. In the case when $\mathcal{X}$ and $\mathcal{Y}$ are separable Hilbert spaces, the condition (ii) of Theorem 1.13 is equivalent to $\beta(A) \leq \alpha(B)$.

Similarly, we can consider the left invertibility of M_C:

THEOREM 1.15. *Let $A \in \mathcal{B}(\mathcal{X})$ and $B \in \mathcal{B}(\mathcal{Y})$ be given operators. The operator matrix M_C is left invertible for $C \in \mathcal{B}(\mathcal{Y}, \mathcal{X})$ if and only if:*

(i) *A is left invertible.*

(ii) $\begin{bmatrix} (I - P_{\mathcal{R}(A)})C \\ B \end{bmatrix}$ *is left invertible.*

With the additional assumption of the regularity of B, we have the following:

THEOREM 1.16. *Let $A \in \mathcal{B}(\mathcal{X})$ and $B \in \mathcal{B}(\mathcal{Y})$ be given operators such that B is regular. The operator matrix M_C is left invertible for some $C \in \mathcal{B}(\mathcal{Y}, \mathcal{X})$ if and only if:*

(i) *A is left invertible.*

(ii) $\mathcal{N}(B) \preceq \mathcal{X}/\mathcal{R}(A)$.

If the conditions (i)–(ii) *are satisfied, the set of all $C \in \mathcal{B}(\mathcal{Y}, \mathcal{X})$ such that M_C is right invertible is described by*

(1.23)

$$S_l(A, B) = \left\{ C \in \mathcal{B}(\mathcal{Y}, \mathcal{X}) \colon \; C = \begin{bmatrix} C_1 & C_2 \\ C_3 & C_4 \end{bmatrix} : \begin{bmatrix} \mathcal{P} \\ \mathcal{N}(B) \end{bmatrix} \to \begin{bmatrix} \mathcal{R}(A) \\ \mathcal{S} \end{bmatrix}, \right.$$
$$\left. C_4 \text{ is left invertible} \right\},$$

where $\mathcal{X} = \mathcal{R}(A) \oplus \mathcal{S}$ and $\mathcal{Y} = \mathcal{P} \oplus \mathcal{N}(B)$.

In the case when $\mathcal{X}$ and $\mathcal{Y}$ are separable Hilbert spaces, the condition (ii) of Theorem 1.16 is equivalent to $\alpha(B) \leq \beta(A)$.

REMARK 1.17. Analogously, as in the case of the point spectrum, when $\mathcal{X}$ and $\mathcal{Y}$ are Hilbert spaces, we can answer Question 3 in the case of right (left) spectrum and find such operator C' in the following cases: when A is a compact or Riesz or polynomial Riesz operator; when B is a normal operator; when the set $P_r = \rho_l(B) \cap \{\lambda \in \mathbb{C} \colon 0 < \beta(A - \lambda I) \leq \alpha(B - \lambda I)\}$ ($P_l = \rho_l(A) \cap \{\lambda \in \mathbb{C} \colon 0 < \alpha(B - \lambda I) \leq \beta(A - \lambda I)\}$) is finite.

1.3. Completions of upper triangular operator matrices to regular operator

In this section, we consider the completion to regularity of an operator matrix

$$\begin{bmatrix} A & ? \\ 0 & B \end{bmatrix} : \begin{bmatrix} \mathcal{X} \\ \mathcal{Y} \end{bmatrix} \to \begin{bmatrix} \mathcal{X} \\ \mathcal{Y} \end{bmatrix},$$

where $\mathcal{X}, \mathcal{Y}$ are Banach spaces and $A \in \mathcal{B}(\mathcal{X})$ and $B \in \mathcal{B}(\mathcal{Y})$ are given operators. We completely describe the set of all $C \in \mathcal{B}(\mathcal{Y}, \mathcal{X})$ such that

$$M_C = \begin{bmatrix} A & C \\ 0 & B \end{bmatrix} : \begin{bmatrix} \mathcal{X} \\ \mathcal{Y} \end{bmatrix} \to \begin{bmatrix} \mathcal{X} \\ \mathcal{Y} \end{bmatrix}$$

is regular. Furthermore, we give a necessary and sufficient condition under which M_C is regular and has an inner inverse of the form $\begin{bmatrix} A^{(1,2)} & X \\ Y & B^{(1,2)} \end{bmatrix}$, where $A^{(1,2)} \in A\{1,2\}$, $B^{(1,2)} \in B\{1,2\}$, and one of the operators $X \in \mathcal{B}(\mathcal{Y}, \mathcal{X})$ and $Y \in \mathcal{B}(\mathcal{X}, \mathcal{Y})$ is zero.

First we should mention the paper [**67**] where the existence of a completion to a closed range upper triangular operator matrix was considered on Hilbert spaces assuming closedness of the ranges of A and B. Notice that in the case of Hilbert spaces closedness of the range of an operator is equivalent with its regularity.

THEOREM 1.18 ([**67**]). *Let $A \in \mathcal{B}(\mathcal{H})$ and $B \in \mathcal{B}(\mathcal{K})$ be given operators. If $\mathcal{R}(A)$ is closed and $\mathcal{R}(B)$ is not closed, then the range of $\begin{bmatrix} A & C \\ 0 & B \end{bmatrix}$ is not closed for every $C \in \mathcal{B}(\mathcal{K}, \mathcal{H})$ if and only if $\dim(\mathcal{N}(A^*))$ is finite.*

In the next theorem, we present a necessary and sufficient condition under which M_C is regular for a given $A \in \mathcal{B}(\mathcal{X})$ and $B \in \mathcal{B}(\mathcal{Y})$. Using its proof, we can completely describe the set $M_C\{1\}$.

THEOREM 1.19. *Let $A \in \mathcal{B}(\mathcal{X})$ and $B \in \mathcal{B}(\mathcal{Y})$ be regular. The operator matrix M_C is regular if and only if $(I - AA^{(1)})C(I - B^{(1)}B)$ is regular, for some $A^{(1)} \in A\{1\}$ and $B^{(1)} \in B\{1\}$.*

PROOF. Since $A \in \mathcal{B}(\mathcal{X})$ and $B \in \mathcal{B}(\mathcal{Y})$ are regular, it follows that $\mathcal{X}$ and $\mathcal{Y}$ have the decompositions

$$\mathcal{X} = \mathcal{T} \oplus \mathcal{N}(A) = \mathcal{R}(A) \oplus \mathcal{S}, \quad \mathcal{Y} = \mathcal{P} \oplus \mathcal{N}(B) = \mathcal{R}(B) \oplus \mathcal{Q}$$

and consequently that A and B are given by

$$A = \begin{bmatrix} A_1 & 0 \\ 0 & 0 \end{bmatrix} : \begin{bmatrix} \mathcal{T} \\ \mathcal{N}(A) \end{bmatrix} \to \begin{bmatrix} \mathcal{R}(A) \\ \mathcal{S} \end{bmatrix}$$

and

$$B = \begin{bmatrix} B_1 & 0 \\ 0 & 0 \end{bmatrix} : \begin{bmatrix} \mathcal{P} \\ \mathcal{N}(B) \end{bmatrix} \to \begin{bmatrix} \mathcal{R}(B) \\ \mathcal{Q} \end{bmatrix},$$

where $A_1 : \mathcal{T} \to \mathcal{R}(A)$ and $B_1 : \mathcal{P} \to \mathcal{R}(B)$ are invertible. There exist $A^{(1)} \in A\{1\}$ and $B^{(1)} \in B\{1\}$ such that $AA^{(1)} = P_{\mathcal{R}(A),\mathcal{S}}$, $A^{(1)}A = P_{\mathcal{T},\mathcal{N}(A)}$, $BB^{(1)} = P_{\mathcal{R}(B),\mathcal{Q}}$, and $B^{(1)}B = P_{\mathcal{P},\mathcal{N}(B)}$. Let C be given by

$$C = \begin{bmatrix} C_1 & C_2 \\ C_3 & C_4 \end{bmatrix} : \begin{bmatrix} \mathcal{P} \\ \mathcal{N}(B) \end{bmatrix} \to \begin{bmatrix} \mathcal{R}(A) \\ \mathcal{S} \end{bmatrix} \tag{1.24}$$

and suppose that M_C is regular. In that case, there exists an operator

$$\begin{bmatrix} X & Y \\ Z & W \end{bmatrix} : \begin{bmatrix} \mathcal{X} \\ \mathcal{Y} \end{bmatrix} \to \begin{bmatrix} \mathcal{X} \\ \mathcal{Y} \end{bmatrix}$$

such that

$$\begin{bmatrix} A & C \\ 0 & B \end{bmatrix} \begin{bmatrix} X & Y \\ Z & W \end{bmatrix} \begin{bmatrix} A & C \\ 0 & B \end{bmatrix} = \begin{bmatrix} A & C \\ 0 & B \end{bmatrix} \tag{1.25}$$

which is equivalent to the following system of the equations:

$$\begin{aligned} &AXA + CZA = A, \\ &BZC + BWB = B, \\ &BZA = 0, \\ &AXC + CZC + AYB + CWB = C. \end{aligned} \tag{1.26}$$

Suppose that X, Y, Z, and W are given by

$$X = \begin{bmatrix} X_1 & X_2 \\ X_3 & X_4 \end{bmatrix} : \begin{bmatrix} \mathcal{R}(A) \\ \mathcal{S} \end{bmatrix} \to \begin{bmatrix} \mathcal{T} \\ \mathcal{N}(A) \end{bmatrix},$$
$$Y = \begin{bmatrix} Y_1 & Y_2 \\ Y_3 & Y_4 \end{bmatrix} : \begin{bmatrix} \mathcal{R}(B) \\ \mathcal{Q} \end{bmatrix} \to \begin{bmatrix} \mathcal{T} \\ \mathcal{N}(A) \end{bmatrix},$$
$$Z = \begin{bmatrix} Z_1 & Z_2 \\ Z_3 & Z_4 \end{bmatrix} : \begin{bmatrix} \mathcal{R}(A) \\ \mathcal{S} \end{bmatrix} \to \begin{bmatrix} \mathcal{P} \\ \mathcal{N}(B) \end{bmatrix}, \tag{1.27}$$
$$W = \begin{bmatrix} W_1 & W_2 \\ W_3 & W_4 \end{bmatrix} : \begin{bmatrix} \mathcal{R}(B) \\ \mathcal{Q} \end{bmatrix} \to \begin{bmatrix} \mathcal{P} \\ \mathcal{N}(B) \end{bmatrix}.$$

We get that the third equation of (1.26), $BZA = 0$, is equivalent with $B_1Z_1A_1 = 0$; i.e., $Z_1 = 0$. Now, using that $Z_1 = 0$, the first equation of (1.26) is equivalent with

$$\begin{aligned} X_1 &= A_1^{-1}(I - C_2Z_3), \\ C_4Z_3 &= 0, \end{aligned} \tag{1.28}$$

while the second one holds if and only if

$$\begin{aligned} W_1 &= (I - Z_2C_3)B_1^{-1}, \\ Z_2C_4 &= 0. \end{aligned} \tag{1.29}$$

Now, the last equation of (1.26) is equivalent with

$$\begin{aligned} &A_1X_2C_3 + C_2Z_4C_3 + A_1Y_1B_1 + C_2W_3B_1 + C_1 = 0, \\ &A_1X_2C_4 + C_2Z_4C_4 = 0, \\ &C_4Z_4C_3 + C_4W_3B_1 = 0, \\ &C_4Z_4C_4 = C_4. \end{aligned} \tag{1.30}$$

From the last equation of (1.30), we conclude that C_4 must be regular. So, we get that the regularity of $(I - AA^{(1)})C(I - B^{(1)}B)$ is a necessary condition for the regularity of M_C. On the other hand, if C_4 is regular, taking $Z_4 = C_4^{(1)}$, $Z_1 = 0$, $Z_2 = 0$, $Z_3 = 0$, $X_1 = A_1^{-1}(I - C_2Z_3)$, $X_2 = -A_1^{-1}C_2C_4^{(1)}$, $Y_1 = -A_1^{-1}(C_1 + A_1X_2C_3 + C_2Z_4C_3 + C_2W_3B_1)B_1^{-1}$, $W_1 = (I - Z_2C_3)B_1^{-1}$, $W_3 = -C_4^{(1)}C_3B_1^{-1}$, where $X_3, X_4, Y_2, Y_3, Y_4, Z_2, Z_3, W_2, W_4$ are arbitrary, using (1.28), (1.29), and (1.30) we can easily check that (1.25) holds. Thus M_C is regular. □

REMARK 1.20. Using the previous theorem, we can completely describe the set $M_C\{1\}$ in the case when M_C is regular. So,

$$\begin{bmatrix} X & Y \\ Z & W \end{bmatrix} : \begin{bmatrix} \mathcal{X} \\ \mathcal{Y} \end{bmatrix} \to \begin{bmatrix} \mathcal{X} \\ \mathcal{Y} \end{bmatrix} \in M_C\{1\}$$

if and only if X, Y, Z, W are given by

$$\begin{aligned} X &= \begin{bmatrix} A_1^{-1}(I - C_2 Z_3) & X_2 \\ X_3 & X_4 \end{bmatrix} : \begin{bmatrix} \mathcal{R}(A) \\ \mathcal{S} \end{bmatrix} \to \begin{bmatrix} \mathcal{T} \\ \mathcal{N}(A) \end{bmatrix}, \\ Y &= \begin{bmatrix} -A_1^{-1} N B_1^{-1} & Y_2 \\ Y_3 & Y_4 \end{bmatrix} : \begin{bmatrix} \mathcal{R}(B) \\ \mathcal{Q} \end{bmatrix} \to \begin{bmatrix} \mathcal{T} \\ \mathcal{N}(A) \end{bmatrix}, \\ Z &= \begin{bmatrix} 0 & Z_2 \\ Z_3 & C_4^{(1)} \end{bmatrix} : \begin{bmatrix} \mathcal{R}(A) \\ \mathcal{S} \end{bmatrix} \to \begin{bmatrix} \mathcal{P} \\ \mathcal{N}(B) \end{bmatrix}, \\ W &= \begin{bmatrix} (I - Z_2 C_3) B_1^{-1} & W_2 \\ -C_4^{(1)} C_3 B_1^{-1} + (I - C_4^{(1)} C_4) S & W_4 \end{bmatrix} : \begin{bmatrix} \mathcal{R}(B) \\ \mathcal{Q} \end{bmatrix} \to \begin{bmatrix} \mathcal{P} \\ \mathcal{N}(B) \end{bmatrix}, \end{aligned} \tag{1.31}$$

where $X_2 = -A_1^{-1} C_2 C_4^{(1)} + M(I - C_4 C_4^{(1)})$, $Z_2 = E(I - C_4 C_4^{(1)})$, $Z_3 = (I - C_4^{(1)} C_4) F$, $N = C_1 + A_1 X_2 C_3 + C_2 Z_4 C_3 + C_2 W_3 B_1$, and the others operators which appear are arbitrary.

In the following theorem we present a necessary and sufficient condition under which M_C is regular and has an inner inverse of the form $\begin{bmatrix} A^{(1,2)} & X \\ Y & B^{(1,2)} \end{bmatrix}$, where $A^{(1,2)} \in A\{1,2\}$, $B^{(1,2)} \in B\{1,2\}$, and one of the operators $X \in \mathcal{B}(\mathcal{Y}, \mathcal{X})$ and $Y \in \mathcal{B}(\mathcal{X}, \mathcal{Y})$ is zero with the assumption that the operator C has the matrix representation given by (1.24).

THEOREM 1.21. *Let $A \in \mathcal{B}(\mathcal{X})$ and $B \in \mathcal{B}(\mathcal{Y})$ be regular. For $C \in \mathcal{B}(\mathcal{Y}, \mathcal{X})$, the operator matrix M_C is regular with an inner inverse of the form $\begin{bmatrix} A^{(1,2)} & X \\ Y & B^{(1,2)} \end{bmatrix}$ where $A^{(1,2)} \in A\{1,2\}$, $B^{(1,2)} \in B\{1,2\}$, and one of the operators $X \in \mathcal{B}(\mathcal{Y}, \mathcal{X})$ and $Y \in \mathcal{B}(\mathcal{X}, \mathcal{Y})$ is zero, if and only if C satisfies one of the following conditions:*

(1) *$C(\mathcal{N}(B)) \subseteq \mathcal{R}(A)$.*

(2) *C_1 is regular, $C_1 C_1^- C_2 = 0$, $C_3 C_1^- C_1 = 0$, $C_3 C_1^- C_2 = -C_4$,*

where the C_i are defined by the representation of C given by

$$C = \begin{bmatrix} C_1 & C_2 \\ C_3 & C_4 \end{bmatrix} : \begin{bmatrix} \mathcal{N}(B) \\ \mathcal{R}(B^{(1,2)}) \end{bmatrix} \to \begin{bmatrix} \mathcal{N}(A^{(1,2)}) \\ \mathcal{R}(A) \end{bmatrix}.$$

PROOF. The fact that for some $C \in \mathcal{B}(\mathcal{Y}, \mathcal{X})$ the operator matrix M_C is regular with an inner inverse of the form $\begin{bmatrix} A^{(1,2)} & X \\ Y & B^{(1,2)} \end{bmatrix}$, for $X \in \mathcal{B}(\mathcal{Y}, \mathcal{X})$ and $Y \in \mathcal{B}(\mathcal{X}, \mathcal{Y})$, is equivalent with the fact that the following system of the equations is solvable:

$$\begin{aligned} &CYA = 0, \\ &AA^{(1,2)}C + CYC + AXB + CB^{(1,2)}B = C, \\ &BYA = 0, \\ &BYC = 0. \end{aligned} \tag{1.32}$$

Since we have an additional condition that one of the operators X and Y is equal to zero, we consider two cases.

First, we consider the case when $Y = 0$. Then the system (1.32) is equivalent to

$$AA^{(1,2)}C + AXB + CB^{(1,2)}B = C. \tag{1.33}$$

Hence, for a given $C \in \mathcal{B}(\mathcal{Y}, \mathcal{X})$ the operator M_C is regular and has an inner inverse of the form $\begin{bmatrix} A^{(1,2)} & X \\ 0 & B^{(1,2)} \end{bmatrix}$, if and only if the following equation is solvable:

$$AXB = C - AA^{(1,2)}C - CB^{(1,2)}B. \tag{1.34}$$

This in turn holds if and only if $(I - AA^{(1,2)})C(I - B^{(1,2)}B) = 0$, which is equivalent with the fact that $C(\mathcal{N}(B)) \subseteq \mathcal{R}(A)$.

In the case when $X = 0$, the system (1.32) is equivalent to

$$\begin{aligned} &CYA = 0, \\ &AA^{(1,2)}C + CYC + CB^{(1,2)}B = C, \\ &BYA = 0, \\ &BYC = 0. \end{aligned} \tag{1.35}$$

Now, it is clear that M_C is regular with an inner inverse of the form

$$\begin{bmatrix} A^{(1,2)} & 0 \\ Y & B^{(1,2)} \end{bmatrix},$$

for each $C \in \mathcal{B}(\mathcal{Y}, \mathcal{X})$ which satisfies the conditions for the solvability of the system (1.35).

Suppose that such C exists. Then C and Y can be represented by

$$\begin{aligned} C &= \begin{bmatrix} C_1 & C_2 \\ C_3 & C_4 \end{bmatrix} : \begin{bmatrix} \mathcal{N}(B) \\ \mathcal{R}(B^{(1,2)}) \end{bmatrix} \to \begin{bmatrix} \mathcal{N}(A^{(1,2)}) \\ \mathcal{R}(A) \end{bmatrix}, \\ Y &= \begin{bmatrix} Y_1 & Y_2 \\ Y_3 & Y_4 \end{bmatrix} : \begin{bmatrix} \mathcal{N}(A^{(1,2)}) \\ \mathcal{R}(A) \end{bmatrix} \to \begin{bmatrix} \mathcal{N}(B) \\ \mathcal{R}(B^{(1,2)}) \end{bmatrix}, \end{aligned} \tag{1.36}$$

for some operators C_i and Y_i, $i = \overline{1,4}$. Since the equation $BYA = 0$ is equivalent with the equation $B^{(1,2)}BYAA^{(1,2)} = 0$ and since

$$\begin{aligned} AA^{(1,2)} &= \begin{bmatrix} 0 & 0 \\ 0 & I \end{bmatrix} : \begin{bmatrix} \mathcal{N}(A^{(1,2)}) \\ \mathcal{R}(A) \end{bmatrix} \to \begin{bmatrix} \mathcal{N}(A^{(1,2)}) \\ \mathcal{R}(A) \end{bmatrix}, \\ B^{(1,2)}B &= \begin{bmatrix} 0 & 0 \\ 0 & I \end{bmatrix} : \begin{bmatrix} \mathcal{N}(B) \\ \mathcal{R}(B^{(1,2)}) \end{bmatrix} \to \begin{bmatrix} \mathcal{N}(B) \\ \mathcal{R}(B^{(1,2)}) \end{bmatrix}, \end{aligned}$$

we conclude that $Y_4 = 0$. Now, $CYA = 0$ is equivalent with $C_1Y_2 = 0$ and $C_3Y_2 = 0$, while $BYC = 0$ is equivalent with $Y_3C_1 = 0$ and $Y_3C_2 = 0$. Now, we are looking for the conditions under which the second equation of the system (1.35) is satisfied. The same computations as above show that the second equation of the system (1.35) is satisfied for some operators C and Y given by (1.36) if and only if operators C_i and Y_i satisfy the following system:

$$C_1Y_1C_1 = C_1, \; C_1Y_1C_2 = 0, \; C_3Y_1C_1 = 0, \; C_3Y_1C_2 = -C_4. \tag{1.37}$$

From the first equation of the system (1.37), it follows that C_1 must be regular and that Y_1 is an inner inverse of C_1. Now, by (1.36) we get that (1.37) is equivalent to (ii). $\square$

REMARK 1.22. Theorem 1.21 is a special case of Theorem 1.19, so each of the conditions (1) and (2) of Theorem 1.21 implies the regularity of $(I - AA^{(1)}) \cdot C(I - B^{(1)}B)$, for some $A^{(1)} \in A\{1\}$ and $B^{(1)} \in B\{1\}$. That this is true for the condition (2) is evident. For the condition (1) this follows from the fact that it is equivalent with the fact that $(I - AA^{(1)})C(I - B^{(1)}B) = 0$.

REMARK 1.23. Obviously, (1.35) can be written as

$$\begin{bmatrix} B \\ C \end{bmatrix} Y \begin{bmatrix} A & C \end{bmatrix} = \begin{bmatrix} 0 & 0 \\ 0 & C - AA^{(1,2)}C - CB^{(1,2)}B \end{bmatrix}. \tag{1.38}$$

Now, it is clear that M_C is regular with an inner inverse of the form

$$\begin{bmatrix} A^{(1,2)} & 0 \\ Y & B^{(1,2)} \end{bmatrix},$$

for each $C \in \mathcal{B}(\mathcal{Y}, \mathcal{X})$ which satisfies the conditions for the solvability of the equation (1.38); i.e.,

$$\begin{bmatrix} B \\ C \end{bmatrix} \begin{bmatrix} B \\ C \end{bmatrix}^{(1)} \begin{bmatrix} 0 & 0 \\ 0 & T \end{bmatrix} \begin{bmatrix} A & C \end{bmatrix}^{(1)} \begin{bmatrix} A & C \end{bmatrix} = \begin{bmatrix} 0 & 0 \\ 0 & T \end{bmatrix},$$

where $T = C - AA^{(1,2)}C - CB^{(1,2)}B$. Using (1.38), we can describe all such Y.

REMARK 1.24. It is interesting to note that by Theorem 1.21, we have that for any regular operator $C \in \mathcal{B}(\mathcal{Y}, \mathcal{X})$ which satisfies $C(I - P_{\mathcal{N}(B)}) = 0$ and $P_{\mathcal{R}(A)}C = 0$, for some projectors $P_{\mathcal{N}(B)}$ and $P_{\mathcal{R}(A)}$, the matrix operator M_C is regular with an inner inverse of the form $\begin{bmatrix} A^{(1,2)} & 0 \\ Y & B^{(1,2)} \end{bmatrix}$.

In the special case of Hilbert spaces, we state the following result concerning the regularity and the form of the Moore-Penrose inverse of M_C. Recall that regularity of an operator on Hilbert space is equivalent with the existence of its Moore-Penrose inverse.

THEOREM 1.25. *Let $A \in \mathcal{B}(\mathcal{H})$ and $B \in \mathcal{B}(\mathcal{K})$ be regular. For $C \in \mathcal{B}(\mathcal{K}, \mathcal{H})$, the operator matrix M_C is regular with the Moore-Penrose inverse of the form $\begin{bmatrix} A^\dagger & X \\ Y & B^\dagger \end{bmatrix}$, where one of the operators $X \in \mathcal{B}(\mathcal{K}, \mathcal{H})$ and $Y \in \mathcal{B}(\mathcal{H}, \mathcal{K})$ is zero if and only if C satisfies one of the following conditions:*

(1) *$\mathcal{R}(C^*) \subseteq \mathcal{R}(B^*)$, $\mathcal{R}(C) \subseteq \mathcal{R}(A)$.*
(2) *C is regular, $\mathcal{R}(C) \subseteq \mathcal{N}(A^*)$, $\mathcal{R}(B^*) \subseteq \mathcal{N}(C)$.*

Furthermore, in the case when $X = 0$, the operator Y must be equal to $C^\dagger$.

PROOF. First, if one of the conditions (1)–(2) is satisfied, then by Theorem 1.21 and Remark 1.22, it follows that M_C is regular, which is equivalent with the existence of the Moore-Penrose inverse. Now, we show that the Moore-Penrose inverse of M_C has the mentioned form if and only if one of the conditions (1)–(2) is satisfied.

First, suppose that for some $C \in \mathcal{B}(\mathcal{Y}, \mathcal{X})$, there exists X such that the operator matrix M_C has the Moore-Penrose inverse of the form $\begin{bmatrix} A^\dagger & X \\ 0 & B^\dagger \end{bmatrix}$. By the four Penrose equations, this is equivalent with the fact that the following system of the equations is solvable:

$$
\begin{aligned}
&AX + CB^\dagger = 0,\\
&A^\dagger C + XB = 0,\\
&AA^\dagger C + AXB + CB^\dagger B = C,\\
&A^\dagger AX + A^\dagger CB^\dagger + XBB^\dagger = X.
\end{aligned} \tag{1.39}
$$

We can check that the system (1.39) has a solution X if and only if $B^\dagger BC^* = C^*$ and $AA^\dagger C = C$ which is equivalent with $\mathcal{R}(C^*) \subseteq \mathcal{R}(B^*)$, $\mathcal{R}(C) \subseteq \mathcal{R}(A)$.

On the other hand, the fact that for some $C \in \mathcal{B}(\mathcal{Y}, \mathcal{X})$, there exists Y such that the operator matrix M_C has the Moore-Penrose inverse of the form $\begin{bmatrix} A^\dagger & 0 \\ Y & B^\dagger \end{bmatrix}$ is equivalent with the fact that the following system of the equations is solvable:

$$
\begin{aligned}
&(CB^\dagger)^* = BY,\ CY = (CY)^*,\ YC = (YC)^*,\\
&(A^\dagger C)^* = YA,\ CYA = 0,\ BYC = 0,\ A^\dagger CB^\dagger = 0,\\
&AA^\dagger C + CYC + CB^\dagger B = C,\\
&YAA^\dagger + YCY + B^\dagger BY = Y.
\end{aligned} \tag{1.40}
$$

Suppose that there exists Y which satisfies (1.40). Then, using that $(CB^\dagger)^* = BY$, we get that $BYC = 0$ is equivalent with $CB^\dagger = 0$ which implies that $B^\dagger BY = 0$. Similarly, using that $(A^\dagger C)^* = YA$, we get that $CYA = 0$ is equivalent with $A^\dagger C = 0$ which implies that $YAA^\dagger = 0$. Now, the system (1.40) has the form:

$$
\begin{aligned}
&BY = 0,\ CY = (CY)^*,\ YC = (YC)^*,\\
&YA = 0,\ A^\dagger C = 0,\ CB^\dagger = 0,\\
&CYC = C,\\
&YCY = Y.
\end{aligned} \tag{1.41}
$$

It is clear that C must be regular and that $A^\dagger C = 0$ and $CB^\dagger = 0$. Furthermore, by (1.41) we get that $Y = C^\dagger$. ☐

1.4. Completions of upper triangular operator matrices to Drazin and generalized Drazin invertible operator

In this section we present certain results related to the Drazin and generalized Drazin completions of M_C as well as some properties of the Drazin and generalized Drazin spectrum of an upper triangular operator matrix. Throughout this section $\mathcal{H}, \mathcal{K}$ are infinite-dimensional separable complex Hilbert spaces. Also $K(0,\delta) = \{\lambda \in \mathbb{C} : |\lambda| < \delta\}$ stands for the open disc with center 0 and radius δ.

Let us first introduce some basic notations and properties of Drazin and generalized Drazin inverses. An operator $A \in \mathcal{B}(\mathcal{H})$ is left Drazin invertible if $\operatorname{asc}(A) < \infty$ and $\mathcal{R}(A^{\operatorname{asc}(A)+1})$ is closed. An operator $A \in \mathcal{B}(\mathcal{H})$ is right Drazin invertible if $\operatorname{dsc}(A) < \infty$ and $\mathcal{R}(A^{\operatorname{dsc}(A)})$ is closed. For interesting results on Drazin and generalized Drazin inverse, see [**42–45, 115, 116**].

The following lemma concerns some equivalent conditions for Drazin invertibility of an operator $A \in \mathcal{B}(\mathcal{X})$, where $\mathcal{X}$ is a Banach space. The equivalence of (i), (ii), and (iii) of the below lemma can be found in [**20, 21**] and the equivalence of (iii), (iv), and (v) is the results of Proposition 38.4 in [**102**].

LEMMA 1.26. *Let $A \in \mathcal{B}(\mathcal{X})$. The following statements are equivalent:*

(i) *A is Drazin invertible.*
(ii) *$A = A_1 \oplus A_2$, where A_1 is invertible and A_2 is nilpotent.*
(iii) *There exists a nonnegative integer k such that $\operatorname{dsc}(A) = \operatorname{asc}(A) = k < \infty$.*
(iv) *There exists a nonnegative integer k such that $\mathcal{X} = \mathcal{R}(A^k) \oplus \mathcal{N}(A^k)$ and*

$$A = \begin{bmatrix} A_{11} & 0 \\ 0 & A_{22} \end{bmatrix} : \mathcal{R}(A^k) \oplus \mathcal{N}(A^k) \to \mathcal{R}(A^k) \oplus \mathcal{N}(A^k),$$

where A_{11} is invertible and A_{22} is nilpotent.
(v) *There exists a nonnegative integer k such that $\mathcal{X} = \mathcal{R}(A^k) \oplus \mathcal{N}(A^k)$.*

Two completely different proofs of the following lemma that will be extensively used throughout this section can be found in [**87**] and [**161**]:

LEMMA 1.27. *For a Banach space $\mathcal{X}$, a given nonnegative integer m, and $A \in \mathcal{B}(\mathcal{X})$, the following conditions are equivalent:*

(i) $\operatorname{dsc}(A) \leq m < \infty$.
(ii) *$\mathcal{N}(A^m) + \mathcal{R}(A^n) = \mathcal{X}$, for every $n \in \mathbb{N}$.*
(iii) *$\mathcal{N}(A^m) + \mathcal{R}(A^n) = \mathcal{X}$, for some $n \in \mathbb{N}$.*

We also need the following result which is proved in [**127, 161**].

LEMMA 1.28. *Let $A \in \mathcal{B}(\mathcal{X})$. We have the following:*

(1) *If $\operatorname{dsc}(A) = m < \infty$, then there exists a constant $\delta > 0$ such that for every $\lambda \in K(0,\delta) \setminus \{0\}$:*
   (i) $\operatorname{dsc}(A - \lambda) = d(A - \lambda) = 0$.
   (ii) $\alpha(A - \lambda) = \dim(\mathcal{N}(A) \cap \mathcal{R}(A^m))$.
(2) *If $\operatorname{asc}(A) = m < \infty$ and $\mathcal{R}(A^{m+k})$ is closed for some $k \geq 1$, then there exists a constant $\delta > 0$ such that for every $\lambda \in K(0,\delta) \setminus \{0\}$:*
   (i) $\operatorname{asc}(A - \lambda) = \alpha(A - \lambda) = 0$.
   (ii) $d(A - \lambda) = \dim\left(\mathcal{R}(A^m)/\mathcal{R}(A^{m+1})\right) = \dim\left(\mathcal{X}/(\mathcal{R}(A) + \mathcal{N}(A^m))\right)$.

The following technical lemma will be used multiple times throughout the section.

LEMMA 1.29. *Suppose $B \in \mathcal{B}(\mathcal{K})$ and p is a positive integer such that $\mathcal{R}(B^p)$ is closed. If B is represented by*

$$(1.42) \qquad B = \begin{bmatrix} 0 & B_1 \\ 0 & B_2 \end{bmatrix} : \begin{bmatrix} \mathcal{N}(B) \cap \mathcal{R}(B^p) \\ (\mathcal{N}(B) \cap \mathcal{R}(B^p))^\perp \end{bmatrix} \to \begin{bmatrix} \mathcal{N}(B) \cap \mathcal{R}(B^p) \\ (\mathcal{N}(B) \cap \mathcal{R}(B^p))^\perp \end{bmatrix},$$

then B_1 and B_2 must satisfy the following conditions:

(i) *The restriction of $B_1 B_2^{p-1}$ on $\mathcal{N}(B_2^p)$ is onto (equivalently: the restriction of B_1 to the subspace $\mathcal{R}(B_2^{p-1}) \cap \mathcal{N}(B_2^p)$ is onto).*
(ii) $\mathcal{R}(B_2^p) \subseteq \mathcal{R}(B^p)$.
(iii) *$\mathcal{R}(B_2^p) \cap \mathcal{N}(B_1) \cap \mathcal{N}(B_2) = \{0\}$ (equivalently: the restriction of B_1 to the subspace $\mathcal{R}(B_2^p) \cap \mathcal{N}(B_2)$ is injective).*

PROOF. Put $\mathcal{S} = \mathcal{N}(B) \cap \mathcal{R}(B^p)$. To see that (i) is true, notice that if $y \in \mathcal{S}$, then $\begin{bmatrix} y \\ 0 \end{bmatrix} = \begin{bmatrix} B_1 B_2^{p-1} x \\ B_2^p x \end{bmatrix}$ for some $x \in \mathcal{S}^\perp$. To see that (ii) is true, notice that for any $x \in \mathcal{S}^\perp$ we have $\begin{bmatrix} 0 \\ B_2^p x \end{bmatrix} = \begin{bmatrix} B_1 B_2^{p-1} x \\ B_2^p x \end{bmatrix} - \begin{bmatrix} B_1 B_2^{p-1} x \\ 0 \end{bmatrix}$ and that by (i) we know that $\begin{bmatrix} B_1 B_2^{p-1} x \\ 0 \end{bmatrix} \in \mathcal{R}(B^p)$. Finally to show (iii), notice that if $y \in \mathcal{R}(B_2^p) \cap \mathcal{N}(B_1) \cap \mathcal{N}(B_2)$, then $y \in \mathcal{S}$ by (ii), and also $y \in \mathcal{S}^\perp$, so $y = 0$. □

In the next result given in [**177**] we obtain certain necessary conditions for the completion of M_C to Drazin invertibility.

LEMMA 1.30. *Let $A \in \mathcal{B}(\mathcal{H})$, $B \in \mathcal{B}(\mathcal{K})$ be given operators. If M_C is Drazin invertible for some $C \in \mathcal{B}(\mathcal{K}, \mathcal{H})$, then* $\operatorname{asc}(A) < \infty$ *and* $\operatorname{dsc}(B) < \infty$.

PROOF. If M_C is Drazin invertible, then there exists a nonnegative integer k such that $\operatorname{asc}(M_C) = \operatorname{dsc}(M_C) = k < \infty$. Hence $M_C^k(\mathcal{H} \oplus \mathcal{K}) = M_C^{k+1}(\mathcal{H} \oplus \mathcal{K})$. In order to prove $\operatorname{dsc}(B) < \infty$, we only need to show that $\mathcal{R}(B^k) \subseteq \mathcal{R}(B^{k+1})$. For any $u \in \mathcal{R}(B^k)$, let $u = B^k y$. Then $M_C^k(0, y) \in \mathcal{R}(M_C^{k+1})$. Thus there exists $(x_0, y_0) \in \mathcal{H} \oplus \mathcal{K}$ such that $(A^{k-1}Cy + A^{k-2}CBy + \cdots + CB^{k-1}y, B^k y) = M_C^k(0, y) = M_C^{k+1}(x_0, y_0) = (A^{k+1}x_0 + A^k C y_0 + A^{k-1}CBy_0 + \cdots + CB^k y_0, B^{k+1} y_0)$, then $u = B^k y = B^{k+1} y_0 \in \mathcal{R}(B^{k+1})$. Thus $\mathcal{R}(B^k) \subseteq \mathcal{R}(B^{k+1})$. This implies $\operatorname{dsc}(B) < \infty$. Since $\mathcal{N}(A^k) \oplus \{0\} \subseteq \mathcal{N}(M_C^k)$ and $\operatorname{asc}(M_C) < \infty$, it follows that $\operatorname{asc}(A) < \infty$. □

The following is a key lemma in the proof of the main result of this section and its motivation is the following: Suppose that $A \in \mathcal{B}(\mathcal{H})$ is a left Drazin invertible operator, $B \in \mathcal{B}(\mathcal{K})$ is an operator with finite descent, and suppose in addition that there exists a constant $\delta > 0$ such that $d(A - \lambda) = \alpha(B - \lambda)$, for every $\lambda \in K(0, \delta) \setminus \{0\}$. Note that if p is any integer with $p \geq \max\{\operatorname{asc}(A), \operatorname{dsc}(B)\}$, then by Lemma 1.28, $\mathcal{R}(A) + \mathcal{N}(A^p) = A^{-\operatorname{asc}(A)}[\mathcal{R}(A^{\operatorname{asc}(A)+1})]$ is a closed subspace of codimension equal to the dimension of the subspace $\mathcal{N}(B) \cap \mathcal{R}(B^p)$. Thus we can fix an invertible operator $J \in \mathcal{B}(\overline{\mathcal{N}(B) \cap \mathcal{R}(B^p)}, (\mathcal{R}(A) + \mathcal{N}(A^p))^\perp)$. Indeed, if $\mathcal{N}(B) \cap \mathcal{R}(B^p)$ is closed, then this is clear. If it is not, then it must be infinite dimensional and so must be the closed subspace $(\mathcal{R}(A) + \mathcal{N}(A^p))^\perp$. But then $\overline{\mathcal{N}(B) \cap \mathcal{R}(B^p)}$ and $(\mathcal{R}(A) + \mathcal{N}(A^p))^\perp$ are both infinite-dimensional separable Hilbert spaces and as such are isomorphic to one another.

LEMMA 1.31. *Let $A \in \mathcal{B}(\mathcal{H})$, $B \in \mathcal{B}(\mathcal{K})$ be given operators such that:*

(i) *A is left Drazin invertible.*
(ii) $\operatorname{dsc}(B) < \infty$.
(iii) *There exists a constant $\delta > 0$ such that $d(A - \lambda) = \alpha(B - \lambda)$, for every $\lambda \in K(0, \delta) \setminus \{0\}$.*

Let $C \in \mathcal{B}(\mathcal{K}, \mathcal{H})$ be given by

$$C = \begin{bmatrix} J & 0 \\ 0 & 0 \end{bmatrix} : \begin{bmatrix} \overline{\mathcal{N}(B) \cap \mathcal{R}(B^p)} \\ (\mathcal{N}(B) \cap \mathcal{R}(B^p))^\perp \end{bmatrix} \to \begin{bmatrix} (\mathcal{R}(A) + \mathcal{N}(A^p))^\perp \\ \mathcal{R}(A) + \mathcal{N}(A^p) \end{bmatrix}, \tag{1.43}$$

where $p \in \mathbb{N}$ is such that $p \geq \max\{\operatorname{asc}(A), \operatorname{dsc}(B)\}$ and $J \in \mathcal{B}(\overline{\mathcal{N}(B) \cap \mathcal{R}(B^p)}$,

$(\mathcal{R}(A)+\mathcal{N}(A^p))^\perp)$ *is any invertible operator. The following are equivalent:*

(i) $\operatorname{dsc}(M_C)\le p$.

(ii) *For any* $x\in\mathcal{H}$ *and* $y\in\mathcal{K}$, *there exist* $x'\in\mathcal{H}$ *and* $y'\in\mathcal{K}$ *such that*

$$A^px = A^{p+1}x' + A^pCy' \tag{1.44}$$

and

$$y - By' \in \mathcal{N}(C)\cap\mathcal{N}(CB)\cap\cdots\cap\mathcal{N}(CB^{p-1})\cap\mathcal{N}(B^p). \tag{1.45}$$

(iii) $\mathcal{K}=\mathcal{R}(B)+\mathcal{N}(C)\cap\mathcal{N}(CB)\cap\mathcal{N}(CB^2)\cap\cdots\cap\mathcal{N}(CB^{p-1})\cap\mathcal{N}(B^p)$.

PROOF. (i) $\Leftrightarrow$ (ii). Since for any $k\in\mathbb{N}$,

$$M_C^k=\begin{bmatrix} A^k & A^{k-1}C+A^{k-2}CB+\cdots+ACB^{k-2}+CB^{k-1}\\ 0 & B^k\end{bmatrix}:\begin{bmatrix}\mathcal{H}\\ \mathcal{K}\end{bmatrix}\to\begin{bmatrix}\mathcal{H}\\ \mathcal{K}\end{bmatrix},$$

it follows that $\operatorname{dsc}(M_C)\le p$ if and only if for any $x\in\mathcal{H}$ and $y\in\mathcal{K}$, there exist $x'\in\mathcal{H}$ and $y'\in\mathcal{K}$ such that

$$\begin{aligned} &A^px+A^{p-1}Cy+A^{p-2}CBy+\cdots+ACB^{p-2}y+CB^{p-1}y\\ &\quad=A^{p+1}x'+A^pCy'+A^{p-1}CBy'+\cdots+ACB^{p-1}y'+CB^py'\\ &\text{and}\\ &B^py=B^{p+1}y'. \end{aligned}\tag{1.46}$$

The case $p=1$ is evident, so suppose that $p>1$. If we suppose that $\operatorname{dsc}(M_C)\le p$, by the second equality in (1.46) we get that $y-By'\in\mathcal{N}(B^p)$. Since $\mathcal{R}(C)\subseteq\mathcal{R}(A)^\perp$, by the first equality in (1.46) we get that $y-By'\in\mathcal{N}(CB^{p-1})$ and

$$\begin{aligned} &A^px+A^{p-1}Cy+A^{p-2}CBy+\cdots+ACB^{p-2}y\\ &\quad=A^{p+1}x'+A^pCy'+A^{p-1}CBy'+\cdots+ACB^{p-1}y'. \end{aligned}\tag{1.47}$$

By (1.47), we have that

$$A^{p-1}x+A^{p-2}Cy+\cdots+CB^{p-2}y-(A^px'+A^{p-1}Cy'+\cdots+CB^{p-1}y')\in\mathcal{N}(A)\subseteq\mathcal{N}(A^p)$$

which implies that $CB^{p-2}y-CB^{p-1}y'\in\mathcal{N}(A^p)+\mathcal{R}(A)$; i.e., $y-By'\in\mathcal{N}(CB^{p-2})$. Continuing in the same manner, we get that (1.45) holds. Now, by (1.46) it follows that (1.44) is also satisfied.

If (ii) holds, then evidently (1.46) is satisfied; i.e., $\operatorname{dsc}(M_C)\le p$.

(ii) $\Rightarrow$ (iii). Evidently (1.45) implies (iii).

(iii) $\Rightarrow$ (ii). Let $x\in\mathcal{H}$ and $y\in\mathcal{K}$ be arbitrary. Then there exists $y_0\in\mathcal{K}$ such that

$$y-By_0\in\mathcal{N}(C)\cap\mathcal{N}(CB)\cap\cdots\cap\mathcal{N}(CB^{p-1})\cap\mathcal{N}(B^p).$$

Let $\mathcal{S}=\overline{\mathcal{R}(A)+\mathcal{N}(A^p)}$. By the definition of the operator C, for given x there exists $y_{00}\in\mathcal{N}(B)\cap\mathcal{R}(B^p)$ such that $(I-P_\mathcal{S})x=Jy_{00}=Cy_{00}$. Since $\mathcal{N}(B)$ is closed we have $By_{00}=0$. Define $y'=P_{\mathcal{N}(B)^\perp}y_0+y_{00}$. Then $By'=By_0$ and $Cy'=Cy_{00}$ which implies that

$$y-By'\in\mathcal{N}(C)\cap\mathcal{N}(CB)\cap\cdots\cap\mathcal{N}(CB^{p-1})\cap\mathcal{N}(B^p)$$

and that

$$(I-P_\mathcal{S})x=Cy'. \tag{1.48}$$

Now, $A^p x = A^p C y' + A^p P_{\mathcal{S}} x$. Since $P_{\mathcal{S}} x \in \mathcal{R}(A) + \mathcal{N}(A^p)$ it follows that $A^p P_{\mathcal{S}} x \in \mathcal{R}(A^{p+1})$, so there exists $x' \in \mathcal{H}$ such that $A^p P_{\mathcal{S}} x = A^{p+1} x'$. Now,

$$A^p x = A^{p+1} x' + A^p C y'. \qquad \square$$

The existence of Drazin invertible completions of M_C was addressed in [**14**] where some sufficient conditions for it were given.

THEOREM 1.32 ([**14**]). *Let $\mathcal{H}$ and $\mathcal{K}$ be separable Hilbert spaces and let $A \in \mathcal{B}(\mathcal{H})$ and $B \in \mathcal{B}(\mathcal{K})$ be given operators such that:*

(i) *A is left Drazin invertible.*
(ii) *B is right Drazin invertible.*
(iii) *There exists a constant $\delta > 0$ such that $d(A-\lambda) = \alpha(B-\lambda)$, for every $\lambda \in K(0,\delta) \setminus \{0\}$.*

Then there exists an operator $C \in \mathcal{B}(\mathcal{K},\mathcal{H})$ such that M_C is Drazin invertible.

The proof of this result presented in [**14**] is not correct, although the result is true. In this section we present a correct proof that is published in [**52**].

Let us first make clear which conditions on the operators A and B are necessary for the existence of some $C \in \mathcal{B}(\mathcal{K},\mathcal{H})$ such that the operator M_C is Drazin invertible. Combining Lemmas 1.28 and 1.30 and Theorem 1.32, we obtain the following result:

THEOREM 1.33. *Let $A \in \mathcal{B}(\mathcal{H})$ and $B \in \mathcal{B}(\mathcal{K})$ be given operators. If there exists an operator $C \in \mathcal{B}(\mathcal{K},\mathcal{H})$ such that M_C is Drazin invertible, then the following hold:*

(i) $\operatorname{asc}(A) < \infty$.
(ii) $\operatorname{dsc}(B) < \infty$.
(iii) *There exists a constant $\delta > 0$ such that $A-\lambda$ is left invertible, $B-\lambda$ is right invertible, and*

$$d(A-\lambda) = \alpha(B-\lambda) = \dim(\mathcal{N}(B) \cap \mathcal{R}(B^{\operatorname{dsc}(B)})),$$

for every $\lambda \in K(0,\delta) \setminus \{0\}$.

In the sequel, we show that the three conditions above together with the assumption that both subspaces $\mathcal{R}(A^{\operatorname{asc}(A)+1})$ and $\mathcal{R}(B^{\operatorname{dsc}(B)})$ are closed (thus meaning that A is left Drazin invertible and B is right Drazin invertible) are actually sufficient for the existence of a Drazin completion of the operator matrix in question. We show that the operator C as defined in (1.43) by the authors of [**14**] indeed does the trick. To properly show that, we first give an equivalent description of when exactly the operator M_C is Drazin invertible for this particular choice of C.

THEOREM 1.34. *Let $A \in \mathcal{B}(\mathcal{H})$, $B \in \mathcal{B}(\mathcal{K})$ be given operators such that:*

(i) *A is left Drazin invertible.*
(ii) $\operatorname{dsc}(B) < \infty$.
(iii) *There exists a constant $\delta > 0$ such that $d(A-\lambda) = \alpha(B-\lambda)$, for every $\lambda \in K(0,\delta) \setminus \{0\}$.*

Then M_C is Drazin invertible for $C \in \mathcal{B}(\mathcal{K},\mathcal{H})$ given by (1.43) if and only if

$$\mathcal{K} = \mathcal{R}(B) + \mathcal{N}(C) \cap \mathcal{N}(CB) \cap \mathcal{N}(CB^2) \cap \cdots \cap \mathcal{N}(CB^{p-1}) \cap \mathcal{N}(B^p).$$

PROOF. First, we prove that such M_C has finite ascent. Let $\begin{bmatrix} x \\ y \end{bmatrix} \in \mathcal{N}(M_C^{p+1})$. Then

$$\left\{ \begin{array}{l} A^{p+1}x + A^pCy + A^{p-1}CBy + \cdots + ACB^{p-1}y + CB^py = 0, \\ B^{p+1}y = 0. \end{array} \right. \tag{1.49}$$

Thus $B^py \in \mathcal{N}(B) \cap \mathcal{R}(B^p)$ and $CB^py \in (\mathcal{R}(A) + \mathcal{N}(A^p))^\perp \subseteq \mathcal{R}(A)^\perp$. It follows that

$$A^{p+1}x + A^pCy + A^{p-1}CBy + \cdots + ACB^{p-1}y = -CB^py = 0.$$

Since J is isometric and $CB^py = JB^py = 0$, then $B^py = 0$. From the first equality in (1.49), we get $A^{p+1}x + A^pCy + A^{p-1}CBy + \cdots + ACB^{p-1}y = 0$; thus, $A^px + A^{p-1}Cy + A^{p-2}CBy + \cdots + CB^{p-1}y \in \mathcal{N}(A)$. This implies that $CB^{p-1}y \in \mathcal{R}(A) + \mathcal{N}(A^p)$; hence $CB^{p-1}y = 0$. Therefore,

$$A^{p-1}x + A^{p-2}Cy + A^{p-3}CBy + \cdots + ACB^{p-3}y + CB^{p-2}y \in \mathcal{N}(A^2).$$

Continuing this process we obtain $Ax + Cy \in \mathcal{N}(A^p)$. Thus $Cy \in \mathcal{R}(A) + \mathcal{N}(A^p)$, which implies that $Cy = 0$ and so $Ax \in \mathcal{N}(A^p)$; that is, $x \in \mathcal{N}(A^{p+1}) = \mathcal{N}(A^p)$. Finally, we get $\begin{bmatrix} x \\ y \end{bmatrix} \in \mathcal{N}(M_C^p)$. Hence, $\operatorname{asc}(M_C) \leq p$. Thus we can conclude that M_C is Drazin invertible if and only if $\operatorname{dsc}(M_C) \leq p$. Now the assertion follows by Lemma 1.31. □

REMARK 1.35. If $B \in \mathcal{B}(\mathcal{K})$ is right Drazin invertible and is given by (1.42), where $p = \operatorname{dsc}(B)$, and if $C \in \mathcal{B}(\mathcal{K}, \mathcal{H})$ is given by (1.43), then

$$\begin{aligned} &\mathcal{N}(C) \cap \mathcal{N}(CB) \cap \mathcal{N}(CB^2) \cap \cdots \cap \mathcal{N}(CB^{p-1}) \cap \mathcal{N}(B^p) \\ &\quad = \mathcal{N}(B_1) \cap \mathcal{N}(B_1B_2) \cap \cdots \cap \mathcal{N}(B_1B_2^{p-1}) \cap \mathcal{N}(B_2^p). \end{aligned} \tag{1.50}$$

Indeed, this is a consequence of the following equalities:

$$\mathcal{N}(C) = (\mathcal{N}(B) \cap \mathcal{R}(B^p))^\perp, \ \mathcal{N}(CB^k) = (\mathcal{N}(B) \cap \mathcal{R}(B^p)) \oplus \mathcal{N}(B_1B_2^{k-1}),$$

the latter of which follows from the representation

$$CB^k = \begin{bmatrix} 0 & JB_1B_2^{k-1} \\ 0 & 0 \end{bmatrix} : \begin{bmatrix} \mathcal{N}(B) \cap \mathcal{R}(B^p) \\ (\mathcal{N}(B) \cap \mathcal{R}(B^p))^\perp \end{bmatrix} \to \begin{bmatrix} (\mathcal{R}(A) + \mathcal{N}(A^p))^\perp \\ \mathcal{R}(A) + \mathcal{N}(A^p) \end{bmatrix}.$$

In the following theorem (see [**52**]), the proof of Theorem 1.32 presented in [**14**] is corrected. Since we make use of Lemma 1.29, in contrast to the previous auxiliary results here we must assume that B is right Drazin invertible.

THEOREM 1.36. *Let $A \in \mathcal{B}(\mathcal{H})$, $B \in \mathcal{B}(\mathcal{K})$ be given operators such that:*

(i) *A is left Drazin invertible.*
(ii) *B is right Drazin invertible.*
(iii) *There exists a constant $\delta > 0$ such that $d(A - \lambda) = \alpha(B - \lambda)$, for every $\lambda \in K(0, \delta) \setminus \{0\}$.*

Then M_C is Drazin invertible for $C \in \mathcal{B}(\mathcal{K}, \mathcal{H})$ given by (1.43).

PROOF. Let B be given by (1.42). Suppose first that $p = 1$. By Theorem 1.34, to prove that M_C is Drazin invertible for $C \in \mathcal{B}(\mathcal{K}, \mathcal{H})$ given by (1.43) it is sufficient to prove that $\mathcal{K} = \mathcal{R}(B) + \mathcal{N}(C) \cap \mathcal{N}(B)$. Since $\operatorname{dsc}(B) = 1$, by Lemma

1.27 it follows that $\mathcal{K} = \mathcal{R}(B) + \mathcal{N}(B)$. Put $\mathcal{S} = \mathcal{N}(B) \cap \mathcal{R}(B)$. As $\mathcal{N}(B) = \mathcal{S} \oplus \mathcal{N}(B_1) \cap \mathcal{N}(B_2)$ and $\mathcal{S} \subseteq \mathcal{R}(B)$, it follows that $\mathcal{K} = \mathcal{R}(B) + \mathcal{N}(B_1) \cap \mathcal{N}(B_2)$. Since $\mathcal{N}(C) \cap \mathcal{N}(B) = \mathcal{N}(B_1) \cap \mathcal{N}(B_2)$, we have $\mathcal{K} = \mathcal{R}(B) + \mathcal{N}(C) \cap \mathcal{N}(B)$.

Now, consider the case when $p > 1$. By Theorem 1.34, we have to prove that

$$\mathcal{K} = \mathcal{R}(B) + \mathcal{N}(C) \cap \mathcal{N}(CB) \cap \mathcal{N}(CB^2) \cap \cdots \cap \mathcal{N}(CB^{p-1}) \cap \mathcal{N}(B^p)$$

which is by (1.50) from the preceding remark equivalent with

$$\mathcal{K} = \mathcal{R}(B) + \mathcal{N}(B_1) \cap \mathcal{N}(B_1B_2) \cap \cdots \cap \mathcal{N}(B_1B_2^{p-1}) \cap \mathcal{N}(B_2^p).$$

Since $\mathcal{K} = \mathcal{R}(B) + \mathcal{N}(B^p)$, which is equivalent with

$$\mathcal{K} = \mathcal{R}(B) + \mathcal{N}(B_1B_2^{p-1}) \cap \mathcal{N}(B_2^p),$$

it is sufficient to prove that

$$\begin{aligned}&\mathcal{N}(B_1B_2^{p-1}) \cap \mathcal{N}(B_2^p)\\ &\quad\subseteq \mathcal{R}(B) + \mathcal{N}(B_1) \cap \mathcal{N}(B_1B_2) \cap \cdots \cap \mathcal{N}(B_1B_2^{p-1}) \cap \mathcal{N}(B_2^p).\end{aligned}$$

Take arbitrary $x \in \mathcal{N}(B_1B_2^{p-1}) \cap \mathcal{N}(B_2^p)$. Now $B_1B_2^{p-2} \in \mathcal{B}((\mathcal{N}(B) \cap \mathcal{R}(B^p))^\perp, \mathcal{N}(B) \cap \mathcal{R}(B^p))$, so $B_1B_2^{p-2}x \in \mathcal{N}(B) \cap \mathcal{R}(B^p)$. Lemma 1.29 says that the operator $B_1B_2^{p-1} \in \mathcal{B}((\mathcal{N}(B) \cap \mathcal{R}(B^p))^\perp, \mathcal{N}(B) \cap \mathcal{R}(B^p))$ maps the subspace $\mathcal{N}(B_2^p)$ onto $\mathcal{N}(B) \cap \mathcal{R}(B^p)$. Hence there exists $y \in \mathcal{N}(B_2^p)$ such that $B_1B_2^{p-2}x = B_1B_2^{p-1}y$. Now, $x - B_2y \in \mathcal{N}(B_1B_2^{p-2}) \cap \mathcal{N}(B_1B_2^{p-1}) \cap \mathcal{N}(B_2^p)$ which together with (ii) of Lemma 1.29 gives that $x \in \mathcal{R}(B) + \mathcal{N}(B_1B_2^{p-2}) \cap \mathcal{N}(B_1B_2^{p-1}) \cap \mathcal{N}(B_2^p)$. We have thus shown that $\mathcal{N}(B_1B_2^{p-1}) \cap \mathcal{N}(B_2^p) \subseteq \mathcal{R}(B) + \mathcal{N}(B_1B_2^{p-2}) \cap \mathcal{N}(B_1B_2^{p-1}) \cap \mathcal{N}(B_2^p)$. Continuing in the same manner we further obtain consecutively

$$\begin{aligned}&\mathcal{N}(B_1B_2^{p-2}) \cap \mathcal{N}(B_1B_2^{p-1}) \cap \mathcal{N}(B_2^p)\\ &\quad\subseteq \mathcal{R}(B) + \mathcal{N}(B_1B_2^{p-3}) \cap \mathcal{N}(B_1B_2^{p-2}) \cap \mathcal{N}(B_1B_2^{p-1}) \cap \mathcal{N}(B_2^p),\end{aligned}$$

..., and finally

$$\begin{aligned}&\mathcal{N}(B_1B_2) \cap \cdots \cap \mathcal{N}(B_1B_2^{p-1}) \cap \mathcal{N}(B_2^p)\\ &\quad\subseteq \mathcal{R}(B) + \mathcal{N}(B_1) \cap \mathcal{N}(B_1B_2) \cap \cdots \cap \mathcal{N}(B_1B_2^{p-1}) \cap \mathcal{N}(B_2^p).\end{aligned}$$

Taking into account all these inclusions, we immediately get (1.51). □

As in the case of ordinary invertibility for which the filling-in-hole problem of 2×2 upper triangular operator matrices was considered in [**98**] and it was showed that if $M_C = \begin{bmatrix} A & C \\ 0 & B \end{bmatrix}$ is an operator acting on the Banach space $\mathcal{X} \oplus \mathcal{Y}$, then

$$\sigma(A) \cup \sigma(B) = \sigma(M_C) \cup W(A, B, C), \tag{1.51}$$

where $W(A, B, C)$ is a subset of $\sigma(A) \cap \sigma(B)$ and a union of certain holes in $\sigma(M_C)$, we have a similar result for the Drazin and generalized Drazin invertibility:

THEOREM 1.37 ([**177**]). *For a given pair (A, B) of operators and for every $C \in \mathcal{B}(\mathcal{Y}, \mathcal{X})$, the following holds:*

$$\sigma_D(A) \cup \sigma_D(B) = \sigma_D(M_C) \cup W_D,$$

where W_D is the union of certain holes in $\sigma_D(M_C)$, which happen to be subsets of $\sigma_D(A) \cap \sigma_D(B)$.

Also, in **[177]** we can find a result that concerns $\bigcap_{C\in\mathcal{B}(Y,X)}\sigma_D(M_C)$:

THEOREM 1.38 ([**177**]). *For a given pair (A, B) of operators, we have:*

(a) $\bigcap_{C\in\mathcal{B}(\mathcal{Y},\mathcal{X})}\sigma_D(M_C)\subseteq\left(\bigcap_{C\in\mathcal{B}(\mathcal{Y},\mathcal{X})}\sigma(M_C)\right)\setminus[\rho_D(A)\cap\rho_D(B)]$.

(b) *In particular, if one of the conditions*

(i) $\sigma_{su}(B)=\sigma(B)$, (iii) $\sigma(A)\cap\sigma(B)=\emptyset$,

(ii) $\operatorname{int}(\sigma_p(B))=\emptyset$, (iv) $\sigma_a(A)=\sigma(A)$

holds, then we have $\bigcap_{C\in\mathcal{B}(\mathcal{Y},\mathcal{X})}\sigma_D(M_C)=\left(\bigcap_{C\in\mathcal{B}(\mathcal{Y},\mathcal{X})}\sigma(M_C)\right)\setminus[\rho_D(A)\cap\rho_D(B)]$.

Notice that if either $\operatorname{acc}(\sigma(B))=\emptyset$ or $\operatorname{int}(\sigma(B))=\emptyset$ holds, then $\operatorname{int}(\sigma_p(B))=\emptyset$, and if either $\operatorname{acc}(\sigma(A))=\emptyset$ or $\operatorname{int}(\sigma(A))=\emptyset$, then $\sigma_a(A)=\sigma(A)$. Using these two facts and Theorem 1.38, we obtain the following result:

COROLLARY 1.39 ([**177**]). For a given pair (A, B) of operators, if any one of the conditions

(i) $\operatorname{acc}(\sigma(B))=\emptyset$, (iii) $\operatorname{acc}(\sigma(A))=\emptyset$,

(ii) $\int\sigma(B)=\emptyset$, (iv) $\operatorname{int}(\sigma(A))=\emptyset$

is satisfied, then we have

$$\bigcap_{C\in\mathcal{B}(\mathcal{Y},\mathcal{X})}\sigma_D(M_C)=\left(\bigcap_{C\in\mathcal{B}(\mathcal{Y},\mathcal{X})}\sigma(M_C)\right)\setminus[\rho_D(A)\cap\rho_D(B)].$$

That the above equality does not always hold even on complex infinite separable dimensional Hilbert spaces can be shown by the following examples:

EXAMPLE 1.40 ([**177**]). Define operators $T, S, C\in\mathcal{B}(l_2)$ by

$$\begin{aligned}T(x_1,x_2,x_3,\dots)&=(0,x_1,x_2,x_3,\dots),\\S(x_1,x_2,x_3,\dots)&=(x_2,x_3,x_4,\dots),\\C(x_1,x_2,x_3,\dots)&=(x_1,0,0,\dots),\end{aligned}$$

for any $(x_1,x_2,x_3,\dots)\in l_2$. Let

$$\begin{aligned}M&=\begin{bmatrix}T&0\\0&S_1\end{bmatrix}:l_2\oplus(l_2\oplus l_2)\to l_2\oplus(l_2\oplus l_2)\\&=\begin{bmatrix}T&0&0\\0&S&0\\0&0&0\end{bmatrix}:l_2\oplus(l_2\oplus l_2)\to l_2\oplus(l_2\oplus l_2),\end{aligned}$$

where $S_1=\begin{bmatrix}S&0\\0&0\end{bmatrix}:l_2\oplus l_2\to l_2\oplus l_2$.

Then we can claim that

$$0\in\bigcap_{Q\in\mathcal{B}(l_2\oplus l_2,l_2)}\sigma(M_Q)\setminus\left(\rho_D(T)\bigcap\rho_D(S_1)\right),\text{ but }0\notin\bigcap_{Q\in\mathcal{B}(l_2\oplus l_2,l_2)}\sigma_D(M_Q).$$

Thus,

$$\bigcap_{Q\in\mathcal{B}(l_2\oplus l_2,l_2)}\sigma_D(M_Q)\neq\left(\bigcap_{Q\in\mathcal{B}(l_2\oplus l_2,l_2)}\sigma(M_Q)\right)\setminus\left(\rho_D(T)\bigcap\rho_D(S_1)\right).$$

In fact,

(i) Since M_Q cannot be surjective for any $Q \in \mathcal{B}(l_2 \oplus l_2, l_2)$, then $M_Q = \begin{bmatrix} T & Q \\ 0 & S_1 \end{bmatrix}$ is not invertible. This implies $0 \in \bigcap_{Q \in \mathcal{B}(l_2 \oplus l_2, l_2)} \sigma(M_Q)$.

(ii) Since $\mathrm{dsc}(T) = \mathrm{asc}(S_1) = \infty$, then neither S_1 nor T is Drazin invertible; that is, $0 \notin (\rho_D(T) \cup \rho_D(S_1))$. Hence $0 \notin \rho_D(T) \cap \rho_D(S_1)$.

(iii) Let $M_{C_1} = \begin{bmatrix} T & C_1 \\ 0 & S_1 \end{bmatrix} : l_2 \oplus (l_2 \oplus l_2) \to l_2 \oplus (l_2 \oplus l_2) = \begin{bmatrix} T & C & 0 \\ 0 & S & 0 \\ 0 & 0 & 0 \end{bmatrix} :$ $l_2 \oplus l_2 \oplus l_2 \to l_2 \oplus l_2 \oplus l_2$, where $C_1 = [C \quad 0] : l_2 \oplus l_2 \to l_2$. Then M_{C_1} is Drazin invertible. Indeed, it is well known that $\begin{bmatrix} T & C \\ 0 & S \end{bmatrix}$ is unitary and hence $\begin{bmatrix} T & C \\ 0 & S \end{bmatrix}$ is invertible. Thus, it follows that M_{C_1} is Drazin invertible from part (iv) of Lemma 1.26. This implies that $0 \notin \bigcap_{Q \in \mathcal{B}(l_2 \oplus l_2, l_2)} \sigma_D(M_Q)$.

In the case of generalized Drazin invertibility we have the following appropriate results:

LEMMA 1.41 ([**178**]). *If any two of operators A, B, and M_C are generalized Drazin invertible, then so is the third.*

As in the cases of ordinary and Drazin invertibility, we have an analogous result for generalized Drazin invertibility:

THEOREM 1.42 ([**178**]). *For a given pair (A, B) of operators, we have*

$$\sigma_{gD}(A) \cup \sigma_{gD}(B) = \sigma_{gD}(M_C) \cup W_{gD}(A, B, C), \tag{1.52}$$

where $W_{gD}(A, B, C)$ is a subset of $\sigma_{gD}(A) \cap \sigma_{gD}(B)$ and a union of certian holes in $\sigma_{gD}(M_C)$.

THEOREM 1.43 ([**178**]). *For given $A \in \mathcal{B}(\mathcal{X})$, $B \in \mathcal{B}(\mathcal{Y})$, and $C \in \mathcal{B}(\mathcal{Y}, \mathcal{X})$,*

$$W(A, B, C) \subseteq W_{gD}(A, B, C),$$

where $W(A, B, C)$ and $W_{gD}(A, B, C)$ are denoted as in equalities (1.51) and (1.52), respectively.

From Theorem 1.42, we obtain immediately the following two results.

COROLLARY 1.44 ([**178**]). If $\sigma_{gD}(A) \cap \sigma_{gD}(B)$ has no interior points, then for every $C \in \mathcal{B}(\mathcal{Y}, \mathcal{X})$,

$$\sigma_{gD}(M_C) = \sigma_{gD}(A) \cup W_{gD}(B). \tag{1.53}$$

In particular, if either $A \in \mathcal{B}(\mathcal{X})$ or $B \in \mathcal{B}(\mathcal{Y})$ is a Riesz operator, then (1.53) holds.

The inclusion in Theorem 1.43 may be strict in general. To see this, consider the following simple example.

EXAMPLE 1.45 ([**178**]). Define operators $A, B_1, C_1 \in \mathcal{B}(l_2)$ as

$$A(x_1, x_2, x_3, \dots) = (0, x_1, x_2, x_3, \dots),$$
$$B_1(x_1, x_2, x_3, \dots) = (x_2, x_3, x_4, \dots),$$
$$C_1(x_1, x_2, x_3, \dots) = (x_1, 0, 0, \dots),$$

for any $(x_1, x_2, x_3, \dots) \in l_2$. Let

$$M_C = \begin{bmatrix} A & C \\ 0 & B \end{bmatrix} : l_2 \oplus (l_2 \oplus l_2) \to l_2 \oplus (l_2 \oplus l_2),$$

i.e.,

$$M_C = \begin{bmatrix} A & C_1 & 0 \\ 0 & B_1 & 0 \\ 0 & 0 & 0 \end{bmatrix} : l_2 \oplus l_2 \oplus l_2 \to l_2 \oplus l_2 \oplus l_2,$$

where

$$B = \begin{bmatrix} B_1 & 0 \\ 0 & 0 \end{bmatrix} : l_2 \oplus l_2 \to l_2 \oplus l_2, \; C = \begin{bmatrix} C_1, 0 \end{bmatrix} : l_2 \oplus l_2 \to l_2.$$

A direct calculation shows the following:

(i) $\sigma(M_C) = \{\lambda \in \mathbb{C} \colon |\lambda| = 1\} \cup \{0\}$, $\sigma(A) = \sigma(B) = \{\lambda \in \mathbb{C} \colon |\lambda| \leq 1\}$.

(ii) $\sigma_{gD}(M_C) = \{\lambda \in \mathbb{C} \colon |\lambda| = 1\}$, $\sigma_{gD}(A) = \sigma_{gD}(B) = \{\lambda \in \mathbb{C} \colon |\lambda| \leq 1\}$.

Accordingly,

$$W(A, B, C) = \{\lambda \in \mathbb{C} \colon 0 < |\lambda| < 1\}, \quad W_{gD}(A, B, C) = \{\lambda \in \mathbb{C} \colon |\lambda| < 1\}.$$

Thus,

$$W(A, B, C) \neq W_{gD}(A, B, C).$$

In spite of Example 1.45, we have the following result:

THEOREM 1.46 ([**178**]). *Let $A \in \mathcal{B}(X)$, $B \in \mathcal{B}(Y)$, and $C \in \mathcal{B}(Y, X)$ be given. If* $\operatorname{iso}(r(W(A, B, C))) = \emptyset$, *then*

$$W(A, B, C) = W_{gD}(A, B, C).$$

It is noteworthy that some interesting results on the completion of an upper triangular matrix to a co-EP matrix can be found in [**180**].

1.5. Completions of upper triangular operator matrices to Fredholm operator

Left (right) semi-Fredholmness of an operator matrix M_C was studied in [**30**]. It is proved that for given operators $A \in \mathcal{B}(\mathcal{H})$ and $B \in \mathcal{B}(\mathcal{K})$ there exists $C \in \mathcal{B}(\mathcal{K}, \mathcal{H})$ such that M_C is left semi-Fredholm if and only if A is left semi-Fredholm and one of the following conditions holds:

1. $\mathcal{R}(B)$ is closed, and one of the conditions $\big(\alpha(B) < \infty,\, \alpha(B) = \beta(A)\big)$ is satisfied.
2. $\mathcal{R}(B)$ is not closed; $\beta(A) = \infty$.

In this section, we present recent results published in [**48**] where, using a method which is completely different than the one used in [**30**], we find different conditions than those given in [**30**] for existence of $C \in \mathcal{B}(\mathcal{K}, \mathcal{H})$ such that M_C is left (right) semi-Fredholm. Furthermore, for the first time we completely describe the set of all such $C \in \mathcal{B}(\mathcal{K}, \mathcal{H})$, for given operators $A \in \mathcal{B}(\mathcal{H})$ and $B \in \mathcal{B}(\mathcal{K})$. The symbol $S_{LF}(A, B)(S_{RF}(A, B))$ stands for the set of all C such that M_C is left (right) semi-Fredholm. If $\mathcal{M}$ is a closed subspace of $\mathcal{H}$, we use the symbol $P_{\mathcal{M}}^{\mathtt{rst}}$ to denote an operator from $\mathcal{B}(\mathcal{H}, \mathcal{M})$ defined by $P_{\mathcal{M}}^{\mathtt{rst}} x = P_{\mathcal{M}} x$, for all $x \in \mathcal{H}$. By $\mathcal{H} = \mathcal{M} \oplus \mathcal{N}$ we shall mean that $\mathcal{M}$ and $\mathcal{N}$ are closed subspaces of $\mathcal{H}$, the direct sum of which is $\mathcal{H}$; in that case by $P_{\mathcal{M},\mathcal{N}}$ we denote the projector onto $\mathcal{M}$ parallel to $\mathcal{N}$, and

$P^{\mathtt{rst}}_{\mathcal{M},\mathcal{N}} \in \mathcal{B}(\mathcal{H},\mathcal{M})$ is defined analogously as $P^{\mathtt{rst}}_{\mathcal{M}}$ above. In this section $\mathcal{H}$ and $\mathcal{K}$ are separable Hilbert spaces.

The following lemma will be auxiliary in proving the main result of this section:

LEMMA 1.47. *Let $A \in \mathcal{B}(\mathcal{H})$, $B \in \mathcal{B}(\mathcal{K})$, and $C \in \mathcal{B}(\mathcal{K},\mathcal{H})$ be given operators. Then M_C is left semi-Fredholm if and only if A and* $\begin{bmatrix} P^{\mathtt{rst}}_{\mathcal{R}(A)^\perp} C \\ B \end{bmatrix} : \mathcal{K} \to \mathcal{R}(A)^\perp \oplus \mathcal{K}$ *are left semi-Fredholm operators.*

PROOF. Without loss of generality, we can suppose that M_C has an operator matrix representation

$$M_C = \begin{bmatrix} A_1 & C_1 \\ 0 & C_2 \\ 0 & B \end{bmatrix} : \mathcal{H} \oplus \mathcal{K} \to \overline{\mathcal{R}(A)} \oplus \overline{\mathcal{R}(A)}^\perp \oplus \mathcal{K},$$

where $A_1 : \mathcal{H} \to \overline{\mathcal{R}(A)}$ is with dense range. If M_C is left semi-Fredholm, then, evidently, A is left semi-Fredholm. Let $(A_1)_r^{-1} : \mathcal{R}(A) \to \mathcal{H}$ be a right inverse of A_1. Then there is an invertible operator

$$V = \begin{bmatrix} I & -(A_1)_r^{-1} C_1 \\ 0 & I \end{bmatrix} : \mathcal{H} \oplus \mathcal{K} \to \mathcal{H} \oplus \mathcal{K}$$

such that

$$M_C V = \begin{bmatrix} A_1 & 0 \\ 0 & C_2 \\ 0 & B \end{bmatrix}. \tag{1.54}$$

By the invertibility of V, we conclude that $M_C V$ is left semi-Fredholm, which implies that

$$\begin{bmatrix} C_2 \\ B \end{bmatrix} : \mathcal{K} \to \mathcal{R}(A)^\perp \oplus \mathcal{K}$$

is left semi-Fredholm, where $C_2 = P_{\mathcal{R}(A)^\perp} C : \mathcal{K} \to \mathcal{R}(A)^\perp$.

Conversely, if A and $\begin{bmatrix} P^{\mathtt{rst}}_{\mathcal{R}(A)^\perp} C \\ B \end{bmatrix} : \mathcal{K} \to \mathcal{R}(A)^\perp \oplus \mathcal{K}$ are left semi-Fredholm, then by (1.54) it follows that M_C is left semi-Fredholm. □

In the following theorem, we give necessary and sufficient conditions for the existence of $C \in \mathcal{B}(\mathcal{K},\mathcal{H})$ such that M_C is left semi-Fredholm and we completely describe the set of all such $C \in \mathcal{B}(\mathcal{K},\mathcal{H})$, for given operators $A \in \mathcal{B}(\mathcal{H})$ and $B \in \mathcal{B}(\mathcal{K})$.

THEOREM 1.48. *Let $A \in \mathcal{B}(\mathcal{H})$ and $B \in \mathcal{B}(\mathcal{K})$ be given operators. Then M_C is left semi-Fredholm for some $C \in \mathcal{B}(\mathcal{K},\mathcal{H})$ if and only if A is left semi-Fredholm and one of the following conditions is satisfied:*

(i) $d(A) = \infty$.

(ii) *B is left semi-Fredholm.*

Furthermore,

$$S_{LF}(A,B) = \{C \in \mathcal{B}(\mathcal{K},\mathcal{H}) \colon \dim[\mathcal{N}(B) \cap \mathcal{N}(P_{\mathcal{R}(A)^\perp} C)] < \infty, \\ \mathcal{R}(B^*) + \mathcal{R}(C^* P_{\mathcal{N}(A^*)}) \text{ is closed}\}.$$

PROOF. Suppose that there exists $C \in \mathcal{B}(\mathcal{K},\mathcal{H})$ such that M_C is left semi-Fredholm. Then by Lemma 1.47, we have that A and

$$\begin{bmatrix} P^{\mathrm{rst}}_{\mathcal{R}(A)^\perp} C \\ B \end{bmatrix} : \mathcal{K} \to \mathcal{R}(A)^\perp \oplus \mathcal{K}$$

are left semi-Fredholm operators. Clearly, if $d(A) = \infty$, then (i) holds; if $d(A) < \infty$, then $P^{\mathrm{rst}}_{\mathcal{R}(A)^\perp} C$ is compact, and so B is left semi-Fredholm. Thus (ii) holds.

Conversely, we first assume that A and B are left semi-Fredholm. Let $C = 0$. Then clearly M_C is left semi-Fredholm. Now we suppose that A is left semi-Fredholm and $d(A) = \infty$. Since $d(A) = \infty$, it follows that there exists a unitary operator $J : \mathcal{K} \to \mathcal{R}(A)^\perp$. Define $C \in \mathcal{B}(\mathcal{K},\mathcal{H})$ by

$$C = \begin{bmatrix} 0 \\ J \end{bmatrix} : \mathcal{K} \to \mathcal{R}(A) \oplus \mathcal{R}(A)^\perp.$$

By Lemma 1.47, M_C is left semi-Fredholm. Since $\begin{bmatrix} P^{\mathrm{rst}}_{\mathcal{R}(A)^\perp} C \\ B \end{bmatrix} : \mathcal{K} \to \mathcal{R}(A)^\perp \oplus \mathcal{K}$ is left semi-Fredholm if and only if

$$\mathcal{N}\left(\begin{bmatrix} P^{\mathrm{rst}}_{\mathcal{R}(A)^\perp} C \\ B \end{bmatrix}\right) = \mathcal{N}(P_{\mathcal{R}(A)^\perp} C) \cap \mathcal{N}(B)$$

is finite dimensional and $\mathcal{R}\left(\begin{bmatrix} P^{\mathrm{rst}}_{\mathcal{R}(A)^\perp} C \\ B \end{bmatrix}\right)$ is closed and since $\mathcal{R}\left(\begin{bmatrix} P^{\mathrm{rst}}_{\mathcal{R}(A)^\perp} C \\ B \end{bmatrix}\right)$ is closed if and only if $\mathcal{R}\left(\begin{bmatrix} C^* P_{\mathcal{N}(A^*)} & B^* \end{bmatrix}\right) = \mathcal{R}(C^* P_{\mathcal{N}(A^*)}) + \mathcal{R}(B^*)$ is closed, by Lemma 1.47 we can infer that

$$\begin{aligned} S_{LF}(A,B) &= \left\{ C \in \mathcal{B}(\mathcal{K},\mathcal{H}) : \begin{bmatrix} P^{\mathrm{rst}}_{\mathcal{R}(A)^\perp} C \\ B \end{bmatrix} : \mathcal{K} \to \mathcal{R}(A)^\perp \oplus \mathcal{K} \text{ is left semi-Fredholm} \right\} \\ &= \{ C \in \mathcal{B}(\mathcal{K},\mathcal{H}) \colon \dim[\mathcal{N}(B) \cap \mathcal{N}(P_{\mathcal{R}(A)^\perp} C)] < \infty, \\ &\qquad\qquad \mathcal{R}(B^*) + \mathcal{R}(C^* P_{\mathcal{N}(A^*)}) \text{ is closed} \}. \end{aligned}$$ □

REMARK 1.49. The main result of the paper [**135**] is a description of the set $\bigcap_{C \in \mathcal{B}(\mathcal{K},\mathcal{H})} \sigma_{le}(M_C)$ given by

$$\bigcap_{C \in \mathcal{B}(\mathcal{K},\mathcal{H})} \sigma_{le}(M_C) = \sigma_{le}(A) \cup \mathcal{F}_l(A,B) \cup \Psi_l(A,B),$$

where

$$\begin{aligned} \mathcal{F}_l(A,B) &= \{\lambda \in \mathbb{C} \colon \mathcal{R}(B-\lambda) \text{ is closed},\ \alpha(A-\lambda) = \infty,\ d(A-\lambda I) < \infty\}, \\ \Psi_l(A,B) &= \{\lambda \in \mathbb{C} \colon \mathcal{R}(B-\lambda) \text{ is not closed},\ d(A-\lambda I) < \infty\}. \end{aligned}$$

As a corollary of Theorem 1.48, we get a description of the set $\bigcap_{C \in \mathcal{B}(\mathcal{K},\mathcal{H})} \sigma_{le}(M_C)$ that is much simpler than the mentioned one given in [**135**]:

COROLLARY 1.50. For given operators $A \in \mathcal{B}(\mathcal{H})$ and $B \in \mathcal{B}(\mathcal{K})$, we have

$$\bigcap_{C \in \mathcal{B}(\mathcal{K},\mathcal{H})} \sigma_{le}(M_C) = \sigma_{le}(A) \cup \left(\sigma_{le}(B) \cap \{\lambda \in \mathbb{C} \colon d(A-\lambda I) < \infty\}\right). \tag{1.55}$$

Similarly, we can obtain the following results concerning the right semi-Fredholmness of M_C.

LEMMA 1.51. *Let $A \in \mathcal{B}(\mathcal{H})$, $B \in \mathcal{B}(\mathcal{K})$, and $C \in \mathcal{B}(\mathcal{K},\mathcal{H})$ be given operators. Then M_C is right semi-Fredholm if and only if B and $\begin{bmatrix} A & C|_{\mathcal{N}(B)} \end{bmatrix} : \mathcal{H}\oplus\mathcal{N}(B) \to \mathcal{H}$ are right semi-Fredholm operators.*

THEOREM 1.52. *Let $A \in \mathcal{B}(\mathcal{H})$ and $B \in \mathcal{B}(\mathcal{K})$ be given operators. Then M_C is right semi-Fredholm for some $C \in \mathcal{B}(\mathcal{K},\mathcal{H})$ if and only if B is right semi-Fredholm and one of the following conditions is satisfied:*

(i) $\alpha(B) = \infty$.

(ii) *A is right semi-Fredholm.*

Furthermore, the set of all C such that M_C is right semi-Fredholm is given by

$$\begin{aligned} S_{RF}(A,B) =& \{C \in \mathcal{B}(\mathcal{K},\mathcal{H})\colon \begin{bmatrix} A & C|_{\mathcal{N}(B)} \end{bmatrix} : \mathcal{H}\oplus\mathcal{N}(B) \to \mathcal{H} \text{ is right semi-Fred.}\} \\ =& \{C \in \mathcal{B}(\mathcal{K},\mathcal{H})\colon \dim[\mathcal{N}(A^*)\cap\mathcal{N}(P_{\mathcal{N}(B)}C^*)] < \infty, \\ & \qquad\qquad \mathcal{R}(A) + \mathcal{R}(CP_{\mathcal{N}(B)}) \text{ is closed}\}. \end{aligned}$$

COROLLARY 1.53. For given operators $A \in \mathcal{B}(\mathcal{H})$ and $B \in \mathcal{B}(\mathcal{K})$, we have

$$\bigcap_{C\in\mathcal{B}(\mathcal{K},\mathcal{H})} \sigma_{re}(M_C) = \sigma_{re}(B) \cup \big(\sigma_{re}(A) \cap \{\lambda \in \mathbb{C}\colon \alpha(A-\lambda I) < \infty\}\big).$$

Finally, we consider the Fredholmness of M_C. Note that the following result does not follow from Theorems 1.48 and 1.52 since the existence of $C \in \mathcal{B}(\mathcal{K},\mathcal{H})$ for which M_C is left semi-Fredholm and the existence of $C \in \mathcal{B}(\mathcal{K},\mathcal{H})$ for which M_C is right semi-Fredholm do not imply the existence of $C \in \mathcal{B}(\mathcal{K},\mathcal{H})$ for which M_C is Fredholm.

THEOREM 1.54. *Let $A \in \mathcal{B}(\mathcal{H})$ and $B \in \mathcal{B}(\mathcal{K})$ be given operators. Then M_C is Fredholm for some $C \in \mathcal{B}(\mathcal{K},\mathcal{H})$ if and only if one of the following conditions is satisfied:*

(i) *A and B are Fredholm.*

(ii) *A is left semi-Fredholm, B is right semi-Fredholm, and $d(A) = \alpha(B) = \infty$.*

Furthermore, if (i) *is satisfied, then $S_F(A,B) = \mathcal{B}(\mathcal{K},\mathcal{H})$, while if* (ii) *holds,*

$$S_F(A,B) = \{C \in \mathcal{B}(\mathcal{K},\mathcal{H})\colon P^{\mathtt{rst}}_{\mathcal{R}(A)^\perp} C|_{\mathcal{N}(B)} : \mathcal{N}(B) \to \mathcal{R}(A)^\perp \text{ is Fredholm}\}.$$

PROOF. Suppose that M_C is Fredholm for some $C \in \mathcal{B}(\mathcal{K},\mathcal{H})$. Then by Theorems 1.48 and 1.52, we get that one of the conditions (i) and (ii) must be satisfied.

Conversely, if (i) holds, then for $C = 0$, we have that M_C is Fredholm. If (ii) holds, then since $\mathcal{N}(B)$ and $\mathcal{R}(A)^\perp$ are infinite dimensional, it follows that there exists a Fredholm operator $S : \mathcal{N}(B) \to \mathcal{R}(A)^\perp$. Define $C \in \mathcal{B}(\mathcal{K},\mathcal{H})$ by

$$\begin{bmatrix} 0 & 0 \\ S & 0 \end{bmatrix} : \mathcal{N}(B)\oplus\mathcal{N}(B)^\perp \to \mathcal{R}(A)\oplus\mathcal{R}(A)^\perp.$$

Then M_C has an operator matrix representation

$$M_C = \begin{bmatrix} A_1 & 0 & 0 \\ 0 & S & 0 \\ 0 & 0 & B_1 \end{bmatrix} : \mathcal{H}\oplus\mathcal{N}(B)\oplus\mathcal{N}(B)^\perp \to \mathcal{R}(A)\oplus\mathcal{R}(A)^\perp\oplus\mathcal{K},$$

where A_1, S, B_1 are Fredholm, and hence M_C is Fredholm.

In the case when one of the conditions (i) and (ii) is satisfied, to describe the set of all C such that M_C is Fredholm, first note that M_C has an operator matrix representation

$$M_C = \begin{bmatrix} A_1 & C_1 & C_2 \\ 0 & C_3 & C_4 \\ 0 & 0 & B_1 \end{bmatrix} : \mathcal{H} \oplus \mathcal{N}(B) \oplus \mathcal{N}(B)^\perp \to \mathcal{R}(A) \oplus \mathcal{R}(A)^\perp \oplus \mathcal{K},$$

where $A_1 : \mathcal{H} \to \mathcal{R}(A)$ is right invertible and $B_1 : \mathcal{N}(B)^\perp \to \mathcal{K}$ is left invertible. Thus there are invertible operators U and V on $\mathcal{H} \oplus \mathcal{K}$ such that

$$UM_CV = \begin{bmatrix} A_1 & 0 & 0 \\ 0 & C_3 & 0 \\ 0 & 0 & B_1 \end{bmatrix}.$$

Obviously, M_C is Fredholm if and only if UM_CV is Fredholm. Now, we conclude that M_C is Fredholm if and only if A is left semi-Fredholm, B is right semi-Fredholm, and $C_3 = P^{\mathtt{rst}}_{\mathcal{R}(A)^\perp} C|_{\mathcal{N}(B)}$ is Fredholm. Thus, the set of all C such that M_C is Fredholm is given by

$$S_F(A,B) = \{C \in \mathcal{B}(\mathcal{K},\mathcal{H})\colon P^{\mathtt{rst}}_{\mathcal{R}(A)^\perp} C|_{\mathcal{N}(B)} : \mathcal{N}(B) \to \mathcal{R}(A)^\perp \text{ is Fredholm}\}.$$

Evidently, when (i) is satisfied, $S_F(A,B) = \mathcal{B}(\mathcal{K},\mathcal{H})$. □

Notice that in Theorem 1.54 the Hilbert spaces $\mathcal{H}$ and $\mathcal{K}$ are isomorphically embedded in $\mathcal{H} \times \mathcal{K}$ as two mutually orthogonal closed subspaces whose direct sum is $\mathcal{H} \times \mathcal{K}$. Here we consider the case when $M_C : \mathcal{H} \to \mathcal{K}$ and $\mathcal{H} = \mathcal{H}_1 \oplus \mathcal{H}_2$, $\mathcal{K} = \mathcal{K}_1 \oplus \mathcal{K}_2$ are topological decompositions which are in general nonorthogonal. Also we do not need the separability condition that is required in Theorem 1.54.

THEOREM 1.55. *Let $\mathcal{H},\mathcal{K}$ be infinite-dimensional Hilbert spaces and let $\mathcal{H} = \mathcal{H}_1 \oplus \mathcal{H}_2$, $\mathcal{K} = \mathcal{K}_1 \oplus \mathcal{K}_2$. An operator matrix $M_C = \begin{bmatrix} A & C \\ 0 & B \end{bmatrix} : \mathcal{H}_1 \oplus \mathcal{H}_2 \to \mathcal{K}_1 \oplus \mathcal{K}_2$ is Fredholm if and only if*

(i) *A is left semi-Fredholm,*
(ii) *B is right semi-Fredholm,*
(iii) *$P^{\mathtt{rst}}_{\mathcal{S},\mathcal{R}(A)} C|_{\mathcal{N}(B)}$ is Fredholm,*

where $\mathcal{K}_1 = \mathcal{R}(A) \oplus \mathcal{S}$.

PROOF. First, note that if M_C is a Fredholm operator, then $\mathcal{R}(A) = \mathcal{R}(M_C|_{\mathcal{H}_1})$ is a closed subspace of $\mathcal{K}_1$ (see Theorem 16.10 from [**148**]). Hence, in the proof of either implication we can suppose that $\mathcal{K}_1 = \mathcal{R}(A) \oplus \mathcal{S}$, for some closed subspace $\mathcal{S}$. So, M_C has the following representation:

$$\begin{bmatrix} A_1 & C_1 & C_2 \\ 0 & C_3 & C_4 \\ 0 & 0 & B_1 \end{bmatrix} : \begin{bmatrix} \mathcal{H}_1 \\ \mathcal{N}(B) \\ \mathcal{T} \end{bmatrix} \to \begin{bmatrix} \mathcal{R}(A) \\ \mathcal{S} \\ \mathcal{K}_2 \end{bmatrix}. \tag{1.56}$$

Let M_C be a Fredholm operator. Since $\mathcal{N}(A) \subseteq \mathcal{N}(M_C)$, we have that A is left semi-Fredholm. Also, since $\mathcal{R}(M_C) \subseteq \mathcal{R}\left(\begin{bmatrix} A & C \end{bmatrix}\right) \oplus \mathcal{R}(B)$ we have that B is right semi-Fredholm. Now, evidently A_1 is right invertible and B_1 is left invertible. Let

$$U = \begin{bmatrix} I & 0 & 0 \\ 0 & I & -C_4(B_1)_l^{-1} \\ 0 & 0 & I \end{bmatrix} : \begin{bmatrix} \mathcal{R}(A) \\ \mathcal{S} \\ \mathcal{K}_2 \end{bmatrix} \to \begin{bmatrix} \mathcal{R}(A) \\ \mathcal{S} \\ \mathcal{K}_2 \end{bmatrix}$$

and

$$V = \begin{bmatrix} I & -(A_1)_r^{-1}C_1 & -(A_1)_r^{-1}C_2 \\ 0 & I & 0 \\ 0 & 0 & I \end{bmatrix} : \begin{bmatrix} \mathcal{H}_1 \\ \mathcal{N}(B) \\ \mathcal{T} \end{bmatrix} \to \begin{bmatrix} \mathcal{H}_1 \\ \mathcal{N}(B) \\ \mathcal{T} \end{bmatrix},$$

where $\mathcal{H}_2 = \mathcal{T} \oplus \mathcal{N}(B)$.

We can check that U, V are invertible and

$$UM_CV = \begin{bmatrix} A_1 & 0 & 0 \\ 0 & C_3 & 0 \\ 0 & 0 & B_1 \end{bmatrix}. \tag{1.57}$$

The operator matrix UM_CV is Fredholm if and only if A_1, C_3, and B_1 are Fredholm operators. Since $C_3 = P^{\mathtt{rst}}_{\mathcal{S},\mathcal{R}(A)}C|_{\mathcal{N}(B)}$, the condition (iii) follows. Hence, all the conditions (i)–(iii) are satisfied.

If we suppose that (i)–(iii) are satisfied, then using (1.56) we can see that M_C is a Fredholm operator. □

REMARK 1.56. The advantage of Theorem 1.55 lies in the fact that the decompositions of the spaces which are used are not orthogonal unlike what is the case in Theorem 1.54 in [**47**] and all other related recently published results. This will particularly be useful in the Section 5.3 for analyzing Fredholmness of a linear combination of idempotents.

REMARK 1.57. It is worth mentioning that Theorem 1.55 is also correct if $\mathcal{H}$ and $\mathcal{K}$ are Banach spaces with the additional assumptions that subspaces $\mathcal{R}(A)$ and $\mathcal{N}(B)$ are complemented.

REMARK 1.58. From the proof of Theorem 1.55 we have that if M_C is Fredholm, then

$$\begin{aligned} \operatorname{ind}(M_C) &= \operatorname{ind}(UM_CV) = \operatorname{ind}(A_1) + \operatorname{ind}(C_3) + \operatorname{ind}(B_1) \\ &= \alpha(A_1) + \operatorname{ind}(C_3) + \beta(B_1) \\ &= \alpha(A) + \operatorname{ind}(C_3) + \beta(B). \end{aligned}$$

Moreover, $\alpha(M_C) = \alpha(A) + \alpha(C_3)$ and $\beta(M_C) = \beta(C_3) + \beta(B)$, where A_1, B_1, and C_3 are defined as in the proof of this theorem. Let us mention that these formulas are correct even when just $\mathcal{R}(A)$ is closed.

The motivation for the results given next was the papers of Y. Li et al. [**136**, **137**] in which the authors give a characterization of the sets

$$\bigcap_{C\in\mathcal{B}^{-1}(\mathcal{K},\mathcal{H})} \sigma_e(M_C) \quad \text{and} \quad \bigcap_{C\in\mathcal{B}^{-1}(\mathcal{K},\mathcal{H})} \sigma_{re}(M_C).$$

They proved that

$$\begin{aligned} &\bigcap_{C\in\mathcal{B}^{-1}(\mathcal{K},\mathcal{H})} \sigma_e(M_C) \\ &\quad = \bigcap_{C\in\mathcal{B}(\mathcal{K},\mathcal{H})} \sigma_e(M_C) \cup \{\lambda \in \mathbb{C} \colon A - \lambda I \text{ is compact or } B - \lambda I \text{ is compact}\} \end{aligned}$$

and

$$\bigcap_{C\in\mathcal{B}^{-1}(\mathcal{K},\mathcal{H})} \sigma_{re}(M_C) = \bigcap_{C\in\mathcal{B}(\mathcal{K},\mathcal{H})} \sigma_{re}(M_C) \cup \{\lambda \in \mathbb{C} \colon A - \lambda I \text{ is compact}\}.$$

Also the motivation for the next results was the papers of X. H. Cao et al. [**28**] and Y. Li et al. [**134**]. In the first one, the authors gave a necessary and sufficient condition for which $M_C \in \mathcal{F}_l^-(\mathcal{H},\mathcal{K})$ (the class of all left semi-Fredholm operators with nonpositive index) for some $C \in \mathcal{B}(\mathcal{K},\mathcal{H})$ and characterize the set $\bigcap_{C\in\mathcal{B}(\mathcal{K},\mathcal{H})}\sigma_{ea}(M_C)$ (the characterization of all such C is given in [**49**]), while in [**134**], the authors characterized the intersection of $\bigcap_{C\in\mathcal{B}_l(\mathcal{K},\mathcal{H})}\sigma_{ea}(M_C)$ and $\bigcap_{C\in\mathcal{B}^{-1}(\mathcal{K},\mathcal{H})}\sigma_{ea}(M_C)$.

In this section, we consider the question of whether or not there exists an invertible operator $C \in \mathcal{B}(\mathcal{K},\mathcal{H})$ such that $C \in S_F(A,B)$ (resp. $S_{RF}(A,B), S_{LF}(A,B)$) and give the characterization for the following sets:

$$\bigcap_{C\in\mathcal{B}^{-1}(\mathcal{K},\mathcal{H})}\sigma_e(M_C),\qquad \bigcap_{C\in\mathcal{B}^{-1}(\mathcal{K},\mathcal{H})}\sigma_{le}(M_C),\qquad \bigcap_{C\in\mathcal{B}^{-1}(\mathcal{K},\mathcal{H})}\sigma_{re}(M_C).$$

Also, for given $A \in \mathcal{B}(\mathcal{H})$ and $B \in \mathcal{B}(\mathcal{K})$, the sets

$$\bigcap_{C\in\mathcal{B}_l(\mathcal{K},\mathcal{H})}\sigma_{le}(M_C),\quad \bigcap_{C\in\mathcal{B}_l(\mathcal{K},\mathcal{H})}\sigma_{re}(M_C),\quad \bigcap_{C\in\mathcal{B}_r(\mathcal{K},\mathcal{H})}\sigma_{le}(M_C),\quad \bigcap_{C\in\mathcal{B}_r(\mathcal{K},\mathcal{H})}\sigma_{re}(M_C)$$

are described.

THEOREM 1.59. *Let $A \in \mathcal{B}(\mathcal{H})$ and $B \in \mathcal{B}(\mathcal{K})$ be given operators. If $S_{RF}(A,B) \neq \emptyset$, then there exists a right invertible operator $C \in S_{RF}(A,B)$. Thus*

$$\bigcap_{C\in\mathcal{B}_r(\mathcal{K},\mathcal{H})}\sigma_{re}(M_C) = \bigcap_{C\in\mathcal{B}(\mathcal{K},\mathcal{H})}\sigma_{re}(M_C).$$

PROOF. Since $S_{RF}(A,B) \neq \emptyset$, it follows from Theorem 1.52 that both A and B are right semi-Fredholm or B is right semi-Fredholm with $\alpha(B) = \infty$. If A and B are right semi-Fredholm, it is easy to check that $S_{RF}(A,B) = \mathcal{B}(\mathcal{K},\mathcal{H})$, and hence there exists a right invertible operator $C \in \mathcal{B}(\mathcal{K},\mathcal{H})$ such that $C \in S_{RF}(A,B)$. Now we suppose that B is right semi-Fredholm with $\alpha(B) = \infty$. Since $\mathcal{N}(B)$ is infinite dimensional, there exists an invertible operator $S \in \mathcal{B}(\mathcal{N}(B),\mathcal{H})$. Define $C \in \mathcal{B}(\mathcal{K},\mathcal{H})$ by

$$C = \begin{bmatrix} S & 0 \end{bmatrix} : \mathcal{N}(B) \oplus \mathcal{N}(B)^\perp \to \mathcal{H}.$$

Evidently, C is right invertible and

$$\begin{bmatrix} A & C|_{\mathcal{N}(B)} \end{bmatrix} = \begin{bmatrix} A & S \end{bmatrix} : \mathcal{H} \oplus \mathcal{N}(B) \to \mathcal{H}$$

is right semi-Fredholm, which implies that $C \in S_{RF}(A,B)$. □

THEOREM 1.60. *Let $A \in \mathcal{B}(\mathcal{H})$ and $B \in \mathcal{B}(\mathcal{K})$ be given operators. Then there exists a left invertible operator $C \in \mathcal{B}(\mathcal{K},\mathcal{H})$ such that $C \in S_{RF}(A,B)$ if and only if A is noncompact and $S_{RF}(A,B) \neq \emptyset$. Therefore*

$$\bigcap_{C\in\mathcal{B}_l(\mathcal{K},\mathcal{H})}\sigma_{re}(M_C) = \bigcap_{C\in\mathcal{B}(\mathcal{K},\mathcal{H})}\sigma_{re}(M_C) \cup \{\lambda \in \mathbb{C}\colon A - \lambda I \textit{ is compact}\}.$$

PROOF. Suppose that there exists a left invertible operator $C \in \mathcal{B}(\mathcal{K},\mathcal{H})$ such that $C \in S_{RF}(A,B)$. Then clearly B is right semi-Fredholm and $S_{RF}(A,B) \neq \emptyset$. Now we prove that A is noncompact. Assume to the contrary that A is compact. Let $C^\dagger$ be the Moore-Penrose inverse of C. Then

$$\begin{bmatrix} I & 0 \\ -BC^\dagger & I \end{bmatrix}\begin{bmatrix} A & C \\ 0 & B \end{bmatrix} = \begin{bmatrix} A & C \\ -BC^\dagger A & 0 \end{bmatrix}$$

is right semi-Fredholm. This, together with the compactness of A, implies that

$$\begin{bmatrix} A & C \\ 0 & 0 \end{bmatrix}$$

is right semi-Fredholm, which is a contradiction.

Conversely, assume that A is noncompact and $S_{RF}(A,B) \neq \emptyset$. By Theorem 1.52 we obtain that both A and B are right semi-Fredholm or B is right semi-Fredholm with $\alpha(B) = \infty$. If A and B are right semi-Fredholm, it is easy to check that $S_{RF}(A,B) = \mathcal{B}(\mathcal{K},\mathcal{H})$, and hence there exists a left invertible operator $C \in \mathcal{B}(\mathcal{K},\mathcal{H})$ such that $C \in S_{RF}(A,B)$. Next suppose that B is right semi-Fredholm with $\alpha(B) = \infty$. Since A is noncompact, there exists an infinite-dimensional closed subspace $\mathcal{M} \subseteq \mathcal{H}$ such that $A|_{\mathcal{M}} : \mathcal{M} \to \mathcal{H}$ is left invertible, which implies $\mathcal{R}(A|_{\mathcal{M}})$ is an infinite-dimensional closed subspace, and thus there exists an invertible operator $C_1 \in \mathcal{B}(\mathcal{K},\mathcal{R}(A|_{\mathcal{M}}))$. On the other hand, since $\alpha(B) = \infty$, there exists a right invertible operator $C_2 \in \mathcal{B}(\mathcal{K},\mathcal{R}(A|_{\mathcal{M}})^{\perp})$ with $\mathcal{N}(C_2)^{\perp} \subseteq \mathcal{N}(B)$. Define $C \in \mathcal{B}(\mathcal{K},\mathcal{H})$ by

$$C = \begin{bmatrix} C_1 \\ C_2 \end{bmatrix} : \mathcal{K} \to \mathcal{R}(A|_{\mathcal{M}}) \oplus \mathcal{R}(A|_{\mathcal{M}})^{\perp}.$$

Note that

$$\begin{bmatrix} C_1^{-1} & 0 \end{bmatrix} \begin{bmatrix} C_1 \\ C_2 \end{bmatrix} = I,$$

and hence C is left invertible. Let us show that M_C is right semi-Fredholm. By the definition of C one can see that $\begin{bmatrix} A & C|_{\mathcal{N}(B)} \end{bmatrix} : \mathcal{H} \oplus \mathcal{N}(B) \to \mathcal{H}$ as an operator from $\mathcal{M} \oplus \mathcal{M}^{\perp} \oplus \mathcal{N}(B)$ to $\mathcal{R}(A|_{\mathcal{M}}) \oplus \mathcal{R}(A|_{\mathcal{M}})^{\perp}$ has the matrix representation

$$\begin{bmatrix} A_1 & A_2 & C_{11} \\ 0 & A_3 & C_{21} \end{bmatrix},$$

where $A_1 : \mathcal{M} \to \mathcal{R}(A|_{\mathcal{M}})$ is invertible and $C_{21} : \mathcal{N}(B) \to \mathcal{R}(A|_{\mathcal{M}})^{\perp}$ is right invertible. Therefore there is an invertible operator U on $\mathcal{H} \oplus \mathcal{K}$ with

$$\begin{bmatrix} A_1 & A_2 & C_{11} \\ 0 & A_3 & C_{21} \end{bmatrix} U = \begin{bmatrix} A_1 & 0 & 0 \\ 0 & A_3 & C_{21} \end{bmatrix}.$$

Clearly, $\begin{bmatrix} A_3 & C_{21} \end{bmatrix} : \mathcal{M}^{\perp} \to \mathcal{R}(A|_{\mathcal{M}})^{\perp}$ is right semi-Fredholm. This, together with the invertibility of A_1, shows that $\begin{bmatrix} A & C|_{\mathcal{N}(B)} \end{bmatrix} : \mathcal{H} \oplus \mathcal{N}(B) \to \mathcal{H}$ is right semi-Fredholm. Now the proof follows from Lemma 1.51. □

Using a similar technique, we obtain the following results:

THEOREM 1.61. *Let $A \in \mathcal{B}(\mathcal{H})$ and $B \in \mathcal{B}(\mathcal{K})$. If $S_{LF}(A,B) \neq \emptyset$, then there exists a left invertible operator $C \in \mathcal{B}(\mathcal{K},\mathcal{H})$ such that $C \in S_{LF}(A,B)$. Thus*

$$\bigcap_{C \in \mathcal{B}_l(\mathcal{K},\mathcal{H})} \sigma_{le}(M_C) = \bigcap_{C \in \mathcal{B}(\mathcal{K},\mathcal{H})} \sigma_{le}(M_C).$$

THEOREM 1.62. *Let $A \in \mathcal{B}(\mathcal{H})$ and $B \in \mathcal{B}(\mathcal{K})$. Then there exists a right invertible operator $C \in \mathcal{B}(\mathcal{K},\mathcal{H})$ such that $C \in S_{LF}(A,B)$ if and only if B is noncompact and $S_{LF}(A,B) \neq \emptyset$. This implies that*

$$\bigcap_{C \in \mathcal{B}_r(\mathcal{K},\mathcal{H})} \sigma_{le}(M_C) = \bigcap_{C \in \mathcal{B}(\mathcal{K},\mathcal{H})} \sigma_{le}(M_C) \cup \{\lambda \in \mathbb{C} : B - \lambda I \text{ is compact}\}.$$

In Theorem 10 from [**137**], it is proved that

$$\bigcap_{C\in\mathcal{B}^{-1}(\mathcal{K},\mathcal{H})}\sigma_e(M_C)=\bigcap_{C\in\mathcal{B}(\mathcal{K},\mathcal{H})}\sigma_e(M_C)\cup\{\lambda\in\mathbb{C}\colon A-\lambda I \text{ or } B-\lambda I \text{ is compact}\}. \tag{1.58}$$

Using the theorem given below, we get that

$$\bigcap_{C\in\mathcal{B}^{-1}(\mathcal{K},\mathcal{H})}\sigma_e(M_C)=\bigcap_{C\in\mathcal{B}(\mathcal{K},\mathcal{H})}\sigma_e(M_C).$$

Hence the set $\{\lambda\in\mathbb{C}\colon A-\lambda I \text{ or } B-\lambda I \text{ is compact}\}$ is redundant in (1.58).

THEOREM 1.63. *Let $A\in\mathcal{B}(\mathcal{H})$ and $B\in\mathcal{B}(\mathcal{K})$. If $S_F(A,B)\neq\emptyset$, then there exists an invertible operator $C\in\mathcal{B}(\mathcal{K},\mathcal{H})$ such that $C\in S_F(A,B)$.*

PROOF. Assume that $S_F(A,B)\neq\emptyset$. By Theorem 1.54, we get that one of the following conditions is satisfied:

(1) A and B are Fredholm.
(2) A is left semi-Fredholm, B is right semi-Fredholm, and $\alpha(B)=d(A)=\infty$.

If (1) holds, then $S_F(A,B)=\mathcal{B}(\mathcal{K},\mathcal{H})$, so there exists an invertible operator $C\in S_F(A,B)$. If (2) holds, then there exists an invertible operator $C_3\in\mathcal{B}(\mathcal{N}(B),\mathcal{R}(A)^\perp)$ since $\alpha(B)=d(A)=\infty$. Also, since A is left semi-Fredholm and B is right semi-Fredholm it follows that $\mathcal{N}(B)^\perp$ and $\mathcal{R}(A)$ are infinite dimensional, which implies that there is an invertible operator $C_2\in\mathcal{B}(\mathcal{N}(B)^\perp,\mathcal{R}(A))$. Define $C\in\mathcal{B}(\mathcal{K},\mathcal{H})$ by

$$C=\begin{bmatrix}0 & C_2\\ C_3 & 0\end{bmatrix}\mathcal{N}(B)\oplus\mathcal{N}(B)^\perp\to\mathcal{R}(A)\oplus\mathcal{R}(A)^\perp.$$

Evidently, C is invertible and $C\in S_F(A,B)$. □

1.6. Completions of upper triangular operator matrices to generalized left and right Weyl operator

Recall that if operator $A\in\mathcal{B}(\mathcal{H},\mathcal{K})$ is such that $\mathcal{R}(A)$ is closed and $\alpha(A)\leq d(A)$, then A is a generalized left Weyl operator. If $A\in\mathcal{B}(\mathcal{H},\mathcal{K})$ is such that $\mathcal{R}(A)$ is closed and $d(A)\leq\alpha(A)$, then A is a generalized right Weyl operator. An operator $A\in\mathcal{B}(\mathcal{H},\mathcal{K})$ is a generalized Weyl operator if it is both generalized right Weyl and generalized left Weyl. The set of all generalized Weyl operators from $\mathcal{B}(\mathcal{H},\mathcal{K})$ is denoted by $W^g(\mathcal{H},\mathcal{K})$, and by $W_{gl}(\mathcal{H},\mathcal{K})$ ($W_{gr}(\mathcal{H},\mathcal{K})$) we denote the subset of $\mathcal{B}(\mathcal{H},\mathcal{K})$ consisting of all generalized left (right) Weyl operators. Notice that in the cases of generalized left (right) Weyl operators, $\alpha(A)$ and $d(A)$ are allowed to be infinity.

For an operator $C\in\mathcal{B}(\mathcal{H})$, the generalized left (right) Weyl spectrum $\sigma^g_{lw}(C)$ ($\sigma^g_{rw}(C)$) is defined by

$$\sigma^g_{lw}(C)(\sigma^g_{rw}(C))=\{\lambda\in\mathbb{C}\colon C-\lambda I\ \text{ is not generalized left (right) Weyl}\}.$$

The generalized Weyl spectrum is defined by

$$\sigma^g_{w}(C)=\{\lambda\in\mathbb{C}\colon C-\lambda I\ \text{ is not generalized Weyl}\}.$$

In this section, for infinite-dimensional complex separable Hilbert spaces $\mathcal{H},\mathcal{K}$ and given operators $A\in\mathcal{B}(\mathcal{H})$, $B\in\mathcal{B}(\mathcal{K})$, we completely describe the sets of all

$C \in \mathcal{B}(\mathcal{K},\mathcal{H})$ such that $M_C = \begin{bmatrix} A & C \\ 0 & B \end{bmatrix}$ is generalized Weyl and the generalized left (right) Weyl operator. We also describe the intersections and unions of the generalized left Weyl spectra

$$\bigcup_{C\in\mathcal{B}(\mathcal{K},\mathcal{H})} \sigma^g_{lw}(M_C), \qquad \bigcap_{C\in\mathcal{B}(\mathcal{K},\mathcal{H})} \sigma^g_{lw}(M_C)$$

and present necessary and sufficient conditions which two operators $A \in \mathcal{B}(\mathcal{H})$ and $B \in \mathcal{B}(\mathcal{K})$ have to satisfy in order for M_C to be a generalized left Weyl operator for each $C \in \mathcal{B}(\mathcal{K},\mathcal{H})$.

In [**130**] the sets $\bigcup_{C\in\mathcal{B}(\mathcal{K},\mathcal{H})} \sigma^g_w(M_C)$ and $\bigcap_{C\in\mathcal{B}(\mathcal{K},\mathcal{H})} \sigma^g_w(M_C)$ were described and some necessary and sufficient conditions for the existence of $C \in \mathcal{B}(\mathcal{K},\mathcal{H})$ such that M_C is generalized Weyl are given, but the set of all such operators C is not described. Here we present results from [**97**] where a description of all $C \in \mathcal{B}(\mathcal{K},\mathcal{H})$ such that M_C is generalized Weyl is given. The symbol $S_{GW}(A,B)$ stands for the set of all such C. Also, we describe the sets $\bigcup_{C\in\mathcal{B}(\mathcal{K},\mathcal{H})} \sigma^g_{lw}(M_C)$ and $\bigcap_{C\in\mathcal{B}(\mathcal{K},\mathcal{H})} \sigma^g_{lw}(M_C)$ for given $A \in \mathcal{B}(\mathcal{H})$ and $B \in \mathcal{B}(\mathcal{K})$ as well as the set of all $C \in \mathcal{B}(\mathcal{K},\mathcal{H})$ such that M_C is generalized left Weyl which is denoted by $S_{GLW}(A,B)$. In an analogous way, similar results can be provided for $\bigcup_{C\in\mathcal{B}(\mathcal{K},\mathcal{H})} \sigma^g_{rw}(M_C)$ and $\bigcap_{C\in\mathcal{B}(\mathcal{K},\mathcal{H})} \sigma^g_{rw}(M_C)$.

Evidently for given $A \in \mathcal{B}(\mathcal{H})$ and $B \in \mathcal{B}(\mathcal{K})$, arbitrary $C \in \mathcal{B}(\mathcal{K},\mathcal{H})$ can be represented by

$$C = \begin{bmatrix} C_1 & C_2 \\ C_3 & C_4 \end{bmatrix} : \begin{bmatrix} \mathcal{N}(B) \\ \mathcal{N}(B)^{\perp} \end{bmatrix} \to \begin{bmatrix} \overline{\mathcal{R}(A)} \\ \mathcal{R}(A)^{\perp} \end{bmatrix}. \tag{1.59}$$

First, we state some auxiliary lemmas which will be used in the proof of the main result.

LEMMA 1.64. *Let $A, T \in \mathcal{B}(\mathcal{H},\mathcal{K})$. If $\mathcal{R}(T)$ is finite dimensional, then the following are equivalent:*

(1) *$\mathcal{R}(A+T)$ is closed.*
(2) *$\mathcal{R}(A)+\mathcal{R}(T)$ is closed.*
(3) *$\mathcal{R}(A)$ is closed.*

LEMMA 1.65. *Let $S \in \mathcal{B}(\mathcal{H})$, $T \in \mathcal{B}(\mathcal{K},\mathcal{H})$, and $R \in \mathcal{B}(\mathcal{H},\mathcal{K})$ be given operators.*

(i) *If $\mathcal{R}(S)$ is nonclosed and $\mathcal{R}(\begin{bmatrix} S & T \end{bmatrix})$ is closed, then $\alpha(\begin{bmatrix} S & T \end{bmatrix}) = \infty$.*
(ii) *If $\mathcal{R}(S)$ is nonclosed and $\mathcal{R}\left(\begin{bmatrix} S \\ R \end{bmatrix}\right)$ is closed, then $d\left(\begin{bmatrix} S \\ R \end{bmatrix}\right) = \infty$.*

PROOF. (i) Suppose that $\mathcal{R}(S)$ is nonclosed, $\mathcal{R}(\begin{bmatrix} S & T \end{bmatrix})$ is closed, and $\alpha(\begin{bmatrix} S & T \end{bmatrix}) < \infty$. Then $\begin{bmatrix} S & T \end{bmatrix}$ is a left semi-Fredholm operator which implies that there exists an operator $\begin{bmatrix} X \\ Y \end{bmatrix} : \mathcal{H} \to \begin{bmatrix} \mathcal{H} \\ \mathcal{K} \end{bmatrix}$ such that

$$\begin{bmatrix} X \\ Y \end{bmatrix} \begin{bmatrix} S & T \end{bmatrix} = I + K,$$

for some compact operator $K \in \mathcal{B}(\mathcal{H} \oplus \mathcal{K})$. Hence, $XS = I_{\mathcal{H}} + K_1$, for some compact operator $K_1 \in \mathcal{B}(\mathcal{H})$, which implies that S is left semi-Fredholm and so $\mathcal{R}(S)$ is closed, which is a contradiction.

(ii) The proof follows by taking adjoints in (i). $\square$

In the following theorem, for given operators $A \in \mathcal{B}(\mathcal{H})$ and $B \in \mathcal{B}(\mathcal{K})$, we present necessary and sufficient conditions for the existence of $C \in \mathcal{B}(\mathcal{K},\mathcal{H})$ such that M_C is a generalized left Weyl operator and we completely describe the set of all such $C \in \mathcal{B}(\mathcal{K},\mathcal{H})$.

THEOREM 1.66. *Let $A \in \mathcal{B}(\mathcal{H})$ and $B \in \mathcal{B}(\mathcal{K})$. There exists $C \in \mathcal{B}(\mathcal{K},\mathcal{H})$ such that M_C is generalized left Weyl if and only if one of the following conditions is satisfied:*

(i) *$\mathcal{R}(A)$ and $\mathcal{R}(B)$ are closed and $\alpha(A)+\alpha(B) \leq d(A)+d(B)$. In this case,*

$$S_{GLW}(A,B) = \{C \in \mathcal{B}(\mathcal{K},\mathcal{H})\colon C \text{ is given by } (1.59), C_3 \text{ has closed range}, \\ \alpha(A)+\alpha(C_3) \leq d(C_3)+d(B)\}.$$

(ii) *$\mathcal{R}(A)$ is closed, $\mathcal{R}(B)$ is nonclosed, and $d(A) = \infty$. In this case,*

$$S_{GLW}(A,B) = \{C \in \mathcal{B}(\mathcal{K},\mathcal{H})\colon \mathcal{R}(B^*) + R(C^*P_{\mathcal{R}(A)^\perp}) \text{ is closed}\}.$$

(iii) *$\mathcal{R}(A)$ is nonclosed, $\mathcal{R}(B)$ is closed, and $\alpha(B) = d(A)+d(B) = \infty$. In this case,*

$$S_{GLW}(A,B) = \{C \in \mathcal{B}(\mathcal{K},\mathcal{H})\colon \mathcal{R}(A)+\mathcal{R}(CP_{\mathcal{N}(B)}) \text{ is closed}, \\ d(B) + \operatorname{codim}(\mathcal{R}(A)+\mathcal{R}(CP_{\mathcal{N}(B)})) = \infty\}.$$

(iv) *$\mathcal{R}(A)$ and $\mathcal{R}(B)$ are nonclosed and $\alpha(B) = d(A) = \infty$. In this case,*

$$S_{GLW}(A,B) = \{C \in \mathcal{B}(\mathcal{K},\mathcal{H})\colon \mathcal{R}(M_C) \text{ is closed}\}.$$

For simplicity, we divide the statement of this theorem into four propositions and prove each of them separately.

PROPOSITION 1.67. Let $A \in \mathcal{B}(\mathcal{H})$ and $B \in \mathcal{B}(\mathcal{K})$ be such that $\mathcal{R}(A)$ and $\mathcal{R}(B)$ are closed. There exists $C \in \mathcal{B}(\mathcal{K},\mathcal{H})$ such that M_C is generalized left Weyl if and only if $\alpha(A)+\alpha(B) \leq d(A)+d(B)$. In this case,

$$S_{GLW}(A,B) = \{C \in \mathcal{B}(\mathcal{K},\mathcal{H})\colon C \text{ is given by } (1.59), C_3 \text{ has closed range}, \\ \alpha(A)+\alpha(C_3) \leq d(C_3)+d(B)\}.$$

PROOF. If $\alpha(A)+\alpha(B) \leq d(A)+d(B)$, then M_0 is a generalized left Weyl operator. Conversely, suppose that there exists $C \in \mathcal{B}(\mathcal{K},\mathcal{H})$ such that M_C is a generalized left Weyl operator and such that C is given by (1.59). Then M_C has a matrix representation

$$M_C = \begin{bmatrix} A_1 & C_1 & C_2 \\ 0 & C_3 & C_4 \\ 0 & 0 & B_1 \end{bmatrix} : \begin{bmatrix} \mathcal{H} \\ \mathcal{N}(B) \\ \mathcal{N}(B)^\perp \end{bmatrix} \to \begin{bmatrix} \mathcal{R}(A) \\ \mathcal{R}(A)^\perp \\ \mathcal{K} \end{bmatrix},$$

where $A_1 : \mathcal{H} \to \mathcal{R}(A)$ is right invertible and $B_1 : \mathcal{N}(B)^\perp \to \mathcal{K}$ is left invertible. Evidently, there exist invertible $U, V \in \mathcal{B}(\mathcal{H}\oplus\mathcal{K})$ such that

$$UM_CV = \begin{bmatrix} A_1 & 0 & 0 \\ 0 & C_3 & 0 \\ 0 & 0 & B_1 \end{bmatrix} : \begin{bmatrix} \mathcal{H} \\ \mathcal{N}(B) \\ \mathcal{N}(B)^\perp \end{bmatrix} \to \begin{bmatrix} \mathcal{R}(A) \\ \mathcal{R}(A)^\perp \\ \mathcal{K} \end{bmatrix}. \tag{1.60}$$

Hence, UM_CV is generalized left Weyl which implies that

$$\alpha(A_1)+\alpha(C_3) \leq d(C_3)+d(B_1). \tag{1.61}$$

Since

$$\alpha(A_1) = \alpha(A),\ \alpha(B) = \alpha(C_3) + \dim(\mathcal{N}(C_3)^\perp),$$
$$d(B_1) = d(B),\ d(A) = d(C_3) + \dim \mathcal{R}(C_3),$$

having in mind that $\dim(\mathcal{N}(C_3)^\perp) = \dim(\mathcal{R}(C_3))$ and (1.61), we get

$$\alpha(A) + \alpha(B) \leq d(A) + d(B).$$

To describe the set of all $C \in \mathcal{B}(\mathcal{K}, \mathcal{H})$ such that M_C is generalized left Weyl, notice that for arbitrary C given by (1.59) there exist invertible $U, V \in \mathcal{B}(\mathcal{H} \oplus \mathcal{K})$ such that UM_CV is given by (1.60). Hence, M_C is generalized left Weyl if and only if UM_CV is generalized left Weyl which is equivalent with the fact that $\mathcal{R}(C_3)$ is closed and that (1.61) holds. □

PROPOSITION 1.68. Let $A \in \mathcal{B}(\mathcal{H})$ and $B \in \mathcal{B}(\mathcal{K})$ be such that $\mathcal{R}(A)$ is closed and $\mathcal{R}(B)$ is nonclosed. There exists $C \in \mathcal{B}(\mathcal{K}, \mathcal{H})$ such that M_C is generalized left Weyl if and only if $d(A) = \infty$. In this case,

$$S_{GLW}(A, B) = \{C \in \mathcal{B}(\mathcal{K}, \mathcal{H})\colon \mathcal{R}(B^*) + \mathcal{R}(C^* P_{\mathcal{R}(A)^\perp}) \text{ is closed}\}.$$

PROOF. Suppose that $d(A) = \infty$. Then M_{C_0} is a generalized left Weyl operator for C_0 given by

$$C_0 = \begin{bmatrix} 0 \\ J \end{bmatrix} : \mathcal{K} \to \mathcal{R}(A) \oplus \mathcal{R}(A)^\perp,$$

where $J : \mathcal{K} \to \mathcal{R}(A)^\perp$ is unitary. Evidently, M_{C_0} is represented by

$$M_{C_0} = \begin{bmatrix} A_1 & 0 \\ 0 & J \\ 0 & B \end{bmatrix} : \mathcal{H} \oplus \mathcal{K} \to \mathcal{R}(A) \oplus \mathcal{R}(A)^\perp \oplus \mathcal{K},$$

where $A_1 : \mathcal{H} \to \mathcal{R}(A)$ is right invertible. Since J is invertible, there exists an invertible operator $U \in \mathcal{B}(\mathcal{H} \oplus \mathcal{K})$ such that

$$UM_{C_0} = \begin{bmatrix} A_1 & 0 \\ 0 & J \\ 0 & 0 \end{bmatrix} : \mathcal{H} \oplus \mathcal{K} \to \mathcal{R}(A) \oplus \mathcal{R}(A)^\perp \oplus \mathcal{K}.$$

Now, it is clear that UM_{C_0} is a generalized left Weyl operator, and so M_{C_0} is a generalized left Weyl operator.

Conversely, suppose that there exists $C \in \mathcal{B}(\mathcal{K}, \mathcal{H})$ such that M_C is generalized left Weyl. Then M_C has a matrix representation

$$M_C = \begin{bmatrix} A_1 & C_1 \\ 0 & C_2 \\ 0 & B \end{bmatrix} : \mathcal{H} \oplus \mathcal{K} \to \mathcal{R}(A) \oplus \mathcal{R}(A)^\perp \oplus \mathcal{K}, \tag{1.62}$$

where $A_1 : \mathcal{H} \to \mathcal{R}(A)$ is right invertible. Thus there exists an invertible operator $V \in \mathcal{B}(\mathcal{H} \oplus \mathcal{K})$ such that

$$M_CV = \begin{bmatrix} A_1 & 0 \\ 0 & C_2 \\ 0 & B \end{bmatrix} : \mathcal{H} \oplus \mathcal{K} \to \mathcal{R}(A) \oplus \mathcal{R}(A)^\perp \oplus \mathcal{K}. \tag{1.63}$$

Now we show that $d(A) = \infty$: Indeed, if $d(A) < \infty$, then $\mathcal{R}(C_2^*)$ is finite dimensional. Since $\mathcal{R}(M_CV)$ is closed, we have that $\mathcal{R}\left(\begin{bmatrix} C_2 \\ B \end{bmatrix}\right)$ is closed. which implies that $\mathcal{R}(B^*) + \mathcal{R}(C_2^*)$ is closed. This, together with $\dim \mathcal{R}(C_2^*) < \infty$, implies that $\mathcal{R}(B)$ is closed. This is a contradiction. Hence, $d(A) = \infty$.

In order to describe the set $S_{GLW}(A, B)$, notice that for arbitrary $C \in \mathcal{B}(\mathcal{K}, \mathcal{H})$, M_C has a form (1.62) and that there exists an invertible operator $V \in \mathcal{B}(\mathcal{H} \oplus \mathcal{K})$ such that $M_C V$ is given by (1.63). Hence, M_C is generalized left Weyl if and only if C is such that $\mathcal{R}\left(\begin{bmatrix} C_2 \\ B \end{bmatrix}\right)$ is closed and such that

$$\alpha(A_1) + \alpha\left(\begin{bmatrix} C_2 \\ B \end{bmatrix}\right) \le d(A_1) + d\left(\begin{bmatrix} C_2 \\ B \end{bmatrix}\right). \tag{1.64}$$

Notice that by Lemma 1.65, we have that for each $C_2 \in \mathcal{B}(\mathcal{K}, \mathcal{R}(A)^{\perp})$ such that $\mathcal{R}\left(\begin{bmatrix} C_2 \\ B \end{bmatrix}\right)$ is closed, it follows that $d\left(\begin{bmatrix} C_2 \\ B \end{bmatrix}\right) = \infty$. Thus,

$$S_{GLW}(A, B) = \{C \in \mathcal{B}(\mathcal{K}, \mathcal{H})\colon \mathcal{R}(B^*) + \mathcal{R}(C^* P_{\mathcal{R}(A)^{\perp}}) \text{ is closed}\}. \qquad \square$$

PROPOSITION 1.69. Let $A \in \mathcal{B}(\mathcal{H})$ and $B \in \mathcal{B}(\mathcal{K})$ be such that $\mathcal{R}(A)$ is non-closed and $\mathcal{R}(B)$ is closed. There exists $C \in \mathcal{B}(\mathcal{K}, \mathcal{H})$ such that M_C is generalized left Weyl if and only if $\alpha(B) = d(A) + d(B) = \infty$. In this case,

$$\begin{aligned} S_{GLW}(A, B) = \{C \in \mathcal{B}(\mathcal{K}, \mathcal{H})\colon & \mathcal{R}(A) + R(CP_{\mathcal{N}(B)}) \text{ is closed}, \\ & d(B) + \operatorname{codim}(\mathcal{R}(A) + \mathcal{R}(CP_{\mathcal{N}(B)})) = \infty\}. \end{aligned}$$

PROOF. Suppose that $\alpha(B) = d(A) + d(B) = \infty$. Then there exists a left invertible operator $C_1 : \mathcal{N}(B) \to \mathcal{H}$ such that $\mathcal{R}(C_1) = \overline{\mathcal{R}(A)}$. We prove that M_C is a generalized left Weyl operator for C given by

$$C = \begin{bmatrix} C_1 & 0 \end{bmatrix} : \mathcal{N}(B) \oplus \mathcal{N}(B)^{\perp} \to \mathcal{H}.$$

Evidently, M_C is represented by

$$M_C = \begin{bmatrix} A & C_1 & 0 \\ 0 & 0 & B_1 \end{bmatrix} : \mathcal{H} \oplus \mathcal{N}(B) \oplus \mathcal{N}(B)^{\perp} \to \mathcal{H} \oplus \mathcal{K},$$

where $B_1 : \mathcal{N}(B)^{\perp} \to \mathcal{K}$ is left invertible and

$$\mathcal{R}(M_C) = (\mathcal{R}(A) + \mathcal{R}(C_1)) \oplus \mathcal{R}(B_1) = \overline{\mathcal{R}(A)} \oplus \mathcal{R}(B).$$

Thus, $\mathcal{R}(M_C)$ is closed and

$$d(M_C) = d(A) + d(B) = \infty;$$

i.e., M_C is a generalized left Weyl operator.

Conversely, suppose that there exists $C \in \mathcal{B}(\mathcal{K}, \mathcal{H})$ such that M_C is generalized left Weyl. It follows that M_C has a matrix representation

$$M_C = \begin{bmatrix} A & C_1 & C_2 \\ 0 & 0 & B_1 \end{bmatrix} : \mathcal{H} \oplus \mathcal{N}(B) \oplus \mathcal{N}(B)^{\perp} \to \mathcal{H} \oplus \mathcal{K}, \tag{1.65}$$

where $B_1 : \mathcal{N}(B)^{\perp} \to \mathcal{K}$ is left invertible and there exists an invertible operator $U \in \mathcal{B}(\mathcal{H} \oplus \mathcal{K})$ such that

$$UM_C = \begin{bmatrix} A & C_1 & 0 \\ 0 & 0 & B_1 \end{bmatrix} : \mathcal{H} \oplus \mathcal{N}(B) \oplus \mathcal{N}(B)^{\perp} \to \mathcal{H} \oplus \mathcal{K}. \tag{1.66}$$

Since UM_C has a closed range, by Lemma 1.64 and the fact that $\mathcal{R}(A)$ is nonclosed we have that $\alpha(B) = \infty$. Also, applying Lemma 1.65 we get that $\alpha([\begin{array}{cc} A & C_1 \end{array}]) = \infty$, which implies that $d(UM_C) = d(B) + d([\begin{array}{cc} A & C_1 \end{array}]) = \infty$. Since $d([\begin{array}{cc} A & C_1 \end{array}]) \leq d(A)$ it follows that $d(A) + d(B) = \infty$.

In order to describe the set $S_{GLW}(A, B)$, notice that for arbitrary $C \in \mathcal{B}(\mathcal{K}, \mathcal{H})$, M_C has a form (1.65) and notice that there exists an invertible operator $U \in \mathcal{B}(\mathcal{H} \oplus \mathcal{K})$ such that UM_C is given by (1.66). Hence, M_C is generalized left Weyl if and only if C is such that $\mathcal{R}\left([\begin{array}{cc} A & C_1 \end{array}]\right)$ is closed and such that

$$\alpha\left([\begin{array}{cc} A & C_1 \end{array}]\right) + \alpha(B_1) \leq d\left([\begin{array}{cc} A & C_1 \end{array}]\right) + d(B_1). \tag{1.67}$$

Notice that if $\mathcal{R}\left([\begin{array}{cc} A & C_1 \end{array}]\right)$ is a closed subspace, then by Lemma 1.65 we have that $\alpha\left([\begin{array}{cc} A & C_1 \end{array}]\right) = \infty$. Hence M_C is a generalized left Weyl operator for $C \in \mathcal{B}(\mathcal{K}, \mathcal{H})$ if and only if $\mathcal{R}\left([\begin{array}{cc} A & C_1 \end{array}]\right)$ is closed and $d\left([\begin{array}{cc} A & C_1 \end{array}]\right) + d(B_1) = \infty$. Obviously $d(B_1) = d(B)$. □

REMARK 1.70. It is interesting to notice that the condition $d(B) + \operatorname{codim}(\mathcal{R}(A) + \mathcal{R}(CP_{\mathcal{N}(B)})) = \infty$ from Proposition 1.69, appearing also in item (iii) of Theorem 1.66, can be replaced by the condition $d(C_3) + d(B) = \infty$, where C_3 is the block-operator defined by (1.59). So, if $A \in \mathcal{B}(\mathcal{H})$ and $B \in \mathcal{B}(\mathcal{K})$ are such that $\mathcal{R}(A)$ is nonclosed and $\mathcal{R}(B)$ is closed, then

$$M_C = \begin{bmatrix} A_1 & C_1 & C_2 \\ 0 & C_3 & C_4 \\ 0 & 0 & B_1 \end{bmatrix} : \begin{bmatrix} \mathcal{H} \\ \mathcal{N}(B) \\ \mathcal{N}(B)^\perp \end{bmatrix} \to \begin{bmatrix} \overline{\mathcal{R}(A)} \\ \mathcal{R}(A)^\perp \\ \mathcal{K} \end{bmatrix},$$

where $A_1 : \mathcal{H} \to \overline{\mathcal{R}(A)}$ is with a dense range and $B_1 : \mathcal{N}(B)^\perp \to \mathcal{K}$ is left invertible. There exists an invertible $U \in \mathcal{B}(\mathcal{H} \oplus \mathcal{K})$ such that

$$UM_C = \begin{bmatrix} A_1 & C_1 & 0 \\ 0 & C_3 & 0 \\ 0 & 0 & B_1 \end{bmatrix} : \begin{bmatrix} \mathcal{H} \\ \mathcal{N}(B) \\ \mathcal{N}(B)^\perp \end{bmatrix} \to \begin{bmatrix} \overline{\mathcal{R}(A)} \\ \mathcal{R}(A)^\perp \\ \mathcal{K} \end{bmatrix}. \tag{1.68}$$

Now, it is evident that $\mathcal{R}(M_C)$ is closed if and only if $\begin{bmatrix} A_1 & C_1 \\ 0 & C_3 \end{bmatrix}$ is closed which is equivalent with the fact that $\mathcal{R}(A) + \mathcal{R}(CP_{\mathcal{N}(B)})$ is closed. Also, in that case

$$d\left(\begin{bmatrix} A_1 & C_1 \\ 0 & C_3 \end{bmatrix}\right) = \alpha\left(\begin{bmatrix} A_1^* & 0 \\ C_1^* & C_3^* \end{bmatrix}\right) = \alpha(C_3^*) = d(C_3),$$

since A_1^* is injective. Hence in this case, the set S_{GLW} can also be described by

$$S_{GLW}(A, B) = \{C \in \mathcal{B}(\mathcal{K}, \mathcal{H})\colon C \text{ is given by (1.59)},\ \mathcal{R}(A) + \mathcal{R}(CP_{\mathcal{N}(B)}) \text{ is closed},\ d(C_3) + d(B) = \infty\}.$$

PROPOSITION 1.71. *Let $A \in \mathcal{B}(\mathcal{H})$ and $B \in \mathcal{B}(\mathcal{K})$ be such that $\mathcal{R}(A)$ and $\mathcal{R}(B)$ are nonclosed. There exists $C \in \mathcal{B}(\mathcal{K}, \mathcal{H})$ such that M_C is generalized left Weyl if and only if $\alpha(B) = d(A) = \infty$. In this case,*

$$S_{GLW}(A, B) = \{C \in \mathcal{B}(\mathcal{K}, \mathcal{H}) : \mathcal{R}(M_C) \text{ is closed}\}.$$

PROOF. Since $\mathcal{R}(A)$ and $\mathcal{R}(B)$ are nonclosed, by Lemma 1.65 we conclude that if $C \in \mathcal{B}(\mathcal{K}, \mathcal{H})$ is such that $\mathcal{R}(M_C)$ is closed, then $\alpha(M_C) = d(M_C) = \infty$. Hence, M_C is generalized left Weyl if and only if $\mathcal{R}(M_C)$ is closed. Now, the proof directly follows by Theorem 2.6 of [**134**]. □

As a corollary of the previous proposition, we get the following description of the set $\bigcap_{C\in\mathcal{B}(\mathcal{K},\mathcal{H})}\sigma^g_{lw}(M_C)$:

COROLLARY 1.72. Let $A\in\mathcal{B}(\mathcal{H})$ and $B\in\mathcal{B}(\mathcal{K})$ be given operators. Then

$$\begin{aligned}\bigcap_{C\in\mathcal{B}(\mathcal{K},\mathcal{H})}\sigma^g_{lw}(M_C) &= \{\lambda\in\mathbb{C}\colon \mathcal{R}(A-\lambda I)\text{ is nonclosed, } \alpha(B-\lambda I)<\infty\}\\ &\cup\{\lambda\in\mathbb{C}\colon \mathcal{R}(B-\lambda I)\text{ is nonclosed, } d(A-\lambda I)<\infty\}\\ &\cup\{\lambda\in\mathbb{C}\colon \mathcal{R}(A-\lambda I)\text{ is nonclosed, } \mathcal{R}(B-\lambda I)\text{ is closed,}\\ &\qquad d(A-\lambda I)+d(B-\lambda I)<\infty\}\\ &\cup\{\lambda\in\mathbb{C}\colon \mathcal{R}(A-\lambda I),\mathcal{R}(B-\lambda I)\text{ are closed,}\\ &\qquad \alpha(A-\lambda I)+\alpha(B-\lambda I)>d(A-\lambda I)+d(B-\lambda I)\}.\end{aligned}\tag{1.69}$$

Using Theorem 1.66, Remark 1.70, and the fact that A is generalized left Weyl if and only if A^* is generalized right Weyl, we can give the description of the set $S_{GW}(A,B)$ which consists of all $C\in\mathcal{B}(\mathcal{K},\mathcal{H})$ such that M_C is generalized Weyl.

THEOREM 1.73. *Let $A\in\mathcal{B}(\mathcal{H})$ and $B\in\mathcal{B}(\mathcal{K})$ be given operators. There exists $C\in\mathcal{B}(\mathcal{K},\mathcal{H})$ such that M_C is generalized Weyl if and only if one of the following conditions is satisfied:*

(i) *$\mathcal{R}(A)$ and $\mathcal{R}(B)$ are closed and $\alpha(A)+\alpha(B)=d(A)+d(B)$. In this case,*

$$S_{GW}(A,B)=\{C\in\mathcal{B}(\mathcal{K},\mathcal{H})\colon C \text{ is given by (1.59)}, C_3 \text{ has closed range,}\\ \alpha(A)+\alpha(C_3)=d(C_3)+d(B)\}.$$

(ii) *$\mathcal{R}(A)$ is closed, $\mathcal{R}(B)$ is nonclosed, and $d(A)=\alpha(A)+\alpha(B)=\infty$. In this case,*

$$S_{GW}(A,B)=\{C\in\mathcal{B}(\mathcal{K},\mathcal{H})\colon C \text{ is given by (1.59)}, \alpha(A)+\alpha(C_3)=\infty,\\ \mathcal{R}(B^*)+R(C^*P_{\mathcal{R}(A)^\perp}) \text{ is closed}\}.$$

(iii) *$\mathcal{R}(A)$ is nonclosed, $\mathcal{R}(B)$ is closed, and $\alpha(B)=d(A)+d(B)=\infty$. In this case,*

$$S_{GW}(A,B)=\{C\in\mathcal{B}(\mathcal{K},\mathcal{H})\colon C \text{ is given by (1.59)}, d(B)+d(C_3)=\infty,\\ \mathcal{R}(A)+\mathcal{R}(CP_{\mathcal{N}(B)}) \text{ is closed}\}.$$

(iv) *$\mathcal{R}(A)$ and $\mathcal{R}(B)$ are nonclosed and $\alpha(B)=d(A)=\infty$. In this case,*

$$S_{GW}(A,B)=\{C\in\mathcal{B}(\mathcal{K},\mathcal{H})\colon \mathcal{R}(M_C) \text{ is closed}\}.$$

PROOF. Since necessary and sufficient conditions for the existence of $C\in\mathcal{B}(\mathcal{K},\mathcal{H})$ such that M_C is generalized Weyl are given in [**130**], we need only prove that the set $S_{GW}(A,B)$ is given as claimed in each of the four possible cases appearing above.

(i) Suppose that $\mathcal{R}(A)$ and $\mathcal{R}(B)$ are closed and $\alpha(A)+\alpha(B)=d(A)+d(B)$. Using Theorem 1.66, we have that $C\in\mathcal{B}(\mathcal{K},\mathcal{H})$ is such that M_C is generalized left Weyl if and only if C is given by (1.59), where C_3 has closed range and

$$\alpha(A)+\alpha(C_3)\le d(C_3)+d(B).$$

Since we are looking for $C\in\mathcal{B}(\mathcal{K},\mathcal{H})$ such that M_C is generalized Weyl, we are asking for which $C\in\mathcal{B}(\mathcal{K},\mathcal{H})$ satisfying the previously mentioned condition is M_C

generalized right Weyl, i.e., $(M_C)^*$ is generalized left Weyl. Since,

$$(M_C)^* = \begin{bmatrix} B^* & C^* \\ 0 & A^* \end{bmatrix} : \begin{bmatrix} \mathcal{K} \\ \mathcal{H} \end{bmatrix} \to \begin{bmatrix} \mathcal{K} \\ \mathcal{H} \end{bmatrix}$$

and, for C given by (1.59), C^* is given by

$$C^* = \begin{bmatrix} C_4^* & C_2^* \\ C_3^* & C_1^* \end{bmatrix} : \begin{bmatrix} \mathcal{N}(A^*) \\ \mathcal{N}(A^*)^\perp \end{bmatrix} \to \begin{bmatrix} \overline{\mathcal{R}(B^*)} \\ \mathcal{R}(B^*)^\perp \end{bmatrix}, \tag{1.70}$$

applying Theorem 1.66 we get that $(M_C)^*$ is a generalized left Weyl operator if and only if $\mathcal{R}(C_3^*)$ is closed and

$$\alpha(B^*) + \alpha(C_3^*) \leq d(C_3^*) + d(A^*)$$

which is equivalent with $\mathcal{R}(C_3)$ being closed and the inequality $d(C_3) + d(B) \leq \alpha(A) + \alpha(C_3)$. Hence, M_C is a generalized Weyl operator if and only if C is given by (1.59), where C_3 has closed range and $\alpha(A) + \alpha(C_3) = d(C_3) + d(B)$.

(ii) Suppose that $\mathcal{R}(A)$ is closed, $\mathcal{R}(B)$ is nonclosed, and $d(A) = \alpha(A)+\alpha(B) = \infty$. Using Theorem 1.66, we have that $C \in \mathcal{B}(\mathcal{K},\mathcal{H})$ is such that M_C is generalized left Weyl if and only if $\mathcal{R}(B^*) + \mathcal{R}(C^*P_{\mathcal{R}(A)^\perp})$ is closed. By item (iii) of Theorem 1.66, using the representations of $(M_C)^*$ given above, we get that $(M_C)^*$ is a generalized left Weyl operator if and only if $\mathcal{R}(B^*) + \mathcal{R}(C^*P_{\mathcal{R}(A)^\perp})$ is closed and $d(A^*) + \operatorname{codim}(\mathcal{R}(B^*) + \mathcal{R}(C^*P_{\mathcal{N}(A^*)})) = \infty$. By Remark 1.70, we have that the last condition is equivalent with $d(C_3^*)+d(A^*) = \infty$; i.e., $\alpha(A)+\alpha(C_3) = \infty$, where C_3 is the block operator in the representation (1.59) of C.

Hence, M_C is a generalized Weyl operator if and only if C is given by (1.59), where $\mathcal{R}(B^*) + \mathcal{R}(C^*P_{\mathcal{R}(A)^\perp})$ is closed and $\alpha(A) + \alpha(C_3) = \infty$.

Items (iii) and (iv) can be proved in a similar manner. □

As a corollary of the previous theorem, we also get the description of the set $\bigcup_{C\in\mathcal{B}(\mathcal{K},\mathcal{H})} \sigma^g_{lw}(M_C)$:

COROLLARY 1.74. For given operators $A \in \mathcal{B}(\mathcal{H})$ and $B \in \mathcal{B}(\mathcal{K})$ we have

$$\begin{aligned}\bigcup_{C\in\mathcal{B}(\mathcal{K},\mathcal{H})} \sigma^g_{lw}(M_C) =& \{\lambda \in \mathbb{C} \colon \mathcal{R}(A-\lambda I) \text{ is not closed}\} \\ &\cup \{\lambda \in \mathbb{C} \colon \mathcal{R}(B-\lambda I) \text{ is not closed}\} \\ &\cup \{\lambda \in \mathbb{C} \colon d(A-\lambda I) = \alpha(B-\lambda I) = \infty\} \\ &\cup \{\lambda \in \mathbb{C} \colon d(A-\lambda I), \alpha(B-\lambda I) < \infty, \\ &\quad \alpha(A-\lambda I) + \alpha(B-\lambda I) > d(A-\lambda I) + d(B-\lambda I)\} \\ &\cup \{\lambda \in \mathbb{C} \colon d(B-\lambda I) < \alpha(B-\lambda I) = \infty\}.\end{aligned} \tag{1.71}$$

In the next theorem we present necessary and sufficient conditions which two operators $A \in \mathcal{B}(\mathcal{H})$ and $B \in \mathcal{B}(\mathcal{K})$ have to satisfy in order for M_C to be a generalized left Weyl operator for each $C \in \mathcal{B}(\mathcal{K},\mathcal{H})$.

THEOREM 1.75. *Let $A \in \mathcal{B}(\mathcal{H})$ and $B \in \mathcal{B}(\mathcal{K})$. Then M_C is a generalized left Weyl operator for each $C \in \mathcal{B}(\mathcal{K},\mathcal{H})$ if and only if $\mathcal{R}(A)$ and $\mathcal{R}(B)$ are closed and one of the following conditions is satisfied:*

(1) $d(A) < \infty$, $\alpha(B) = \infty$, $d(B) = \infty$.
(2) $d(A) = \infty$, $\alpha(B) < \infty$.
(3) $d(A), \alpha(B) < \infty$, $\alpha(A) + \alpha(B) \leq d(A) + d(B)$.

PROOF. Suppose that M_C is a generalized left Weyl operator for each $C \in \mathcal{B}(\mathcal{K},\mathcal{H})$. If at least one of $\mathcal{R}(A)$ and $\mathcal{R}(B)$ is not closed, we have that M_0 is not a generalized left Weyl operator since its range is not closed. So, it follows that $\mathcal{R}(A)$ and $\mathcal{R}(B)$ are closed subspaces.

Notice that for arbitrary $C \in \mathcal{B}(\mathcal{K},\mathcal{H})$, M_C is given by

$$M_C = \begin{bmatrix} 0 & A_1 & C_1 & C_2 \\ 0 & 0 & C_3 & C_4 \\ 0 & 0 & 0 & B_1 \\ 0 & 0 & 0 & 0 \end{bmatrix} : \begin{bmatrix} \mathcal{N}(A) \\ \mathcal{N}(A)^{\perp} \\ \mathcal{N}(B) \\ \mathcal{N}(B)^{\perp} \end{bmatrix} \to \begin{bmatrix} \mathcal{R}(A) \\ \mathcal{R}(A)^{\perp} \\ \mathcal{R}(B) \\ \mathcal{R}(B)^{\perp} \end{bmatrix}, \tag{1.72}$$

where A_1, B_1 are invertible operators and that there exist invertible U, $V \in \mathcal{B}(\mathcal{H} \oplus \mathcal{K})$ such that

$$UM_CV = \begin{bmatrix} 0 & A_1 & 0 & 0 \\ 0 & 0 & C_3 & 0 \\ 0 & 0 & 0 & B_1 \\ 0 & 0 & 0 & 0 \end{bmatrix} : \begin{bmatrix} \mathcal{N}(A) \\ \mathcal{N}(A)^{\perp} \\ \mathcal{N}(B) \\ \mathcal{N}(B)^{\perp} \end{bmatrix} \to \begin{bmatrix} \mathcal{R}(A) \\ \mathcal{R}(A)^{\perp} \\ \mathcal{R}(B) \\ \mathcal{R}(B)^{\perp} \end{bmatrix}. \tag{1.73}$$

So, for any $C_3 \in \mathcal{B}(\mathcal{N}(B), \mathcal{R}(A)^{\perp})$, we have that $\mathcal{R}(C_3)$ is closed and

$$\alpha(A) + \alpha(C_3) \le d(B) + d(C_3).$$

Hence at least one of $d(A)$ and $\alpha(B)$ is finite. So, we consider all possible cases (there are three in total) when at least one of $d(A)$ and $\alpha(B)$ is finite.

Suppose first that $d(A) < \infty$ and $\alpha(B) = \infty$. Since for any $C_3 \in \mathcal{B}(\mathcal{N}(B), \mathcal{R}(A)^{\perp})$ it follows that $\alpha(C_3) = \infty$ and since there exists $C_3 \in \mathcal{B}(\mathcal{N}(B), \mathcal{R}(A)^{\perp})$ such that $d(C_3) = 0$, we conclude that $\alpha(M_C) \le d(M_C)$, for each $C \in \mathcal{B}(\mathcal{K},\mathcal{H})$ if and only if $d(B) = \infty$.

If $d(A) = \infty$ and $\alpha(B) < \infty$, then for any $C_3 \in \mathcal{B}(\mathcal{N}(B), \mathcal{R}(A)^{\perp})$, we have that $d(C_3) = \infty$, so $\alpha(M_C) \le d(M_C)$ is satisfied for any $C \in \mathcal{B}(\mathcal{K},\mathcal{H})$.

If $d(A), \alpha(B) < \infty$, then for any $C_3 \in \mathcal{B}(\mathcal{N}(B), \mathcal{R}(A)^{\perp})$, we have that $\alpha(B) - \alpha(C_3) = d(A) - d(C_3)$, so $\alpha(M_C) \le d(M_C)$ if and only if $\alpha(A)+\alpha(B) \le d(A)+d(B)$.

The converse implication can be proved in the same manner. □

REMARK 1.76. Throughout this section we have used the following fact: For given operators $A \in \mathcal{B}(\mathcal{H})$ and $B \in \mathcal{B}(\mathcal{K})$ in the each of three cases

1. $\mathcal{R}(A)$ and $\mathcal{R}(B)$ are closed,
2. $\mathcal{R}(A)$ is closed, and $\mathcal{R}(B)$ is nonclosed,
3. $\mathcal{R}(A)$ is nonclosed, and $\mathcal{R}(B)$ is closed,

we have that $\mathcal{R}(M_C)$ is closed if and only if the respective condition below is satisfied:

(1) $\mathcal{R}(C_3)$ is closed,
(2) $\mathcal{R}(B^*) + \mathcal{R}(C^* P_{\mathcal{R}(A)^{\perp}})$ is closed,
(3) $\mathcal{R}(A) + R(CP_{\mathcal{N}(B)})$ is closed.

1.7. Completions of upper triangular operator matrices to Browder operator

Recall that an operator A is called Browder if it is Fredholm of finite ascent and descent. The Browder spectrum $\sigma_b(A)$ of A is defined by

$$\sigma_b(A) = \{\lambda \in \mathbb{C} \colon A - \lambda I \text{ is not Browder}\}.$$

Let $\mathcal{F}_+^-(\mathcal{H})$ be the class of all left semi-Fredholm operators with nonpositive index. Let the Browder essential approximate point spectrum for any $A \in \mathcal{B}(\mathcal{H})$, be defined by

$$\sigma_{ab}(A) = \{\lambda \in \mathbb{C}\colon A - \lambda I \notin \mathcal{F}(\mathcal{H})_+^- \text{ or } \operatorname{asc}(A - \lambda I) = \infty\}.$$

Let $A \in \mathcal{B}(\mathcal{H})$ and $B \in \mathcal{B}(\mathcal{K})$ be given. In this section, we present results related to Browder completions of upper triangular operator matrices. Certain sufficient conditions for the existence of $C \in \mathcal{B}(\mathcal{K}, \mathcal{H})$ such that M_C is a Browder operator in the case when $\mathcal{H}$ and $\mathcal{K}$ are separable Hilbert spaces are given in [**31**] and read as follows.

THEOREM 1.77 ([**31**]). *Let $A \in \mathcal{B}(\mathcal{H})$ be left semi-Fredholm with finite ascent, and let $B \in \mathcal{B}(\mathcal{K})$ be right semi-Fredholm with finite descent. If $d(A) = \alpha(B) = \infty$, then there exists $C \in \mathcal{B}(\mathcal{K}, \mathcal{H})$ such that M_C is Browder.*

PROOF. Let $p = \max\{\operatorname{asc}(A), \operatorname{dsc}(B)\}$. From the fact that $d(A) = \infty$ and $\alpha(A^p) < \infty$, we know that $\dim(\mathcal{R}(A) + \mathcal{N}(A^p))^\perp = \infty$. Also using the fact that $d((B^*)^p) = \alpha(B^p) = \infty$ and $\alpha((B^*)^p) = d(B^p) < \infty$, we know $\dim(\mathcal{R}(B^*) + \mathcal{N}((B^*)^p))^\perp = \dim(\mathcal{R}(B^*)^\perp \cap \mathcal{N}((B^*)^p)^\perp) = \dim(\mathcal{N}(B) \cap \mathcal{R}(B^p)) = \infty$. Since $\mathcal{N}(B) \cap \mathcal{R}(B^p)$ and $(\mathcal{R}(A) + \mathcal{N}(A^p))^\perp$ are separable, there exists a linear operator T with domain $\mathcal{N}(B) \cap \mathcal{R}(B^p)$ and range $(\mathcal{R}(A) + \mathcal{N}(A^p))^\perp$ such that $\|Ty\| = \|y\|$ for every $y \in \mathcal{N}(B) \cap \mathcal{R}(B^p)$. Define an operator $C : \mathcal{K} \to \mathcal{H}$ by

$$C = \begin{bmatrix} T & 0 \\ 0 & 0 \end{bmatrix} : \begin{bmatrix} \mathcal{N}(B) \cap \mathcal{R}(B^p) \\ \mathcal{R}(B^*) + \mathcal{N}((B^*)^p) \end{bmatrix} \to \begin{bmatrix} (\mathcal{R}(A) + \mathcal{N}(A^p))^\perp \\ \mathcal{R}(A) + \mathcal{N}(A^p) \end{bmatrix}.$$

Then we claim that M_C is Browder. We divide the proof into four steps.

(I) $\alpha(M_C) < \infty$.

Let $\begin{bmatrix} x_0 \\ y_0 \end{bmatrix} \in \mathcal{N}(M_C)$. Then $Ax_0 + Cy_0 = 0$ and $By_0 = 0$. Thus $Ax_0 = -Cy_0 \in \mathcal{R}(A) \cap (\mathcal{R}(A) + \mathcal{N}(A^p))^\perp \subseteq \mathcal{R}(A) \cap \mathcal{R}(A)^\perp$ and hence $Ax_0 = 0$ and $Cy_0 = 0$. Let $y_0 = y_1 + y_2$, where $y_1 \in \mathcal{N}(B) \cap \mathcal{R}(B^p)$ and $y_2 \in \mathcal{R}(B^*) + \mathcal{N}((B^*)^p)$. Then $Cy_0 = Ty_1 = 0$ implies that $y_1 = 0$, which means that $y_0 \in \mathcal{R}(B^*) + \mathcal{N}((B^*)^p)$. Thus $y_0 \in (\mathcal{R}(B^*) + \mathcal{N}((B^*)^p)) \cap \mathcal{N}(B)$. Now, we get that $\mathcal{N}(M_C) \subseteq \mathcal{N}(A) \oplus (\mathcal{R}(B^*) + \mathcal{N}((B^*)^p)) \cap \mathcal{N}(B)$.

We claim that $\dim(\mathcal{R}(B^*) + \mathcal{N}((B^*)^p)) \cap \mathcal{N}(B) < \infty$. To the contrary, we assume that $(\mathcal{R}(B^*) + \mathcal{N}((B^*)^p)) \cap \mathcal{N}(B)$ is infinite dimensional. Let $\{u_n\}_{n=1}^\infty$ be the orthonormal sequence in $(\mathcal{R}(B^*) + \mathcal{N}((B^*)^p)) \cap \mathcal{N}(B)$, and suppose $u_n = w_n + v_n$, where $w_n \in \mathcal{R}(B^*)$ and $v_n \in \mathcal{N}((B^*)^p)$. Then $\{v_n\}_{n=1}^\infty \subseteq \mathcal{N}((B^*)^p)$ is linear independent. In fact, for any $n \in \mathbb{N}$, if $a_1v_1 + a_2v_2 + \cdots + a_nv_n = 0$, then $a_1(u_1 - w_1) + a_2(u_2 - w_2) + \cdots + a_n(u_n - w_n) = 0$, and hence $a_1u_1 + a_2u_2 + \cdots + a_nu_n = a_1w_1 + a_3w_2 + \cdots + a_nw_n \in \mathcal{R}(B^*) \cap \mathcal{N}(B)$. Since, $\mathcal{R}(B^*) = \mathcal{N}(B)^\perp$, it follows that $\mathcal{R}(B^*) \cap \mathcal{N}(B) = \mathcal{N}(B)^\perp \cap \mathcal{N}(B) = \{0\}$. Then $a_1u_1 + a_2u_2 + \cdots + a_nu_n = 0$. Thus $a_1 = a_2 = \cdots = a_n = 0$; therefore $\{v_n\}_{i=1}^n \subseteq \mathcal{N}((B^*)^p)$ is linear independent for any $n \in \mathbb{N}$. This induces that $\dim(\mathcal{N}((B^*)^p)) = \infty$. It is in contradiction to the fact that $\dim(\mathcal{N}((B^*)^p)) = d(B^p) < \infty$.

From the preceding proof, we get that

$$\alpha(M_C) \leq \alpha(A) + \dim\left((\mathcal{R}(B^*) + \mathcal{N}((B^*)^p)) \cap \mathcal{N}(B)\right) < \infty.$$

(II) $\mathcal{R}(M_C)$ is closed.

Suppose that $M_C \begin{bmatrix} x_n \\ y_n \end{bmatrix} \to \begin{bmatrix} u_0 \\ v_0 \end{bmatrix}$, $n \to \infty$. Then $Ax_n + Cy_n \to u_0$ and $By_n \to v_0$, $n \to \infty$. Thus $\{Ax_n\}_{n=1}^{\infty}$ and $\{Cy_n\}_{n=1}^{\infty}$ are Cauchy sequences. Write $y_n = u_n + v_n$, where $u_n \in \mathcal{N}(B) \cap R(B^p)$ and $v_n \in \mathcal{R}(B^*) + \mathcal{N}((B^*)^p)$. Then $Cy_n = Tu_n$, which implies that $\{u_n\}_{n=1}^{\infty}$ is a Cauchy sequence. Also $\{Bv_n\}$ is a Cauchy sequence since $By_n = Bv_n$. Using the fact that $\mathcal{R}(B^*)$ is closed and $\dim(\mathcal{N}((B^*)^p)) < \infty$, we know that $\mathcal{R}(B^*) + \mathcal{N}((B^*)^p)$ is closed. There is orthogonal decomposition $\mathcal{R}(B^*) + \mathcal{N}((B^*)^p) = \mathcal{N}(B)^{\perp} \oplus M$ and $M \subseteq \mathcal{N}(B)$. Let $v_n = w_n + z_n$, where $w_n \in \mathcal{N}(B)^{\perp}$ and $z_n \in M$. Then $Bv_n = Bw_n$, and hence $\{Bw_n\}_{n=1}^{\infty}$ is a Cauchy sequence. We claim that $\{w_n\}$ is a Cauchy sequence. Indeed, since $Bv_n = Bw_n \to v_0$, $n \to \infty$, and $\mathcal{R}(B)$ is closed, it follows that there exists $z_0 \in \mathcal{N}(B)^{\perp}$ such that $Bv_n = Bw_n \to Bz_0 = v_0$, $n \to \infty$, and hence $B(w_n - z_0) \to 0$, $n \to \infty$. Since $B|_{\mathcal{N}(B)^{\perp}}$ is invertible, we get that $w_n - z_0 \to 0$, $n \to \infty$. Then $\{w_n\}$ is a Cauchy sequence.

Let $y'_n = u_n + w_n$. Then $\{y'_n\}_{n=1}^{\infty}$ is a Cauchy sequence. Suppose $y'_n \to y_0$, $n \to \infty$. Then $Cy_n = Cy'_n \to Cy_0$, $n \to \infty$ and $By_n = By'_n \to By_0$, $n \to \infty$. Suppose $Ax_n \to Ax_0$, $n \to \infty$. We get that $M_C \begin{bmatrix} x_0 \\ y_0 \end{bmatrix} = \begin{bmatrix} u_0 \\ v_0 \end{bmatrix}$. This proves that $\mathcal{R}(M_C)$ is closed.

The preceding proof tells us that M_C is upper semi-Fredholm.

(III) M_C has finite ascent.

We only need to prove that $\mathcal{N}(M_C^p) = \mathcal{N}(M_C^{p+1})$. Let $\begin{bmatrix} x \\ y \end{bmatrix} \in \mathcal{N}(M_C^{p+1})$. Then

$$\begin{cases} A^{p+1}x + A^pCy + A^{p-1}CBy + \cdots + ACB^{p-1}y + CB^py = 0, \\ B^{p+1}y = 0. \end{cases}$$

Thus $B^py \in \mathcal{N}(B) \cap \mathcal{R}(B^p)$ and $CB^py \in (\mathcal{R}(A) + \mathcal{N}(A^p))^{\perp} \subseteq \mathcal{R}(A)^{\perp}$. It induces that $A^{p+1}x + A^pCy + A^{p-1}CBy + \cdots + ACB^{p-1}y = -CB^py \in \mathcal{R}(A) \cap (\mathcal{R}(A) + \mathcal{N}(A^p))^{\perp} \subseteq \mathcal{R}(A) \cap \mathcal{R}(A)^{\perp} = \{0\}$. Then

$$\begin{aligned} &A^{p+1}x + A^pCy + A^{p-1}CBy + \cdots + ACB^{p-1}y = 0, \\ &CB^py = 0. \end{aligned}$$

Using the definition of operator C, we know that $CB^py = TB^py = 0$. This implies that $B^py = 0$. Since $A^{p+1}x + A^pCy + A^{p-1}CBy + \cdots + ACB^{p-1}y = A(A^px + A^{p-1}Cy + A^{p-2}CBy + \cdots + ACB^{p-2}y + CB^{p-1}y) = 0$, it follows that $A^px + A^{p-1}Cy + A^{p-2}CBy + \cdots + ACB^{p-2}y + CB^{p-1}y \in \mathcal{N}(A)$. Let $A^px + A^{p-1}Cy + A^{p-2}CBy + \cdots + ACB^{p-2}y + CB^{p-1}y = x_1 \in \mathcal{N}(A)$. Then

$$A^px + A^{p-1}Cy + A^{p-2}CBy + \cdots + ACB^{p-2}y - x_1 + CB^{p-1}y = 0.$$

It follows that $A^px + A^{p-1}Cy + A^{p-2}CBy + \cdots + ACB^{p-2}y - x_1 = -CB^{p-1}y \in (\mathcal{R}(A) + \mathcal{N}(A)) \cap (\mathcal{R}(A) + \mathcal{N}(A^p))^{\perp} \subseteq (\mathcal{R}(A) + \mathcal{N}(A^p)) \cap (\mathcal{R}(A) + \mathcal{N}(A^p))^{\perp} = \{0\}$. Then $A^px + A^{p-1}Cy + A^{p-2}CBy + \cdots + ACB^{p-2}y - x_1 = -CB^{p-1}y = 0$. Hence

$$\begin{cases} A^px + A^{p-1}Cy + A^{p-2}CBy + \cdots + ACB^{p-2}y = x_1, \\ CB^{p-1}y = 0. \end{cases}$$

From $A^p x + A^{p-1}Cy + A^{p-2}CBy + \cdots + ACB^{p-2}y = x_1$, we know that $A^{p-1}x + A^{p-2}Cy + A^{p-3}CBy + \cdots + CB^{p-2}y \in \mathcal{N}(A^2)$. Let $A^{p-1}x + A^{p-2}Cy + A^{p-3}CBy + \cdots + CB^{p-2}y = x_2$, where $x_2 \in \mathcal{N}(A^2)$. Then

$$A^{p-1}x + A^{p-2}Cy + A^{p-3}CBy + \cdots + ACB^{p-3}y - x_2 + CB^{p-2}y = 0.$$

This induces $A^{p-1}x + A^{p-2}Cy + A^{p-3}CBy + \cdots + ACB^{p-3}y - x_2 = -CB^{p-2}y \in (\mathcal{R}(A) + \mathcal{N}(A^2)) \cap (\mathcal{R}(A) + \mathcal{N}(A^p))^\perp \subseteq (\mathcal{R}(A) + \mathcal{N}(A^p)) \cap (\mathcal{R}(A) + \mathcal{N}(A^p))^\perp = \{0\}$. So

$$\begin{cases} A^{p-1}x + A^{p-2}Cy + A^{p-3}CBy + \cdots + ACB^{p-3}y = x_2, \\ CB^{p-2}y = 0. \end{cases}$$

Continuing this process, we can get that

$$\begin{cases} A^2x + ACy = x_{p-1}, \\ CBy = 0, \end{cases}$$

where $x_{p-1} \in \mathcal{N}(A^{p-1})$. Then there exists $x_p \in \mathcal{N}(A^p)$ such that

$$Ax + Cy - x_p = 0.$$

Thus $Ax - x_p = -Cy \in (\mathcal{R}(A) + \mathcal{N}(A^p)) \cap (\mathcal{R}(A) + \mathcal{N}(A^p))^\perp = \{0\}$; hence $Ax = x_p$ and $Cy = 0$. It follows that $x \in \mathcal{N}(A^{p+1}) = \mathcal{N}(A^p)$. This induces that $\begin{bmatrix} x \\ y \end{bmatrix} \in \mathcal{N}(M_C^p)$, which means that $\mathcal{N}(M^p) = \mathcal{N}(M^{p+1})$. Thus M_C has finite ascent.

(IV) M_C has finite descent.

Since $\mathcal{R}(M_C)$ is closed, we know that $\operatorname{dsc}(M_C) = \operatorname{asc}(M_C^*)$. Then we need to prove that M_C^* has finite ascent, which can be done as in (III). □

In the case when $A \in \mathcal{B}(\mathcal{H})$ is Fredholm with finite ascent and $B \in \mathcal{B}(\mathcal{K})$ is Fredholm with finite descent, we have the following result:

THEOREM 1.78 ([**31**]). *Suppose $A \in \mathcal{B}(\mathcal{H})$ is Fredholm with* $\operatorname{asc}(A) < \infty$ *and $B \in \mathcal{B}(\mathcal{K})$ is Fredholm with* $\operatorname{dsc}(B) < \infty$. *If $\alpha(A) + \alpha(B) = d(A) + d(B)$, then there exists $C \in \mathcal{B}(\mathcal{K}, \mathcal{H})$ such that M_C is Browder.*

PROOF. By the given conditions we know that M_C is Weyl for every $C \in \mathcal{B}(\mathcal{K}, \mathcal{H})$. It is well known that if T is Weyl with finite ascent or with finite descent, T is Browder. We only need to prove that there is $C \in \mathcal{B}(\mathcal{K}, \mathcal{H})$ such that $\operatorname{asc}(M_C) < \infty$ or $\operatorname{asc}(M_C^*) < \infty$. Let $p = \max\{\operatorname{asc}(A), \operatorname{dsc}(B)\}$. There are two cases to consider.

Case 1. Suppose $\dim(\mathcal{N}(B) \cap \mathcal{R}(B^p)) \leq \dim(\mathcal{R}(A) + \mathcal{N}(A^p))^\perp$.

Let $M \subseteq (\mathcal{R}(A) + \mathcal{N}(A^p))^\perp$ be such that $\dim(\mathcal{N}(B) \cap \mathcal{R}(B^p)) = \dim M < \infty$. Then there exists a linear operator T with domain $\mathcal{N}(B) \cap \mathcal{R}(B^p)$ and range M such that $\|Ty\| = \|y\|$ for every $y \in (\mathcal{N}(B) \cap \mathcal{R}(B^p))$. Define an operator $C : \mathcal{K} \to \mathcal{H}$ by

$$C = \begin{bmatrix} T & 0 \\ 0 & 0 \end{bmatrix} : \begin{bmatrix} \mathcal{N}(B) \cap \mathcal{R}(B^p) \\ \mathcal{R}(B^*) + \mathcal{N}((B^*)^p) \end{bmatrix} \to \begin{bmatrix} (\mathcal{R}(A) + \mathcal{N}(A^p))^\perp \\ \mathcal{R}(A) + \mathcal{N}(A^p) \end{bmatrix}.$$

Adopting a technique in Theorem 1.77 ((III) of the proof), we get that M_C is Browder.

Case 2. Suppose $\dim(\mathcal{N}(B) \cap \mathcal{R}(B^p)) \geq \dim(\mathcal{R}(A) + \mathcal{N}(A^p))^\perp$.

Let $M \subseteq \mathcal{N}(B) + \mathcal{R}(B^p)$ be such that $\dim(\mathcal{R}(A) + \mathcal{N}(A^p))^\perp = \dim M < \infty$. Then there exists a linear operator T with domain $(\mathcal{R}(A) + \mathcal{N}(A^p))^\perp$ and range

M such that $\|Ty\| = \|y\|$ for every $y \in (\mathcal{R}(A) + \mathcal{N}(A^p))^\perp$. Define an operator $C_1 : \mathcal{H} \to \mathcal{K}$ by

$$C_1 = \begin{bmatrix} T & 0 \\ 0 & 0 \end{bmatrix} : \begin{bmatrix} (\mathcal{R}(A) + \mathcal{N}(A^p))^\perp \\ \mathcal{R}(A) + \mathcal{N}(A^p) \end{bmatrix} \to \begin{bmatrix} \mathcal{N}(B) \cap \mathcal{R}(B^p) \\ \mathcal{R}(B^*) + \mathcal{N}((B^*)^p) \end{bmatrix}.$$

Let $M_{C_1^*} = \begin{bmatrix} A & C_1^* \\ 0 & B \end{bmatrix}$. Then $M_{C_1^*}^*$ has finite ascent. Thus $M_{C_1^*}$ has finite descent.

From the preceding proof, there is $C \in \mathcal{B}(\mathcal{K}, \mathcal{H})$ such that M_C is Browder. □

If M_C is Browder for some $C \in \mathcal{B}(\mathcal{K}, \mathcal{H})$, then $A \in \mathcal{B}(\mathcal{H})$ is left semi-Fredholm with finite ascent, $B \in \mathcal{B}(\mathcal{K})$ is right semi-Fredholm with finite descent, and $d(A) < \infty$ if and only if $\alpha(B) < \infty$. If $d(A) = \alpha(B) = \infty$, then $\alpha(A) + \alpha(B) = d(A) + d(B)$. But if $d(A) < \infty$, both A and B are Fredholm. Since M_C is Weyl, it follows from the decomposition

$$M_C = \begin{bmatrix} I & 0 \\ 0 & B \end{bmatrix} \begin{bmatrix} I & C \\ 0 & I \end{bmatrix} \begin{bmatrix} A & 0 \\ 0 & I \end{bmatrix}$$

that $\operatorname{ind}(A) + \operatorname{ind}(B) = 0$; that is, $\alpha(A) + \alpha(B) = d(A) + d(B)$. So, it follows:

THEOREM 1.79 ([**31**]). *If M_C is Browder for some $C \in \mathcal{B}(\mathcal{K}, \mathcal{H})$, then A is left semi-Fredholm with finite ascent, B is right semi-Fredholm with finite descent, and $\alpha(A) + \alpha(B) = d(A) + d(B)$.*

From Theorems 1.77–1.79, we get the following:

COROLLARY 1.80 ([**31**]). A 2×2 operator matrix $M_C = \begin{bmatrix} A & C \\ 0 & B \end{bmatrix}$ is Browder for some $C \in \mathcal{B}(\mathcal{K}, \mathcal{H})$ if and only if $A \in \mathcal{B}(\mathcal{H})$ is left semi-Fredholm with finite ascent, $B \in \mathcal{B}(\mathcal{K})$ is right semi-Fredholm with finite descent, and $\alpha(A) + \alpha(B) = d(A) + d(B)$.

In [**62**], the set $\bigcap_{C \in \mathcal{B}(\mathcal{K}, \mathcal{H})} \sigma_b(M_C)$ was investigated and it was shown that

$$\bigcap_{C \in \mathcal{B}(\mathcal{K}, \mathcal{H})} \sigma_b(M_C) \subseteq \sigma_{le}(A) \cup \sigma_{re}(B) \cup W(A, B) \cup W_1(A, B),$$

where

$$W(A, B) = \{\lambda \in \mathbb{C} \colon \mathcal{N}(B - \lambda I) \text{ and } X/\overline{\mathcal{R}(A - \lambda I)} \text{ are not isomorphic up to a finite-dimensional subspace}\},$$

$$W_1(A, B) = \{\lambda \in \mathbb{C} \colon (A - \lambda I)^D \text{ does not exist or } (B - \lambda I)^D \text{ does not exist}\}.$$

In the next corollary, the set $\bigcap_{C \in \mathcal{B}(\mathcal{K}, \mathcal{H})} \sigma_b(M_C)$ is completely characterized.

COROLLARY 1.81. For a given pair (A, B) of operators we have

$$\bigcap_{C \in B(K,H)} \sigma_b(M_C) = \{\lambda \in \mathbb{C} \colon \alpha(A - \lambda I) + \alpha(B - \lambda I) \neq d(A - \lambda I) + d(B - \lambda I)\} \cup \sigma_{ab}(A) \cup \sigma_{ab}(B^*).$$

As we mentioned before, in the case of ordinary invertibility (see [**98**]), it was shown that the passage from $\sigma(A) \cup \sigma(B)$ to $\sigma(M_C)$ is accomplished by removing a certain open subset of $\sigma(A) \cap \sigma(B)$ from the former; that is, there is equality

$$\sigma(A) \cup \sigma(B) = \sigma(M_C) \cup W_1(A, B, C),$$

where $W_1(A,B,C)$ is the union of certain of the holes in $\sigma(M_C)$, which happen to be the subsets of $\sigma(A)\cap\sigma(B)$. The passage from the $\sigma_b(A)\cup\sigma_b(B)$ to $\sigma_b(M_C)$ is more delicate as will be shown in the next result:

THEOREM 1.82 ([**31**]). *For a given pair (A,B) of operators we have that for every $C\in\mathcal{B}(\mathcal{K},\mathcal{H})$,*

$$\eta(\sigma_b(A)\cup\sigma_b(B))=\eta(\sigma_b(M_C)),$$

where $\eta(\cdot)$ denotes the polynomially convex hull. More precisely,

$$\sigma_b(A)\cup\sigma_b(B)=\sigma_b(M_C)\cup W_3(A,B,C),$$

where $W_3(A,B,C)$ is the union of certain of the holes in $\sigma_b(M_C)$, which happen to be subsets of $\sigma_{ab}(A^*)\cap\sigma_{ab}(B)$.

COROLLARY 1.83 ([**31**]). If $\sigma_{ab}(A^*)\cap\sigma_{ab}(B)$ has no interior points, then:

(1) $\sigma_b(M_C)=\sigma_b(A)\cup\sigma_b(B)$ for every $C\in\mathcal{B}(\mathcal{K},\mathcal{H})$.

(2) $\sigma_w(M_C)=\sigma_w(M_0)$ for every $C\in\mathcal{B}(\mathcal{K},\mathcal{H})$.

H. Weyl [**165**] has shown that every hermitian operator $A\in\mathcal{B}(\mathcal{H})$ satisfies the equality

$$\sigma(A)\backslash\sigma_w(A)=\pi_{00}(A), \tag{1.74}$$

where $\pi_{00}(A)=\{\lambda\in \text{iso}\,\sigma(A):0<\dim\mathcal{N}(A-\lambda I)<\infty\}$. Today we say that Weyl's theorem holds for $A\in\mathcal{B}(\mathcal{H})$ if A satisfies the equality (1.74). Weyl's theorem has been extended from hermitian operators to hyponormal operators and to Toeplitz operator by L. Coburn [**40**], to several classes of operators including seminormal operators by S. Berberian [**18, 19**], and to a few classes of Banach space operators [**149**]. But Weyl's theorem may fail for upper triangular operator matrices. Weyl's theorem for upper triangular operator matrices is more delicate in comparison with the diagonal matrices. An operator $A\in\mathcal{B}(\mathcal{H})$ is called isoloid if every isolated point of $\sigma(A)$ is an eigenvalue of A (see [**19**]).

THEOREM 1.84 ([**31**]). *Suppose $\sigma_{ab}(A^*)\cap\sigma_{ab}(B)$ has no interior points and suppose A is an isoloid operator for which Weyl's theorem holds. If Weyl's theorem holds for M_0, then for every $C\in\mathcal{B}(\mathcal{K},\mathcal{H})$, Weyl's theorem holds for M_C.*

The condition "$\sigma_{ab}(A^*)\cap\sigma_{ab}(B)$ has no interior points" is essential in Theorem 1.84. For example if A, B, and C are operators on $l_2\otimes l_2$ defined by

$$A=U\otimes I \text{ and } B=U^*\otimes I,\quad C=\begin{bmatrix}0&&&\\&I-UU^*&&\\&&I-UU^*&\\&&&\ddots\end{bmatrix},$$

where U is the unilateral shift on l_2, then:

(a) $\sigma_{ab}(A^*)\cap\sigma_{ab}(B)=\mathcal{D}$ has interior points, where $\mathcal{D}$ denotes the closed unit disk.

(b) $\sigma(A)=\sigma_w(A)=\mathcal{D}$ and $\pi_{00}(A)=\emptyset$ which says that A is isoloid and Weyl's theorem holds for A.

Also a straightforward calculation shows that

$$\sigma(M_0) = \sigma_w(M_0) = \sigma(M_C) = \mathcal{D}, \quad \sigma_w(M_C) = \mathcal{T}, \quad \text{and} \quad \pi_{00}(M_0) = \pi_{00}(M_C) = \emptyset,$$

where $\mathcal{T}$ denotes the closed unit disk circle, which implies that Weyl's theorem holds for M_0 but fails for M_C.

In the following example we show that the "isoloid" condition is essential in Theorem 1.84.

EXAMPLE 1.85. Let $A, B, C \in \mathcal{B}(l_2)$ be defined by

$$\begin{aligned} A(x_1, x_2, x_3, \dots) &= \left(0, x_1, 0, \frac{1}{2}x_2, 0, \frac{1}{3}x_3, 0, \dots\right), \\ B(x_1, x_2, x_3, \dots) &= (0, x_2, 0, x_4, 0, x_6, \dots), \\ C(x_1, x_2, x_3, \dots) &= (0, 0, x_2, 0, x_3, 0, x_4, \dots), \end{aligned}$$

where $x = (x_1, x_2, x_3, \dots) \in l_2$. Then:

(a) $\sigma_{ab}(A^*) \cap \sigma_{ab}(B) = \{0\}$ has no interior points.

(b) $\sigma(A) = \sigma_w(A) = \{0\}$ and $\pi_{00}(A) = \emptyset$, which means that Weyl's theorem holds for A, but A is not isoloid.

(c) $\sigma(M_0) = \sigma_w(M_0) = \{0, 1\}$, $\pi_{00}(M_0) = \emptyset$, so Weyl's theorem holds for M_0.

But since $\sigma(M_C) = \sigma_w(M_C) = \{0, 1\}$ and $\pi_{00}(M_C) = \{0\}$, it follows that Weyl's theorem fails for M_C.

Also the condition "Weyl's theorem holds for A" cannot be dropped in Theorem 1.84 as will be shown in the following example:

EXAMPLE 1.86. Let $A, B, C \in \mathcal{B}(l_2)$ be defined by

$$\begin{aligned} A(x_1, x_2, x_3, \dots) &= \left(0, 0, 0, \frac{1}{2}x_2, 0, \frac{1}{3}x_3, 0, \frac{1}{4}x_4, \dots\right), \\ B(x_1, x_2, x_3, \dots) &= (0, x_2, 0, x_4, 0, x_6, \dots), \\ C(x_1, x_2, x_3, \dots) &= (x_1, 0, x_2, 0, x_3, 0, \dots). \end{aligned}$$

Then Weyl's theorem fails for A and the other conditions in Theorem 1.84 hold, but since $\sigma(M_C) = \sigma_w(M_C) = \{0, 1\}$ and $\pi_{00}(M_C) = \{0\}$, we know that Weyl's theorem fails for M_C.

In the following example we present a pair of operators A and B such that Weyl's theorem holds for M_C independently of the choice of C.

EXAMPLE 1.87. Let A, $B \in \mathcal{B}(l_2)$ be defined by

$$\begin{aligned} A(x_1, x_2, x_3, \dots) &= (x_2, x_4, x_6, \dots), \\ B(x_1, x_2, x_3, \dots) &= (0, x_1, 0, x_2, 0, x_3, 0, \dots), \end{aligned}$$

where $x = (x_1, x_2, x_3, \dots) \in l_2$. Then:

(1) $\sigma(A) = \sigma_w(A) = \mathcal{D}$ and $\pi_{00}(A) = \emptyset$. Then A is isoloid and Weyl's theorem holds for A.

(2) $\sigma_{ab}(A^*) = \sigma_{ab}(B) = \partial\mathcal{D}$. So $\sigma_{ab}(A^*) \cap \sigma_{ab}(B)$ has no interior points.

(3) $\sigma(M_0) = \sigma_w(M_0) = \mathcal{D}$ and $\pi_{00}(M_0) = \emptyset$. Then Weyl's theorem holds for M_0.

Then for every $C \in \mathcal{B}(l_2)$, Weyl's theorem holds for M_C.

THEOREM 1.88 ([**176**]). *For a given pair* (A, B) *of operators, we have*

$$\bigcap_{C\in\mathcal{B}(\mathcal{K},\mathcal{H})} \sigma_b(M_C) \subseteq \sigma_{le}(A) \cup \sigma_{re}(B) \cup W(A,B) \cup \sigma_D(A) \cup \sigma_D(B).$$

In particular, under one of the conditions

(i) $\sigma_{su}(B) = \sigma(B)$,
(ii) $\operatorname{int}\sigma_p(B) = \emptyset$,
(iii) $\sigma(A) \cap \sigma(B) = \emptyset$,
(iv) $\sigma_a(A) = \sigma(A)$,

we have

$$\bigcap_{C\in\mathcal{B}(\mathcal{K},\mathcal{H})} \sigma_b(M_C) = \sigma_{le}(A) \cup \sigma_{re}(B) \cup W(A,B) \cup \sigma_D(A) \cup \sigma_D(B),$$

where $W(A,B) = \{\lambda \in \mathbb{C} \colon \mathcal{N}(B-\lambda I)$ *and* $\mathcal{H}/\overline{\mathcal{R}(A-\lambda I)}$ *are not isomorphic up to a finite-dimensional subspace*$\}$.

COROLLARY 1.89 ([**176**]). For a given pair (A, B) of operators, if any one of the conditions

(i) $\operatorname{acc}(\sigma(B)) = \emptyset$,
(ii) $\operatorname{int}(\sigma(B)) = \emptyset$,
(iii) $\operatorname{acc}(\sigma(A)) = \emptyset$,
(iv) $\operatorname{int}(\sigma(A)) = \emptyset$

is satisfied, then we have

$$\bigcap_{C\in\mathcal{B}(\mathcal{K},\mathcal{H})} \sigma_b(M_C) = \sigma_{le}(A) \cup \sigma_{re}(B) \cup W(A,B) \cup \sigma_D(A) \cup \sigma_D(B).$$

The results of Theorem 1.88 and Corollary 1.89 are similar to that of Theorem 3.9 in [**62**].

In the following theorem a connection between the Browder spectrum and the Drazin spectrum is given.

THEOREM 1.90 ([**176**]). *If* $\sigma_b(M_C) = \sigma_b(A) \cup \sigma_b(B)$, *then*

$$\sigma_D(M_C) = \sigma_D(A) \cup \sigma_D(B).$$

Before explaining the main theorem, we give a counterexample to say that there exist $A \in \mathcal{B}(\mathcal{H})$, $B \in \mathcal{B}(\mathcal{K})$, $C \in \mathcal{B}(\mathcal{K},\mathcal{H})$, such that neither

$$\bigcap_{C\in\mathcal{B}(\mathcal{K},\mathcal{H})} \sigma_b(M_C) = \left(\bigcap_{C\in\mathcal{B}(\mathcal{K},\mathcal{H})} \sigma(M_C) \right) \setminus (\rho_b(A) \cap \rho_b(B))$$

nor $W_3(A,B,C) = W_1(A,B,C)$ holds.

EXAMPLE 1.91. Let X_n be a complex finite-dimensional Hilbert space with $\dim X_n = n < \infty$. Define operators $T, S, C \in \mathcal{B}(l_2)$ by

$$T(x_1, x_2, x_3, \dots) = (0, x_1, x_2, x_3, \dots) \text{ for any } (x_1, x_2, x_3, \dots) \in l_2$$

and $S = T^*$, $C = I - TS$. Let

$$M_{C_1} = \begin{bmatrix} T & C_1 \\ 0 & S' \end{bmatrix} : l_2 \oplus (l_2 \oplus X_n) \to l_2 \oplus (l_2 \oplus X_n),$$

i.e.,

$$M_{C_1} = \begin{bmatrix} T & C & 0 \\ 0 & S & 0 \\ 0 & 0 & 0 \end{bmatrix} : l_2 \oplus l_2 \oplus X_n \to l_2 \oplus l_2 \oplus X_n,$$

where $C_1 = \begin{bmatrix} C & 0 \end{bmatrix} : l_2 \oplus X_n \to l_2$, $S' = \begin{bmatrix} S & 0 \\ 0 & 0 \end{bmatrix} : l_2 \oplus X_n \to l_2 \oplus X_n$.

Then we can claim that
$0 \in \left(\bigcap_{Q\in\mathcal{B}(l_2\oplus X_n,l_2)} \sigma(M_Q)\right) \backslash (\rho_b(T)\cap\rho_b(S'))$ but $0 \notin \bigcap_{Q\in\mathcal{B}(l_2\oplus X_n,l_2)} \sigma_b(M_Q)$.
Thus,

$$\bigcap_{Q\in\mathcal{B}(l_2\oplus X_n,l_2)} \sigma_b(M_Q) \neq \left(\bigcap_{Q\in\mathcal{B}(l_2\oplus X_n,l_2)} \sigma(M_Q)\right) \backslash (\rho_b(T) \cap \rho_b(S')).$$

In fact:

(i) Since $M_Q = \begin{bmatrix} T & Q \\ 0 & S' \end{bmatrix}$ cannot be surjective for any $Q \in \mathcal{B}(l_2 \oplus X_n, l_2)$, then M_Q is not invertible. Thus $0 \in \bigcap_{Q\in\mathcal{B}(l_2\oplus X_n,l_2)} \sigma(M_Q)$.

(ii) Since $\mathrm{dsc}(T) = \mathrm{asc}(S') = \infty$, then neither S' nor T is a Browder operator. Hence $0 \notin \rho_b(T) \cap \rho_b(S')$.

(iii) It well known that $\begin{bmatrix} T & C \\ 0 & S \end{bmatrix}$ is unitary, so it is invertible. Moreover, since $\dim X_n = n < \infty$, then M_{C_1} is a Drazin invertible Fredholm operator. Therefore $0 \notin \bigcap_{Q\in\mathcal{B}(l_2\oplus X_n,l_2)} \sigma_b(M_Q)$.

From (iii), we can get that $\sigma(M_{C_1}) = \{\lambda \in \mathbb{C}\colon |\lambda| = 1\} \cup \{0\}$ and $\sigma_b(M_{C_1}) = \{\lambda \in \mathbb{C}\colon |\lambda| = 1\}$. It is easy to prove that

$$\sigma(T) = \sigma_b(T) = \sigma(S') = \sigma_b(S') = \{\lambda \in \mathbb{C}\colon |\lambda| \leq 1\}.$$

Thus

$$\{\lambda \in \mathbb{C}\colon 0 < |\lambda| < 1\} = W_1(T, S', C_1) \neq W_3(T, S', C_1) = \{\lambda \in \mathbb{C}\colon |\lambda| < 1\}$$

and

$$\{0\} = \mathrm{iso}\,\sigma(M_{C_1}) \neq \mathrm{iso}(\sigma(T) \cup \sigma(S')) = \emptyset.$$

1.8. Completions of upper triangular matrices to Kato nonsingular operator

Vast research can be found in the literature concerned with the study of bounded linear operators with closed range. Their significance stems from the many applications they have, e.g., in the spectral study of differential operators or in the context of perturbation theory, but also from the important role they play when it comes to purely theoretical considerations. By an old theorem of Banach, for such operators the equation $Tx = y$ is solvable if for all bounded linear functionals g there is the implication $gT = 0 \Rightarrow gy = 0$. In Hilbert spaces, these are precisely the regular operators (those possessing the Moore-Penrose inverse). S. Goldberg (see [**84**]), among others, considered the set $\sigma_c(T)$ consisting of all $\lambda \in \mathbb{C}$ such that $\mathcal{R}(T - \lambda)$ is not closed. However, $\sigma_c(T)$ fails to have some basic properties expected from of a decent spectrum. For example, it may be empty and does not satisfy the spectral mapping theorem (not even for polynomials). On the other hand, M. Mbekhta (see [**140**]–[**145**]) defines $\sigma_h(T)$ as the set of all $\lambda \in \mathbb{C}$ such that there are a neighborhood U of 0 in $\mathbb{C}$ and a holomorphic function $H : U \to \mathcal{X}$ such that $T - \lambda = (T - \lambda)H(\lambda)(T - \lambda)$ for all $\lambda \in U$, and he shows that, in a Hilbert space, this set is always nonempty, closed, and subject to the spectral mapping theorem.

A useful characterization of closed range operators was given by T. Kato (see [**114**]) who introduced the reduced minimum modulus

$$\gamma(T) = \inf\{\|Tx\| \colon \operatorname{dist}(x, \mathcal{N}(T)) = 1\}$$

and proved that $\mathcal{R}(T)$ is closed if and only if $\gamma(T) > 0$. Related to this is the study of the function $z \to \gamma(T - z)$ which, although not continuous in general, has nice continuity properties. For example we have the following result.

THEOREM 1.92 ([**146**]). *Let $\mathcal{X}$ be a Banach space and let $T \in \mathcal{B}(\mathcal{X})$ be with closed range. The following conditions are equivalent:*

(1) *The function $z \to \gamma(T - z)$ is continuous at $z = 0$.*
(2) $\mathcal{N}(T) \subseteq \bigcap_{n\in\mathbb{N}} \mathcal{R}(T^n)$.
(3) $\bigcup_{n\in\mathbb{N}} \mathcal{N}(T^n) \subseteq \mathcal{R}(T)$.
(4) $\bigcup_{n\in\mathbb{N}} \mathcal{N}(T^n) \subseteq \bigcap_{n\in\mathbb{N}} \mathcal{R}(T^n)$.

The equivalence between (2), (3), and (4) is of a purely algebraic nature and does not require the assumption of closedness of the range. An operator $T \in \mathcal{B}(\mathcal{X})$ satisfying the condition (2) of the previous theorem is said to be hyperexact (for more details see [**101**]). Various equivalents of hyperexactness and related concepts can be found in the excellent paper of R. E. Harte [**100**]. Hyperexact operators with closed range are called Kato nonsingular operators. For many interesting properties of this class of operators see [**3, 73, 148**].

There are only a few results that are concerned with completion problems to Kato operators. M. Barraa et al. [**14**] considered the completion problem to a Kato nonsingular M_C and gave some sufficient conditions for the existence of such operators C. These conditions always assumed, among other things, Kato nonsigularity of A or injectivity of B. Later Y. N. Dou et al. [**67**] generalized some of their results. In this section, we completely solve the problem of completion to Kato nonsingularity in each of the following seven cases:

Case 1: if one of the operators A or B is Kato nonsingular,
Case 2: if B is injective,
Case 3: if A is with dense range,
Case 4: if B is with finite ascent,
Case 5: if A is with finite descent,
Case 6: if $0 \notin \operatorname{int}(\sigma_p(B))$,
Case 7: if $0 \notin \operatorname{int}(\sigma_{cp}(A))$.

Throughout the section $\mathcal{H}, \mathcal{K}$ are infinite-dimensional separable complex Hilbert spaces. For $A \in \mathcal{B}(\mathcal{H})$ and $B \in \mathcal{B}(\mathcal{K})$, we use the following notation:

$$S_K(A, B) = \{C \in \mathcal{B}(\mathcal{K}, \mathcal{H}) \colon M_C \text{ is Kato nonsingular}\}.$$

We also set $\mathcal{R}^\infty(A) = \bigcap_{n\in\mathbb{N}} \mathcal{R}(A^n)$.

We begin by listing the results that will be made use of later in the section. The next two are well-known useful results.

LEMMA 1.93. *If $\mathcal{X}$ is a closed and $\mathcal{X}_0$ a finite-dimensional subspace of a Banach space, then $\mathcal{X} + \mathcal{X}_0$ is a closed subspace.*

Using elementary techniques one readily obtains the result below.

LEMMA 1.94. *Let $A \in \mathcal{B}(\mathcal{H}_1, \mathcal{H}_2)$, $B \in \mathcal{B}(\mathcal{K}_1, \mathcal{K}_2)$, and $C \in \mathcal{B}(\mathcal{K}_1, \mathcal{H}_2)$ be given operators. If $\overline{\mathcal{R}(A)} = \mathcal{H}_2$ and $\mathcal{R}(M_C)$ is closed, then $\mathcal{R}(B)$ is closed.*

Throughout this section we make extensive use of a variant of a particular result from [**67**] on the completion problem to closed range operator matrices. One readily checks that both the statement and the proof of Theorem 2.5 from [**67**] remain valid if $A \in \mathcal{B}(\mathcal{H}_1, \mathcal{H}_2)$ and $B \in \mathcal{B}(\mathcal{K}_1, \mathcal{K}_2)$, for Hilbert spaces $\mathcal{H}_1, \mathcal{H}_2, \mathcal{K}_1$, and $\mathcal{K}_2$, where neither $\mathcal{H}_1$ and $\mathcal{H}_2$ nor $\mathcal{K}_1$ and $\mathcal{K}_2$ are assumed to be the same. More precisely, we have the following result.

THEOREM 1.95. *Let $A \in \mathcal{B}(\mathcal{H}_1, \mathcal{H}_2)$ and $B \in \mathcal{B}(\mathcal{K}_1, \mathcal{K}_2)$. If $\mathcal{R}(A)$ is not closed and $\mathcal{R}(B)$ is closed, then $\mathcal{R}(M_C)$ is closed for some $C \in \mathcal{B}(\mathcal{K}_1, \mathcal{H}_2)$ if and only if $\alpha(B) = \infty$.*

Now, we list some results concerning properties of Kato nonsingular operators, the proofs of which can be found in [**148**].

LEMMA 1.96. *If $A \in \mathcal{B}(\mathcal{H})$ is (hyperexact) Kato nonsingular, then so is A^n for all $n \in \mathbb{N}$. A is Kato nonsingular if and only if A^* is such.*

THEOREM 1.97. *If T is Kato nonsingular, then $T - z$ is Kato nonsingular for all z in a neighborhood of $0 \in \mathbb{C}$.*

LEMMA 1.98 ([**147**]). *Let $T \in \mathcal{B}(\mathcal{H})$ be Kato nonsingular and set $U = \{z \in \mathbb{C} : |z| < \gamma(T)\}$. Then for every $\lambda \in U$ and $x \in \mathcal{N}(T - \lambda)$ there exists an analytic function $f : U \to \mathcal{H}$ such that $(T - z)f(z) = 0$ for all $z \in U$ and $f(\lambda) = x$.*

The next proposition plays an important role in the proof of the main result.

PROPOSITION 1.99. If M_C is hyperexact for some $C \in \mathcal{B}(\mathcal{K}, \mathcal{H})$, then

$$\operatorname{codim}_{\mathcal{N}(B^k)} \left(\mathcal{R}(B^n) \cap \mathcal{N}(B^k)\right) \leq \beta(A^k), \tag{1.75}$$

for all $n, k \in \mathbb{N}$.

PROOF. In view of Lemma 1.96, if we prove the statement above for $k = 1$, then we are done, as $M_C^m = \begin{bmatrix} A^m & * \\ 0 & B^m \end{bmatrix}$ and

$$\operatorname{codim}_{\mathcal{N}(B^m)} \left(\mathcal{R}(B^n) \cap \mathcal{N}(B^m)\right) \leq \operatorname{codim}_{\mathcal{N}(B^m)} \left(\mathcal{R}(B^{n+1}) \cap \mathcal{N}(B^m)\right),$$

for all $n, m \in \mathbb{N}$.

Let $P_{\mathcal{K}} \in \mathcal{B}(\mathcal{H} \oplus \mathcal{K}, \mathcal{K})$ be the orthogonal projector onto $\mathcal{K}$ and set $\mathcal{S} = P_{\mathcal{K}}[\mathcal{N}(M_C)]$. If $y \in \mathcal{S}$, then $\begin{bmatrix} x \\ y \end{bmatrix} \in \mathcal{N}(M_C)$ for some $x \in \mathcal{H}$, so for each $n \in \mathbb{N}$ we have

$$\begin{bmatrix} x \\ y \end{bmatrix} = M_C^n \begin{bmatrix} p_n \\ q_n \end{bmatrix} = \begin{bmatrix} A^n p_n + X_n q_n \\ B^n q_n \end{bmatrix},$$

for some $\begin{bmatrix} p_n \\ q_n \end{bmatrix} \in \mathcal{H} \oplus \mathcal{K}$, where $X_n = \sum_{i=1}^{n} A^{n-i} C B^{i-1}$. Hence $\mathcal{S} \subseteq \mathcal{R}(B^n)$, for all $n \in \mathbb{N}$. Actually, $\mathcal{S} \subseteq \mathcal{R}(B^n) \cap \mathcal{N}(B)$, so

$$\operatorname{codim}_{\mathcal{N}(B)} \left(\mathcal{R}(B^n) \cap \mathcal{N}(B)\right) \leq \operatorname{codim}_{\mathcal{N}(B)} \mathcal{S}.$$

Pick subspaces $\mathcal{S}_0 \subseteq \mathcal{N}(B)$ and $\mathcal{L} \subseteq \mathcal{H}$ such that $\mathcal{S} \oplus \mathcal{S}_0 = \mathcal{N}(B)$ and $\mathcal{R}(A) \oplus \mathcal{L} = \mathcal{H}$ and let $P_{\mathcal{L}} : \mathcal{H} \to \mathcal{H}$ be the idempotent linear operator with range $\mathcal{L}$ and null space

$\mathcal{R}(A)$. Then $P_{\mathcal{L}}C$ takes $\mathcal{S}_0$ injectively into $\mathcal{L}$. Indeed, let $y_0 \in \mathcal{S}_0$ be nonzero. Then $y_0 \in \mathcal{N}(B)$ but $y_0 \notin \mathcal{S} = \mathcal{N}(B) \cap C^{-1}[\mathcal{R}(A)]$, so $Cy_0 \notin \mathcal{R}(A)$; i.e., $P_{\mathcal{L}}Cy_0 \neq 0$. This proves that $\operatorname{codim}_{\mathcal{N}(B)}\mathcal{S} = \dim \mathcal{S}_0 \leq \dim \mathcal{L} = \beta(A)$. □

Using Lemma 1.96 and taking adjoints in Proposition 1.99 we directly obtain the following result.

PROPOSITION 1.100. If M_C is Kato nonsingular for some $C \in \mathcal{B}(\mathcal{K}, \mathcal{H})$, then

$$\operatorname{codim}_{\mathcal{N}((A^*)^k)} \left(\mathcal{R}\left((A^*)^n\right) \cap \mathcal{N}\left((A^*)^k\right)\right) \leq \beta\left((B^*)^k\right), \tag{1.76}$$

for all $n, k \in \mathbb{N}$.

Now, we consider completion to Kato nonsingular operator in the following seven cases:

Case 1: A or B is Kato nonsingular. In this case we address the question of completion of M_C to Kato nonsingularity under the additional assumption that one of the operators A or B is Kato nonsingular. As we will see in Cases 3 and 4 below, for certain important classes of operators (such as the compact ones, those with finite descent, quasinilpotent operators, ...) Kato nonsingularity of the operator $A \in \mathcal{B}(\mathcal{H})$ is a necessary condition for the existence of some $C \in \mathcal{B}(\mathcal{K}, \mathcal{H})$ such that M_C is Kato nonsingular.

Some particular cases of the completion problem to Kato nonsingularity in the case A is Kato nonsingular were already considered in [**14**] and [**67**]. The following two theorems, which we often use in the sequel, were proved:

THEOREM 1.101 ([**14**]). *Let $A \in \mathcal{B}(\mathcal{H})$ and $B \in \mathcal{B}(\mathcal{K})$ be given operators. If A is Kato nonsingular, $\mathcal{R}(B)$ is closed, and $\alpha(B) \leq d(A)$, then there exists $C \in \mathcal{B}(\mathcal{K}, \mathcal{H})$ such that M_C is Kato nonsingular.*

In the sequel we present a proof of the following theorem given in [**67**] which will be substantially shortened and at the same time will describe a way to produce operators in $S_K(A, B)$.

THEOREM 1.102 ([**67**]). *Let $A \in \mathcal{B}(\mathcal{H})$ be Kato nonsingular and let $B \in \mathcal{B}(\mathcal{K})$ be an arbitrary operator. If $d(A) = \infty$, then there exists $C \in \mathcal{B}(\mathcal{K}, \mathcal{H})$ such that M_C is Kato nonsingular.*

PROOF. If $d(A) = \infty$, there exists $C_3 \in \mathcal{B}(\mathcal{N}(B), \mathcal{R}(A)^\perp)$ left invertible such that $d(C_3) = \infty$. For such C_3 we get that there exists $C_4 \in \mathcal{B}(\mathcal{N}(B)^\perp, \mathcal{R}(A)^\perp)$ such that $S = \begin{bmatrix} C_3 & C_4 \\ 0 & B_1 \end{bmatrix}$ has a closed range, where $B = \begin{bmatrix} 0 & B_1 \end{bmatrix} : \mathcal{N}(B) \oplus \mathcal{N}(B)^\perp \to \mathcal{K}$ (if $\mathcal{R}(B)$ is closed take $C_4 = 0$; otherwise, use Theorem 1.95). Now, for $C \in \mathcal{B}(\mathcal{K}, \mathcal{H})$ given by

$$C = \begin{bmatrix} 0 & 0 \\ C_3 & C_4 \end{bmatrix} : \begin{bmatrix} \mathcal{N}(B)^\perp \\ \mathcal{N}(B) \end{bmatrix} \to \begin{bmatrix} \mathcal{R}(A) \\ \mathcal{R}(A)^\perp \end{bmatrix}, \tag{1.77}$$

M_C has a matrix representation

$$M_C = \begin{bmatrix} A_1 & 0 & 0 & 0 \\ 0 & 0 & C_3 & C_4 \\ 0 & 0 & 0 & 0 \\ 0 & 0 & 0 & B_1 \end{bmatrix} : \begin{bmatrix} \mathcal{N}(A)^\perp \\ \mathcal{N}(A) \\ \mathcal{N}(B) \\ \mathcal{N}(B)^\perp \end{bmatrix} \to \begin{bmatrix} \mathcal{R}(A) \\ \mathcal{R}(A)^\perp \\ \mathcal{R}(B)^\perp \\ \overline{\mathcal{R}(B)} \end{bmatrix}.$$

Since $\mathcal{R}(A)$ and $\mathcal{R}(S)$ are closed it follows that $\mathcal{R}(M_C)$ is closed. Also, by the injectivity of B_1 and C_3, $\mathcal{N}(S) = \{0\}$. So, $\mathcal{N}(M_C) = \mathcal{N}(A) \oplus \mathcal{N}(S) = \mathcal{N}(A) \subseteq \mathcal{R}(A^n) \subseteq \mathcal{R}(M_C^n)$. Hence, M_C is Kato nonsingular. □

It is interesting to notice that for C_3 and C_4 chosen as in the proof above and for arbitrary C_1 and C_2 from the appropriate spaces, the operator $C \in \mathcal{B}(\mathcal{K}, \mathcal{H})$ given by

$$C = \begin{bmatrix} C_1 & C_2 \\ C_3 & C_4 \end{bmatrix} : \begin{bmatrix} \mathcal{N}(B)^{\perp} \\ \mathcal{N}(B) \end{bmatrix} \to \begin{bmatrix} \overline{\mathcal{R}(A)} \\ \mathcal{R}(A)^{\perp} \end{bmatrix} \tag{1.78}$$

belongs to $S_K(A, B)$.

In the two theorems below, which are the main result of this part, we completely answer the question posed.

THEOREM 1.103. *Let $A \in \mathcal{B}(\mathcal{H})$ be Kato nonsingular and let $B \in \mathcal{B}(\mathcal{K})$. Then M_C is Kato nonsingular for some $C \in \mathcal{B}(\mathcal{K}, \mathcal{H})$ if and only if one of the following two conditions is satisfied:*

(1) $d(A) = \infty$.
(2) $d(A) < \infty$, *$\mathcal{R}(B^n)$ is closed for all $n \in \mathbb{N}$ and* (1.75) *holds for all $n, k \in \mathbb{N}$.*

THEOREM 1.104. *Let $B \in \mathcal{B}(\mathcal{K})$ be Kato nonsingular and let $A \in \mathcal{B}(\mathcal{H})$. Then M_C is Kato nonsingular for some $C \in \mathcal{B}(\mathcal{K}, \mathcal{H})$ if and only if one of the following two conditions is satisfied:*

(1) $\alpha(B) = \infty$.
(2) $\alpha(B) < \infty$, *$\mathcal{R}(A^n)$ is closed for all $n \in \mathbb{N}$ and* (1.76) *holds for all $n, k \in \mathbb{N}$.*

The second theorem obviously follows by taking adjoints in Theorem 1.103 and using Lemma 1.96.

By virtue of Theorem 1.102, to prove Theorem 1.103 it remains to find necessary and sufficient conditions for M_C to be Kato nonsingular for some $C \in \mathcal{B}(\mathcal{K}, \mathcal{H})$, under the assumption that A is Kato nonsingular and $d(A) < \infty$. It turns out that this case is much more complicated to deal with. We first prove an auxiliary result in which some sufficient conditions are given for the set $S_K(A, B)$ to be nonempty.

PROPOSITION 1.105. Let $A \in \mathcal{B}(\mathcal{H})$ and $B \in \mathcal{B}(\mathcal{K})$ be given operators. If A is Kato nonsingular, $d(A) < \infty$, $\mathcal{R}(B)$ is closed, and (1.75) holds for all $n, k \in \mathbb{N}$, then there is some $C \in \mathcal{B}(\mathcal{K}, \mathcal{H})$ such that M_C is Kato nonsingular provided that either

(i) $\mathcal{R}^{\infty}(B)$ is closed or
(ii) $\operatorname{codim}_{\mathcal{N}(B)} (\mathcal{R}(B) \cap \mathcal{N}(B)) = d(A)$.

PROOF. For each $k \in \mathbb{N}$ set $S_k = \mathcal{R}^{\infty}(B) \cap \mathcal{N}(B^k)$; we have that

$$\operatorname{codim}_{\mathcal{N}(B^k)} \left(\mathcal{R}(B^n) \cap \mathcal{N}(B^k)\right) \leq \beta(A^k) \leq k\beta(A) < \infty$$

for all $n \in \mathbb{N}$, by Lemma 1.93, so there must be some $m_k \in \mathbb{N}$ such that $S_k = \mathcal{R}(B^{m_k}) \cap \mathcal{N}(B^k) = \mathcal{R}(B^l) \cap \mathcal{N}(B^k)$ for all $l \geq m_k$.

We claim that for each $k \in \mathbb{N}$ and $x \in S_k$ there is some $x_1 \in S_{k+1}$ such that $x = Bx_1$. Indeed, $x \in S_k$ implies that $x = B^{m_{k+1}+1}x_2$ for some $x_2 \in \mathcal{K}$. This means that $Bx_1 = x$ for $x_1 = B^{m_{k+1}}x_2 \in \mathcal{N}(B^{k+1})$. But $x_1 \in \mathcal{R}(B^{m_{k+1}}) \cap \mathcal{N}(B^{k+1}) = S_{k+1}$.

Take a finite-dimensional subspace $\mathcal{X}$ of $\mathcal{N}(B)$ such that $S_1 \oplus \mathcal{X} = \mathcal{N}(B)$. Since $\dim \mathcal{X} \le d(A)$ by assumption, we can fix a left invertible $C_0 \in \mathcal{B}(\mathcal{X}, \mathcal{R}(A)^\perp)$. If (i) holds, let $\mathcal{X}_1 = \mathcal{R}^\infty(B)$, and if (ii) holds, let $\mathcal{X}_1 = \mathcal{R}(B)$. Note that if (ii) holds, then it must be $S_1 = \mathcal{R}(B) \cap \mathcal{N}(B)$. In any case $\mathcal{X}_1 \cap \mathcal{X} = \{0\}$ so $\mathcal{X}_1$ and $\mathcal{X}$ are two closed complementary subspaces of the closed subspace $\mathcal{Y} = \mathcal{X}_1 \oplus \mathcal{X} = \mathcal{X}_1 + \mathcal{N}(B)$ of $\mathcal{K}$ (recall that $\dim \mathcal{X} < \infty$). Thus we can define a $C \in \mathcal{B}(\mathcal{K}, \mathcal{H})$ by

$$C = \begin{bmatrix} 0 & 0 & 0 \\ 0 & 0 & C_0 \end{bmatrix} : \begin{bmatrix} \mathcal{Y}^\perp \\ \mathcal{X}_1 \\ \mathcal{X} \end{bmatrix} \to \begin{bmatrix} \mathcal{R}(A) \\ \mathcal{R}(A)^\perp \end{bmatrix}.$$

We show that M_C is Kato nonsingular. First use left invertibility of C_0 to verify that $\mathcal{N}(M_C) = \mathcal{N}(A) \oplus S_1$. Next, for a given $n \in \mathbb{N}$ we show that $\mathcal{N}(M_C) \subseteq \mathcal{R}(M_C^n)$. So let $\begin{bmatrix} x \\ y_1 \end{bmatrix} \in \mathcal{N}(A) \oplus S_1$. By an earlier observation we can pick recursively $y_i \in S_i$ for $2 \le i \le n+1$ such that $By_i = y_{i-1}$. Since $y_i \in \mathcal{R}^\infty(B) \subseteq \mathcal{X}_1$ we have $Cy_i = 0$, for $1 \le i \le n+1$. A is Kato nonsingular and $x \in \mathcal{N}(A)$ so $x = A^n p$ for some $p \in \mathcal{H}$. Thus from $y_1 = B^n y_{n+1}$ and $\left(\sum_{i+j=n-1} A^i C B^j\right) y_{n+1} = 0$ we see that

$$\begin{bmatrix} x \\ y_1 \end{bmatrix} = M_C^n \begin{bmatrix} p \\ y_{n+1} \end{bmatrix}.$$

Since C is a finite-rank operator, we have that $\mathcal{R}(M_C)$ is closed if and only if $\mathcal{R}(M_0)$ is closed (Lemma 1.64), and the latter indeed is such, both $\mathcal{R}(A)$ and $\mathcal{R}(B)$ being closed by assumption. □

Now we can complete the proof of Theorem 1.103.

THEOREM 1.106. *Let $B \in \mathcal{B}(\mathcal{K})$ and let $A \in \mathcal{B}(\mathcal{H})$ be Kato nonsingular and $d(A) < \infty$. Then M_C is Kato nonsingular for some $C \in \mathcal{B}(\mathcal{K}, \mathcal{H})$ if and only if $\mathcal{R}(B^n)$ is closed for all $n \in \mathbb{N}$ and* (1.75) *holds for all $n, k \in \mathbb{N}$.*

PROOF. By Proposition 1.105 and Proposition 1.99, we only need to show that $\mathcal{R}(B^n)$ is closed for all $n \in \mathbb{N}$ provided that A is Kato nonsingular, $d(A) < \infty$, and that M_C is Kato nonsingular for some C.

Given $n \in \mathbb{N}$ we have $(M_C^*)^n = \begin{bmatrix} B^{*n} & * \\ 0 & A^{*n} \end{bmatrix}$. By Lemma 1.96 we know that the ranges of both $(M_C^*)^n$ and A^{*n} are closed. Should $\mathcal{R}(B^{*n})$ not be closed, Theorem 1.95 would then imply that $\infty = \alpha(A^{*n}) = d(A^n) \le n \cdot d(A) < \infty$, a contradiction. □

Cases 2–5: B injective or with finite ascent; A with dense range or with finite descent. In this case for given operators $A \in \mathcal{B}(\mathcal{H})$ and $B \in \mathcal{B}(\mathcal{K})$ we discuss the completion problem of M_C under the hypothesis that B is with finite ascent or under the more restrictive one that it is an injective operator. In both of these cases we solve the problem completely.

First we deal with the case when B is with finite ascent. We note that in this case the closedness of the range of A turns out to be a necessary condition. Furthermore, A must be a Kato nonsingular operator.

We begin by establishing the algebraic result given below.

THEOREM 1.107. *Let $A \in \mathcal{B}(\mathcal{H})$ and $B \in \mathcal{B}(\mathcal{K})$ be such that* $\operatorname{asc}(B) < \infty$. *If $C \in \mathcal{B}(\mathcal{K}, \mathcal{H})$ is such that M_C is hyperexact, then:*

(1) *A is hyperexact.*
(2) *$\mathcal{N}(M_C^k) = \mathcal{N}(A^k)$ for all $k \in \mathbb{N}$.*
(3) *$\alpha(B) \le \beta(A)$.*

PROOF. Let $k_0 = \operatorname{asc}(B)$.

(1) Let $x \in \mathcal{N}(A)$ and $m \in \mathbb{N}$. Since $\begin{bmatrix} x \\ 0 \end{bmatrix} \in \mathcal{N}(M_C)$ and M_C is Kato nonsingular, there are $p \in \mathcal{H}$, $q \in \mathcal{K}$ such that

$$\begin{cases} x = A^{m+k_0}p + \left(\sum_{i=0}^{m+k_0-1} A^{m+k_0-1-i}CB^i\right) q, \\ 0 = B^{m+k_0}q\,. \end{cases}$$

From here it follows that $q \in \mathcal{N}(B^{m+k_0}) = \mathcal{N}(B^{k_0})$ and thus

$$x = A^m \left(A^{k_0}p + \left(\sum_{i=0}^{k_0-1} A^{k_0-1-i}CB^i \right) q \right) \in \mathcal{R}(A^m)\,.$$

(2) This is proved by induction. For $\begin{bmatrix} x \\ y \end{bmatrix} \in \mathcal{N}(M_C)$ it follows that $\begin{bmatrix} x \\ y \end{bmatrix} \in \mathcal{R}(M_C^n)$ for every $n \in \mathbb{N}$ which implies that $\begin{bmatrix} x \\ y \end{bmatrix} = M_C^{k_0} \begin{bmatrix} p \\ q \end{bmatrix}$ for some $p \in \mathcal{H}$, $q \in \mathcal{K}$. So $q \in \mathcal{N}(B^{k_0+1}) = \mathcal{N}(B^{k_0})$ and thus $y = B^{k_0}q = 0$. But this implies that $Ax = 0$; i.e., $\mathcal{N}(M_C) = \mathcal{N}(A)$.

Assume now that $\mathcal{N}(M_C^k) = \mathcal{N}(A^k)$ for some k and let $\begin{bmatrix} x \\ y \end{bmatrix} \in \mathcal{N}(M_C^{k+1})$. Then

$$\begin{bmatrix} Ax + Cy \\ By \end{bmatrix} \in \mathcal{N}(M_C^k) = \mathcal{N}(A^k)\,,$$

so $By = 0$ and $Ax + Cy \in \mathcal{N}(A^k) \subseteq \mathcal{R}(A)$ (by item (1)). From this we have

$$\begin{bmatrix} A & C \\ 0 & B \end{bmatrix} \begin{bmatrix} x_1 \\ y \end{bmatrix} = \begin{bmatrix} 0 \\ 0 \end{bmatrix},$$

for some $x_1 \in \mathcal{H}$; i.e., $\begin{bmatrix} x_1 \\ y \end{bmatrix} \in \mathcal{N}(M_C) = \mathcal{N}(A)$. Hence $y = 0$ and consequently $x \in \mathcal{N}(A^{k+1})$. We have thus shown $\mathcal{N}(M_C^{k+1}) = \mathcal{N}(A^{k+1})$.

(3) Let $\mathcal{S}$, $\mathcal{S}_0$, $\mathcal{L}$, and $P_{\mathcal{L}}$ be as in the proof of Proposition 1.99; we know that $P_{\mathcal{L}}C$ maps $\mathcal{S}_0$ injectively into $\mathcal{L}$. By (1) we have $\mathcal{S} = \{0\}$, so $\mathcal{S}_0 = \mathcal{N}(B)$ and thus $\alpha(B) = \dim(\mathcal{N}(B)) \le \dim \mathcal{L} = \beta(A)$. □

In the next result we show that if M_C is Kato nonsingular for some $C \in \mathcal{B}(\mathcal{K}, \mathcal{H})$ with the additional condition that $\mathcal{N}(M_C) = \mathcal{N}(A)$, then $\mathcal{R}(A)$ must be closed.

PROPOSITION 1.108. Let $A \in \mathcal{B}(\mathcal{H})$, $B \in \mathcal{B}(\mathcal{K})$ be such that there exists $C \in \mathcal{B}(\mathcal{K}, \mathcal{H})$ such that M_C is Kato nonsingular and $\mathcal{N}(M_C) = \mathcal{N}(A)$. Then $\mathcal{R}(A)$ must be closed.

PROOF. Suppose that M_C is Kato nonsingular for some $C \in \mathcal{B}(\mathcal{K}, \mathcal{H})$ given by (1.78). In that case, M_C has a matrix representation

$$M_C = \begin{bmatrix} A_1 & 0 & C_1 & C_2 \\ 0 & 0 & C_3 & C_4 \\ 0 & 0 & 0 & 0 \\ 0 & 0 & 0 & B_1 \end{bmatrix} : \begin{bmatrix} \mathcal{N}(A)^\perp \\ \mathcal{N}(A) \\ \mathcal{N}(B) \\ \mathcal{N}(B)^\perp \end{bmatrix} \to \begin{bmatrix} \overline{\mathcal{R}(A)} \\ \mathcal{R}(A)^\perp \\ \mathcal{R}(B)^\perp \\ \overline{\mathcal{R}(B)} \end{bmatrix},$$

where A_1, B_1 are injections with dense ranges.

By Lemma 1.94, using the fact that B_1^* has a dense range, it follows that $S = \begin{bmatrix} A_1 & C_1 \\ 0 & C_3 \end{bmatrix}$ has a closed range. Since $\mathcal{N}(M_C) = \mathcal{N}(A)$, we conclude that S is injective. Hence S is left invertible, so A_1 is left invertible; i.e., $\mathcal{R}(A)$ is closed. $\square$

In the following theorem, we resolve the completion problem in the case B is with finite ascent. In addition to that, we give a description of the set $S_K(A, B)$ whenever it is nonempty.

THEOREM 1.109. *Let $A \in \mathcal{B}(\mathcal{H})$ and $B \in \mathcal{B}(\mathcal{K})$ be such that* $\operatorname{asc}(B) < \infty$. *Then M_C is Kato nonsingular for some $C \in \mathcal{B}(\mathcal{K}, \mathcal{H})$ if and only if one of the following conditions is satisfied:*

(i) *A is Kato nonsingular, $\mathcal{R}(B)$ is closed, and $\alpha(B) \le d(A)$. In this case,*

$$S_K(A, B) = \left\{ C \in \mathcal{B}(\mathcal{K}, \mathcal{H}) \colon C = \begin{bmatrix} C_1 & C_2 \\ C_3 & C_4 \end{bmatrix} : \begin{bmatrix} \mathcal{N}(B) \\ \mathcal{N}(B)^\perp \end{bmatrix} \to \begin{bmatrix} \mathcal{R}(A) \\ \mathcal{R}(A)^\perp \end{bmatrix}, \right.$$
$$\left. C_3 \text{ is left invertible} \right\}.$$

(ii) *A is Kato nonsingular, $\mathcal{R}(B)$ is not closed, and $d(A) = \infty$. In this case,*

$$S_K(A, B) = \left\{ C \in \mathcal{B}(\mathcal{K}, \mathcal{H}) \colon C = \begin{bmatrix} C_1 & C_2 \\ C_3 & C_4 \end{bmatrix} : \begin{bmatrix} \mathcal{N}(B) \\ \mathcal{N}(B)^\perp \end{bmatrix} \to \begin{bmatrix} \mathcal{R}(A) \\ \mathcal{R}(A)^\perp \end{bmatrix}, \right.$$
$$\left. C_3 \text{ is left invertible, } \mathcal{R}(B^*) + \mathcal{R}(C_4^*(I - P_{\mathcal{R}(C_3)})) \text{ is closed} \right\}.$$

PROOF. By Proposition 1.108 and Theorems 1.95, 1.101, 1.102, and 1.107 it immediately follows that M_C is Kato nonsingular for some $C \in \mathcal{B}(\mathcal{K}, \mathcal{H})$ if and only if (i) or (ii) holds. It remains to describe the set $S_K(A, B)$ when one of the conditions (i) or (ii) is satisfied.

Assume for the moment that M_C is Kato nonsingular for some $C \in \mathcal{B}(\mathcal{K}, \mathcal{H})$ given by (1.78). We know that $\mathcal{N}(M_C) = \mathcal{N}(A)$ as well as that $\mathcal{R}(A)$ is closed. In that case, M_C has a matrix representation

$$M_C = \begin{bmatrix} A_1 & 0 & C_1 & C_2 \\ 0 & 0 & C_3 & C_4 \\ 0 & 0 & 0 & 0 \\ 0 & 0 & 0 & B_1 \end{bmatrix} : \begin{bmatrix} \mathcal{N}(A)^\perp \\ \mathcal{N}(A) \\ \mathcal{N}(B) \\ \mathcal{N}(B)^\perp \end{bmatrix} \to \begin{bmatrix} \mathcal{R}(A) \\ \mathcal{R}(A)^\perp \\ \mathcal{R}(B)^\perp \\ \overline{\mathcal{R}(B)} \end{bmatrix},$$

where A_1 is invertible and B_1 is an injection with dense range.

By $\mathcal{N}(M_C) = \mathcal{N}(A)$, we have that $S = \begin{bmatrix} A_1 & C_1 \\ 0 & C_3 \end{bmatrix}$ is injective which, using the invertibility (surjectivity to be precise) of A_1, implies that C_3 is injective.

Now, using twice Lemma 1.94, we first conclude that $P = \begin{bmatrix} C_3 & C_4 \\ 0 & B_1 \end{bmatrix}$ has closed range and then that $\mathcal{R}(C_3)$ is closed. So, C_3 must be left invertible.

From the discussion above it follows that:

(a) If M_C is Kato nonsingular for some $C \in \mathcal{B}(\mathcal{K}, \mathcal{H})$ given by (1.78), then S, C_3, and P are left invertible.

Also, the following fact can easily be verified:

(b) If $\mathcal{R}(A)$ is closed, then $\mathcal{R}(M_C)$ is closed if and only if $\mathcal{R}(P)$ is closed.

Suppose (i) holds: Denote the set from the right side of the equality in (i) by $\mathcal{S}_1$. By (a), we immediately have $S_K(A,B) \subseteq \mathcal{S}_1$. On the other hand, take $C \in \mathcal{S}_1$. Since C_3 is injective it can be checked that $\mathcal{N}(M_C) = \mathcal{N}(A)$ which by hyperexactness of A implies that M_C is hyperrexact. Since A_1 and B_1 are invertible and C_3 is left invertible, it is not difficult to verify that $\mathcal{R}(M_C)$ is closed (use left invertibility of A_1, B_1, and C_3). Hence, $C \in S_K(A,B)$. Suppose (ii) holds: Let us denote the set from the right side of the equality in (ii) by $\mathcal{S}_2$ and let $C \in \mathcal{B}(\mathcal{K}, \mathcal{H})$ be such that M_C is Kato nonsingular. Then by (a), we have that C_3 and P are left invertible. By Theorem 1.15, it follows that $\begin{bmatrix} (I - P_{\mathcal{R}(C_3)})C_4 \\ B_1 \end{bmatrix}$ is left invertible. As B_1 is injective, this is equivalent to saying that $\mathcal{R}\left(\begin{bmatrix} C_4^*(I - P_{\mathcal{R}(C_3)}) & B_1^* \end{bmatrix}\right)$ is closed. Hence $C \in \mathcal{S}_2$. Conversely, let $C \in \mathcal{S}_2$. By injectivity of C_3 we have $\mathcal{N}(M_C) = \mathcal{N}(A)$, so M_C is hyperexact since A is. By Theorem 2.5 of [**47**] and injectivity of B_1, we have that P is left invertible. Thus $\mathcal{R}(P)$ is closed. From (b) it now follows that $\mathcal{R}(M_C)$ is closed. Hence, $C \in S_K(A,B)$. □

In the special case when B is a nilpotent operator, we have the following result:

COROLLARY 1.110. Let $A \in \mathcal{B}(\mathcal{H})$ and let $B \in \mathcal{B}(\mathcal{K})$ be nilpotent. Then M_C is Kato nonsingular for some $C \in \mathcal{B}(\mathcal{K}, \mathcal{H})$ if and only if A is Kato nonsingular and $d(A) = \infty$.

PROOF. By Theorem 1.109 we only need to show that if M_C is Kato nonsingular for some C, then $d(A) = \infty$. But then, as A is Kato nonsingular, so is $M_C^k = \begin{bmatrix} A^k & * \\ 0 & 0 \end{bmatrix}$, where $k \in \mathbb{N}$ is such that $B^k = 0$. Using Theorem 1.107 once again we conclude that $k\, d(A) \geq d(A^k) \geq \min\{\dim(\mathcal{N}(B^k)), \infty\} = \infty$. □

Dually to Theorem 1.109 one has the following result.

THEOREM 1.111. *Let* $A \in \mathcal{B}(\mathcal{H})$ *and* $B \in \mathcal{B}(\mathcal{K})$ *be such that* $\operatorname{dsc}(A) < \infty$. *Then* M_C *is Kato nonsingular for some* $C \in \mathcal{B}(\mathcal{K}, \mathcal{H})$ *if and only if one of the following conditions is satisfied:*

(i) *B is Kato nonsingular, $\mathcal{R}(A)$ is closed, and $d(A) \leq \alpha(B)$. In this case,*

$$S_K(A,B) = \left\{ C \in \mathcal{B}(\mathcal{K}, \mathcal{H})\colon C = \begin{bmatrix} C_1 & C_2 \\ C_3 & C_4 \end{bmatrix} : \begin{bmatrix} \mathcal{N}(B) \\ \mathcal{N}(B)^\perp \end{bmatrix} \to \begin{bmatrix} \mathcal{R}(A) \\ \mathcal{R}(A)^\perp \end{bmatrix}, \right.$$
$$\left. C_3 \textit{ is right invertible} \right\}.$$

(ii) *B is Kato nonsingular, $\mathcal{R}(A)$ is not closed, and $\alpha(B) = \infty$. In this case,*

$$S_K(A,B) = \left\{ C \in \mathcal{B}(\mathcal{K},\mathcal{H})\colon C = \begin{bmatrix} C_1 & C_2 \\ C_3 & C_4 \end{bmatrix} : \begin{bmatrix} \mathcal{N}(B) \\ \mathcal{N}(B)^\perp \end{bmatrix} \to \begin{bmatrix} \overline{\mathcal{R}(A)} \\ \mathcal{R}(A)^\perp \end{bmatrix}, \right.$$

$$\left. C_3 \text{ is right invertible, } \mathcal{R}(A) + \mathcal{R}(C_1(I - P_{\mathcal{R}(C_3^*)})) \text{ is closed} \right\}.$$

PROOF. Notice that $\mathrm{dsc}(A) < \infty$ implies that $\mathrm{asc}(A^*) < \infty$, so the proof follows directly by Theorem 1.109 and Lemma 1.96. □

The problem of existence of $C \in \mathcal{B}(\mathcal{K},\mathcal{H})$ such that M_C is Kato nonsingular in the case when B is injective was almost completely covered by Theorems 2.3 and 2.4 from [**14**]. Below we complete these results and additionally provide a description of the set $S_K(A,B)$.

THEOREM 1.112. *Let $A \in \mathcal{B}(\mathcal{H})$ and $B \in \mathcal{B}(\mathcal{K})$ be an injective operator. Then M_C is Kato nonsingular for some $C \in \mathcal{B}(\mathcal{K},\mathcal{H})$ if and only if one of the following conditions is satisfied:*

(i) *A is Kato nonsingular and $\mathcal{R}(B)$ are closed. In this case,*

$$S_K(A,B) = \mathcal{B}(\mathcal{K},\mathcal{H}).$$

(ii) *A is Kato nonsingular, $\mathcal{R}(B)$ is nonclosed, and $d(A) = \infty$. In this case,*

$$S_K(A,B) = \{C \in \mathcal{B}(\mathcal{K},\mathcal{H})\colon \mathcal{R}(B^*) + \mathcal{R}(C^* P_{\mathcal{R}(A)^\perp}) \text{ is closed}\}. \tag{1.79}$$

PROOF. From Theorems 2.3 and 2.4 from [**14**] and Theorem 1.109 it follows that M_C is Kato nonsingular for some $C \in \mathcal{B}(\mathcal{K},\mathcal{H})$ if and only if (i) or (ii) holds. Also that $S_K(A,B) = \mathcal{B}(\mathcal{K},\mathcal{H})$, provided (i) holds, was proved in Theorem 2.4 from [**14**]. We only need to establish the equality (1.79) if the conditions in (ii) are satisfied. But this directly follows from Theorem 1.109 since $\mathcal{N}(B) = \{0\}$ implies that $C_3 = 0$. □

Taking adjoints in the previous theorem, we find necessary and sufficient conditions for the existence of $C \in \mathcal{B}(\mathcal{K},\mathcal{H})$ such that M_C is Kato nonsingular in the case when A is with dense range.

THEOREM 1.113. *Let $A \in \mathcal{B}(\mathcal{H})$ be with dense range and $B \in \mathcal{B}(\mathcal{K})$. Then M_C is Kato nonsingular for some $C \in \mathcal{B}(\mathcal{K},\mathcal{H})$ if and only if one of the following conditions is satisfied:*

(i) *B is Kato nonsingular and $\mathcal{R}(A)$ is closed. In this case,*

$$S_K(A,B) = \mathcal{B}(\mathcal{K},\mathcal{H}).$$

(ii) *B is Kato nonsingular, $\mathcal{R}(A)$ is nonclosed, and $\alpha(B) = \infty$. In this case,*

$$S_K(A,B) = \{C \in \mathcal{B}(\mathcal{K},\mathcal{H})\colon \mathcal{R}(A) + \mathcal{R}(C P_{\mathcal{N}(B)}) \text{ is closed}\}.$$

Cases 6–7: $0 \notin \mathrm{int}(\sigma_p(B))$ or $0 \notin \mathrm{int}(\sigma_{cp}(A))$. We have seen that when $\mathrm{asc}(B) < \infty$ or $\mathrm{dsc}(A) < \infty$, one of the operators of A and B must be Kato nonsingular in order for some $C \in \mathcal{B}(\mathcal{K},\mathcal{H})$ to exist such that M_C is Kato nonsingular. However it can happen that M_C is Kato nonsingular without $\mathcal{R}(B)$ or $\mathcal{R}(A)$ being closed (see Example 3 of [**107**]). In what follows we describe yet one more situation when, for the completion problem to Kato nonsingularity to have an affirmative answer, it is necessary that one of the operators A or B is Kato nonsingular.

PROPOSITION 1.114. Let $A \in \mathcal{B}(\mathcal{H})$ and $B \in \mathcal{B}(\mathcal{K})$. If M_C is Kato nonsingular for some $C \in \mathcal{B}(\mathcal{K},\mathcal{H})$ and $\operatorname{acc}\big(\mathbb{C}\backslash\sigma_p(B)\big)\cap K(0,\gamma(M_C)) \neq \emptyset$, then $\mathcal{N}(A) = \mathcal{N}(M_C)$.

PROOF. We know that $\mathcal{N}(A) \subseteq \mathcal{N}(M_C)$. Towards a contradiction suppose there is some $\begin{bmatrix} x \\ y \end{bmatrix} \in \mathcal{N}(M_C) \setminus \mathcal{N}(A)$. Then

$$\begin{cases} Ax + Cy = 0, \\ By = 0, \end{cases}$$

and $y \neq 0$. Put $r = \gamma(M_C)$. By Lemma 1.98 there is an analytic function $f : K(0,r) \to \mathcal{H}\oplus\mathcal{K}$ such that $f(0) = \begin{bmatrix} x \\ y \end{bmatrix}$ and $f(\lambda) \in \mathcal{N}(M_C - \lambda)$ for all $\lambda \in K(0,r)$. The assumption that $\operatorname{acc}\big(\mathbb{C}\backslash\sigma_p(B)\big)\cap K(0,\gamma(M_C)) \neq \emptyset$ implies that for the analytic function $z \mapsto P_{\mathcal{K}}f(z)$, $z \in K(0,r)$, we have $P_{\mathcal{K}}f(z) = 0$ for infinitely many points z belonging to a closed disc contained in the open connected set $K(0,r)$. Thus $P_{\mathcal{K}}f(z) = 0$ for all $z \in K(0,r)$. But then $y = P_{\mathcal{K}}f(0) = 0$, a contradiction. □

For our next result we need the following characterization of Kato nonsingularity from [**147**]:

THEOREM 1.115 ([**147**]). *Let $T \in \mathcal{B}(\mathcal{H})$ be an operator with closed range. The following conditons are equivalent:*

(1) *T is Kato nonsingular.*
(2) *$\mathcal{N}(T) \subseteq \overline{\bigvee_{z\neq 0} \mathcal{N}(T - z)}$.*

THEOREM 1.116. *Let $A \in \mathcal{B}(\mathcal{H})$ and $B \in \mathcal{B}(\mathcal{K})$. If M_C is Kato nonsingular for some $C \in \mathcal{B}(\mathcal{K},\mathcal{H})$ and $0 \notin \operatorname{int}(\sigma_p(B))$, then:*

(1) *$\mathcal{N}(A^k) = \mathcal{N}(M_C^k)$ for all $k \in \mathbb{N}$.*
(2) *A is Kato nonsingular.*

PROOF. Observe that if $0 \notin \sigma_p(B)$, then both conclusions follow from Theorem 1.107 and Proposition 1.108. Thus in the sequel we assume that $0 \in \sigma_p(B)$.

(1) Since $\sigma_p(q(B)) = \{q(\lambda) \colon \lambda \in \sigma_p(B)\}$, for all polynomials q, we have $0 \notin \operatorname{int}(\sigma_p(B^k))$ for all $k \in \mathbb{N}$. Thus, using Lemma 1.96, we need only prove the assertion under (1) for $k = 1$. But this case is covered by Proposition 1.114 since $0 \in \sigma_p(B) \setminus \operatorname{int}(\sigma_p(B))$ and $0 \in K(0,\gamma(M_C))$.

(2) By Proposition 1.108 and what we have just proved under (1) we already know that $\mathcal{R}(A)$ must be closed.

Put $r = \gamma(M_C)$. By Theorems 1.92 and 1.97, there is some $\epsilon \in (0, r/2)$ such that $M_C - zI$ is Kato nonsingular and $\gamma(M_C - zI) > r/2$ for all $z \in K(0,\epsilon)$.

Now we prove that $\mathcal{N}(M_C - zI) = \mathcal{N}(A - zI)$ for all $z \in K(0,\epsilon)$. To see this fix $z \in K(0,\epsilon)$. Here $0 \in \sigma_p(B) \setminus \operatorname{int}(\sigma_p(B))$ gives

$$-z \in \sigma_p(B - zI) \setminus \operatorname{int}(\sigma_p(B - zI)).$$

It follows that

$$-z \in \operatorname{acc}(\mathbb{C} \setminus \sigma_p(B - zI)) \cap K(0, r/2),$$

which in turn, by Kato nonsingularity of $M_C - zI$ and Proposition 1.114, yields $\mathcal{N}(M_C - zI) = \mathcal{N}(A - zI)$, since $r/2 < \gamma(M_C - zI)$.

Now let $x \in \mathcal{N}(A)$ be arbitrary. Since $x \in \mathcal{N}(M_C)$ by Lemma 1.98 there is an analytic function $f : K(0,\epsilon) \to \mathcal{H} \oplus \mathcal{K}$ with $f(0) = x$ and $f(z) \in \mathcal{N}(M_C - zI)$ for all $z \in K(0,\epsilon)$. We have

$$x = f(0) = \frac{1}{2\pi i}\int_{|z|=\epsilon/2} \frac{f(z)}{z}dz \in \overline{\bigvee_{|z|=\epsilon/2} \mathcal{N}(M_C - zI)}.$$

Since $\mathcal{N}(M_C - zI) = \mathcal{N}(A - zI)$, we thus have $x \in \overline{\bigvee_{z\neq 0} \mathcal{N}(A - zI)}$, and consequently A must be Kato nonsingular by Theorem 1.115. □

Combining Theorems 1.103 and 1.116 the reader can easily give the answer to the completion problem considered, in the case $0 \notin \operatorname{int}(\sigma_p(B))$.

COROLLARY 1.117. Let $A \in \mathcal{B}(\mathcal{H})$ and $B \in \mathcal{B}(\mathcal{K})$. If A is not Kato nonsingular, then there is no $C \in \mathcal{B}(\mathcal{K},\mathcal{H})$ such that M_C is Kato nonsingular whenever $\operatorname{int}\big(\sigma_p(B)\big) = \emptyset$, thus in particular in each of the following cases:

- B is a compact operator,
- B is a quasinilpotent operator,
- B is an algebraic operator.

Easily, we can conclude the following:

THEOREM 1.118. *Let $A \in \mathcal{B}(\mathcal{H})$ and $B \in \mathcal{B}(\mathcal{K})$. If M_C is Kato nonsingular for some $C \in \mathcal{B}(\mathcal{K},\mathcal{H})$ and $0 \notin \operatorname{int}(\sigma_{cp}(A))$, then B must be Kato nonsingular.*

1.9. Fredholm consistency of upper triangular operator matrices

Let $\mathcal{S}(\mathcal{H})$ denote a subset of $\mathcal{B}(\mathcal{H})$. An operator $A \in \mathcal{B}(\mathcal{H})$ is said to be $\mathcal{S}(\mathcal{H})$ consistent, or consistent in $\mathcal{S}(\mathcal{H})$, if

$$AB \in \mathcal{S}(\mathcal{H}) \Leftrightarrow BA \in \mathcal{S}(\mathcal{H}),$$

for every $B \in \mathcal{B}(\mathcal{H})$.

A characterization of operators consistent in invertibility (CI operators) and a characterization of CI operators invariant under compact perturbations are given in [**86**]. It is worth mentioning the very interesting paper of D. S. Djordjević et al. [**63**] where the property of being $\mathcal{S}(\mathcal{H})$ consistent is related with the single-valued extension property (SVEP) for a variety of choices of the subset $\mathcal{S}(\mathcal{H})$ of $\mathcal{B}(\mathcal{H})$. Recall that $A \in \mathcal{B}(\mathcal{H})$ has the SVEP at a point $\lambda_0 \in \mathbb{C}$ if for every open neighborhood U_λ of λ_0 the only analytic function $f : U_\lambda \to \mathcal{H}$ satisfying $(A-\lambda)f(\lambda) = 0$ for every $\lambda \in U_\lambda$ is the function $f = 0$. Evidently, A has the SVEP at every point of the resolvent set $\rho(A)$ of A. The SVEP provides a simple sufficient condition for determining $\mathcal{S}(\mathcal{H})$ consistent operators $A \in \mathcal{B}(\mathcal{H})$ for a variety of choices of the subset $\mathcal{S}(\mathcal{H})$ of $\mathcal{B}(\mathcal{H})$.

The work presented in this section is a continuation of the work of G. Hai et al. [**91**] in which the problem of completions of the upper triangular operator matrix

$$\begin{bmatrix} A & ? \\ 0 & B \end{bmatrix} : \begin{bmatrix} \mathcal{H} \\ \mathcal{K} \end{bmatrix} \to \begin{bmatrix} \mathcal{H} \\ \mathcal{K} \end{bmatrix} \tag{1.80}$$

to consistent invertibility was considered, where $A \in \mathcal{B}(\mathcal{H})$ and $B \in \mathcal{B}(\mathcal{K})$ are given operators on separable Hilbert spaces.

In this section we present the results published in [**51**] that concern the problem of completion of the upper triangular operator matrix given by (1.80) to Fredholm consistent (FC). Let us recall that for $A \in \mathcal{B}(\mathcal{H})$, by $\sigma_{FC}(A)$ we denote

$$\sigma_{FC}(A) = \{\lambda \in \mathbb{C}\colon A - \lambda I \text{ is not Fredholm consistent}\}$$

and

$$\rho_{FC}(A) = \mathbb{C} \setminus \sigma_{FC}(A).$$

We answer the following three questions which repeatedly arise:

Question 1: Is there an operator $C \in \mathcal{B}(\mathcal{K}, \mathcal{H})$ such that M_C is FC?

Question 2: $\bigcap_{C\in\mathcal{B}(\mathcal{K},\mathcal{H})} \sigma_{FC}(M_C) =?$

Question 3: Is there an operator $C' \in \mathcal{B}(\mathcal{K}, \mathcal{H})$ such that

$$\sigma_{FC}(M_{C'}) = \bigcap_{C\in\mathcal{B}(\mathcal{Y},\mathcal{X})} \sigma_{FC}(M_C)?$$

Furthermore, for given operators $A \in \mathcal{B}(\mathcal{H})$ and $B \in \mathcal{B}(\mathcal{K})$, we describe the set

$$\bigcup_{C\in\mathcal{B}(\mathcal{Y},\mathcal{X})} \sigma_{FC}(M_C)$$

and characterize the set of all $C \in \mathcal{B}(\mathcal{K}, \mathcal{H})$ such that the operator matrix $M_C = \begin{bmatrix} A & C \\ 0 & B \end{bmatrix}$ is Fredholm consistent.

Recall that for given $A \in \mathcal{B}(\mathcal{H})$ and $B \in \mathcal{B}(\mathcal{K})$, the set of all $C \in \mathcal{B}(\mathcal{K}, \mathcal{H})$ such that M_C is Fredholm, we denote by $S_F(A, B)$, respectively.

First we state a well-known result which gives necessary and sufficient conditions for Fredholm consistency of an operator $A \in \mathcal{B}(\mathcal{H})$:

THEOREM 1.119. *Let $A \in \mathcal{B}(\mathcal{H})$. Then A is FC if and only if one of the following conditions is satisfied:*

(i) $A \in \mathcal{F}(\mathcal{H})$.
(ii) *$\mathcal{R}(A)$ is closed, $\alpha(A) = d(A) = \infty$.*
(iii) *$\mathcal{R}(A)$ is nonclosed.*

Throughout the section for given operators $A \in \mathcal{B}(\mathcal{H})$ and $B \in \mathcal{B}(\mathcal{K})$, we suppose that an arbitrary $C \in \mathcal{B}(\mathcal{K}, \mathcal{H})$ is given by

$$C = \begin{bmatrix} C_1 & C_2 \\ C_3 & C_4 \end{bmatrix} : \begin{bmatrix} \mathcal{N}(B)^{\perp} \\ \mathcal{N}(B) \end{bmatrix} \to \begin{bmatrix} \overline{\mathcal{R}(A)} \\ \mathcal{R}(A)^{\perp} \end{bmatrix}. \tag{1.81}$$

We now begin by giving necessary and sufficient conditions such that M_C given by (1.1) is FC:

THEOREM 1.120. *Let $A \in \mathcal{B}(\mathcal{H})$, $B \in \mathcal{B}(\mathcal{K})$, and $C \in \mathcal{B}(\mathcal{K}, \mathcal{H})$ be given operators and let C be given by (1.81). The operator matrix M_C is FC if and only if one of the following conditions is satisfied:*

(i) $A, B \in \mathcal{F}(\mathcal{H})$.
(ii) $A \in \mathcal{F}_l(\mathcal{H})$, $B \in \mathcal{F}_r(\mathcal{K})$, $\alpha(B) = d(A) = \infty$, $C_4 \in \mathcal{F}(\mathcal{N}(B), \mathcal{R}(A)^{\perp})$.
(iii) *$\mathcal{R}(A)$, $\mathcal{R}(B)$ are closed and one of the following conditions holds:*
  (a) *$\mathcal{R}(C_4)$ is closed and $\alpha(A) + \alpha(C_4) = d(B) + d(C_4) = \infty$.*
  (b) *$\mathcal{R}(C_4)$ is nonclosed.*

(iv) $\mathcal{R}(A)$ *is closed,* $\mathcal{R}(B)$ *is not closed, and one of the following conditions holds:*
  (c) $\mathcal{R}(C_4)$ *is closed,* $d(C_4) = \infty$, $\mathcal{R}(B^*) + \mathcal{R}(C_3^* P_{\mathcal{R}(C_4)^\perp}) = \overline{\mathcal{R}(B^*)}$, $\alpha(A) + \alpha(C_4) = \infty$.
  (d) $\mathcal{R}(C_4)$ *is closed,* $d(C_4) < \infty$.
  (e) $\mathcal{R}(C_4)$ *is closed,* $\mathcal{R}(B^*) + \mathcal{R}(C_3^* P_{\mathcal{R}(C_4)^\perp}) \neq \overline{\mathcal{R}(B^*)}$.
  (f) $\mathcal{R}(C_4)$ *is not closed.*

(v) $\mathcal{R}(A)$ *is nonclosed,* $\mathcal{R}(B)$ *is closed, and one of the following conditions holds:*
  (g) $\mathcal{R}(C_4)$ *is closed,* $\alpha(C_4) = \infty$, $\mathcal{R}(A) + \mathcal{R}(C_2 P_{\mathcal{N}(C_4)}) = \overline{\mathcal{R}(A)}$, $d(B) + d(C_4) = \infty$.
  (h) $\mathcal{R}(C_4)$ *is closed,* $\alpha(C_4) < \infty$.
  (i) $\mathcal{R}(C_4)$ *is closed,* $\mathcal{R}(A) + \mathcal{R}(C_2 P_{\mathcal{N}(C_4)}) \neq \overline{\mathcal{R}(A)}$.
  (j) $\mathcal{R}(C_4)$ *is not closed.*

(vi) $\mathcal{R}(A)$, $\mathcal{R}(B)$ *are nonclosed subspaces.*

PROOF. We have that M_C is Fredholm consistent if and only if one of the conditions (i)–(iii) from Theorem 1.119 holds. So, we first look for necessary and sufficient conditions which the operators A, B, and C have to satisfy in order for (i) from Theorem 1.119 to hold for M_C, i.e., so that $M_C \in \mathcal{F}(\mathcal{H} \oplus \mathcal{K})$. The answer to this question is completely given in Theorem 1.54. Hence, from this theorem, it follows that $M_C \in \mathcal{F}(\mathcal{H} \oplus \mathcal{K})$ if and only if $A \in \mathcal{F}(\mathcal{H})$, $B \in \mathcal{F}(\mathcal{K})$ or $A \in \mathcal{F}_l(\mathcal{H})$, $B \in \mathcal{F}_r(\mathcal{K})$, $\alpha(B) = d(A) = \infty$, and C given by (1.81) is such that $C_4 \in \mathcal{F}(\mathcal{N}(B), \mathcal{R}(A)^\perp)$. Now we are looking for necessary and sufficient conditions for the operators A, B, and C for one of the conditions (ii) or (iii) from Theorem 1.119 to hold true for M_C. We consider four cases depending, on the closedness of the ranges of operators A and B:

Case 1: $\mathcal{R}(A)$, $\mathcal{R}(B)$ are closed. Then M_C has a matrix representation

$$M_C = \begin{bmatrix} A_1 & 0 & C_1 & C_2 \\ 0 & 0 & C_3 & C_4 \\ 0 & 0 & B_1 & 0 \\ 0 & 0 & 0 & 0 \end{bmatrix} : \begin{bmatrix} \mathcal{N}(A)^\perp \\ \mathcal{N}(A) \\ \mathcal{N}(B)^\perp \\ \mathcal{N}(B) \end{bmatrix} \to \begin{bmatrix} \mathcal{R}(A) \\ \mathcal{R}(A)^\perp \\ \mathcal{R}(B) \\ \mathcal{R}(B)^\perp \end{bmatrix},$$

where A_1, B_1 are invertible. We can verify that $\mathcal{R}(M_C)$ is closed if and only if $\mathcal{R}(C_4)$ is closed. Also, we have that

$$\alpha(M_C) = \alpha(A) + \alpha\left(\begin{bmatrix} A_1 & C_2 \\ 0 & C_4 \end{bmatrix}\right) = \alpha(A) + \alpha(C_4) \tag{1.82}$$

and

$$\alpha(M_C^*) = \alpha(B^*) + \alpha\left(\begin{bmatrix} C_3^* & B_1^* \\ C_4^* & 0 \end{bmatrix}\right) = \alpha(B^*) + \alpha(C_4^*). \tag{1.83}$$

So, in this case we get that $\mathcal{R}(M_C)$ is nonclosed if and only if $\mathcal{R}(C_4)$ is nonclosed while $\mathcal{R}(M_C)$ is closed and $\alpha(M_C) = d(M_C) = \infty$ if and only if $\mathcal{R}(C_4)$ is closed and $\alpha(A) + \alpha(C_4) = d(B) + d(C_4) = \infty$.

Case 2: $\mathcal{R}(A)$ is closed, $\mathcal{R}(B)$ is nonclosed. In this case, we are looking for necessary and sufficient conditions for M_C to satisfy (ii) or (iii) from Theorem 1.119 (evidently (i) cannot be satisfied). M_C has the matrix representation

$$M_C = \begin{bmatrix} A_1 & 0 & C_1 & C_2 \\ 0 & 0 & C_3 & C_4 \\ 0 & 0 & B_1 & 0 \\ 0 & 0 & 0 & 0 \end{bmatrix} : \begin{bmatrix} \mathcal{N}(A)^\perp \\ \mathcal{N}(A) \\ \mathcal{N}(B)^\perp \\ \mathcal{N}(B) \end{bmatrix} \to \begin{bmatrix} \mathcal{R}(A) \\ \mathcal{R}(A)^\perp \\ \overline{\mathcal{R}(B)} \\ \mathcal{R}(B)^\perp \end{bmatrix},$$

where A_1 is invertible and B_1 is injective with dense range. It can be checked that $\mathcal{R}(M_C)$ is closed if and only if $M_1 = \begin{bmatrix} C_3 & C_4 \\ B_1 & 0 \end{bmatrix}$ has closed range. Using Theorems 2.5 and 2.6 from [**67**], we have that there exists $C_3 \in \mathcal{B}(\mathcal{N}(B)^\perp, R(A)^\perp)$ such that $\mathcal{R}(M_1)$ is closed if and only if $C_4 \in \mathcal{B}(\mathcal{N}(B), R(A)^\perp)$ has closed range and $\alpha(C_4^*) = \infty$. In order to describe all C_3 such that $\mathcal{R}(M_1)$ is closed, suppose that C_4 is such that $\mathcal{R}(C_4)$ is closed and $\alpha(C_4^*) = \infty$. In that case M_1 can be represented by

$$M_1 = \begin{bmatrix} C_{31} & C_{41} & 0 \\ C_{32} & 0 & 0 \\ B_1 & 0 & 0 \end{bmatrix} : \begin{bmatrix} \mathcal{N}(B)^\perp \\ \mathcal{N}(C_4)^\perp \\ \mathcal{N}(C_4) \end{bmatrix} \to \begin{bmatrix} \mathcal{R}(C_4) \\ \mathcal{R}(C_4)^\perp \\ \mathcal{R}(B)^\perp \end{bmatrix}, \tag{1.84}$$

where C_{41} is invertible. Evidently, $\mathcal{R}(M_1)$ is closed if and only if $\mathcal{R}(B_1^*)+\mathcal{R}(C_{32}^*)$ is closed. Since $C_{32} = P^{\mathtt{rst}}_{\mathcal{R}(C_4)^\perp} C_3$ and $\mathcal{R}(B_1^*) = \mathcal{R}(B^*)$, the last condition is equivalent with $\mathcal{R}(B^*)+\mathcal{R}(C_3^* P_{\mathcal{R}(C_4)^\perp})$ being closed, i.e., with $\mathcal{R}(B^*)+\mathcal{R}(C_3^* P_{\mathcal{R}(C_4)^\perp}) = \overline{\mathcal{R}(B^*)}$.

By Lemma 1.65, if $\mathcal{R}(M_C)$ is closed, then since $\mathcal{R}(B)$ is not closed, we get that $d(M_C) = \infty$. Finally, since $\alpha(M_C) = \alpha(A) + \alpha(C_4)$, we can conclude that $\mathcal{R}(M_C)$ is closed and $\alpha(M_C) = d(M_C) = \infty$ if and only if C_4 has closed range, $d(C_4) = \infty$, $\mathcal{R}(B^*) + \mathcal{R}(C_3^* P_{\mathcal{R}(C_4)^\perp}) = \overline{\mathcal{R}(B^*)}$, and $\alpha(A) + \alpha(C_4) = \infty$.

Now, we consider the case when $\mathcal{R}(M_C)$ is not closed. From the discussion above, $\mathcal{R}(M_C)$ is not closed if and only if $\mathcal{R}(M_1)$ is not closed. Using Theorems 2.5 and 2.6 from [**67**], we conclude that $\mathcal{R}(M_1)$ is not closed precisely in one of the following cases:

1. $\mathcal{R}(C_4)$ is closed, $d(C_4) < \infty$.
2. $\mathcal{R}(C_4)$ is closed, $\mathcal{R}(B^*) + \mathcal{R}(C_3^* P_{\mathcal{R}(C_4)^\perp}) \neq \overline{\mathcal{R}(B^*)}$.
3. $\mathcal{R}(C_4)$ is not closed.

Case 3: $\mathcal{R}(A)$ is nonclosed, $\mathcal{R}(B)$ is closed. This case is analogous to Case 2.

Case 4: $\mathcal{R}(A)$, $\mathcal{R}(B)$ are nonclosed. In this case for any $C \in \mathcal{B}(\mathcal{K}, \mathcal{H})$, we have that M_C satisfies conditions from (ii) or (iii) of Theorem 1.119: If $\mathcal{R}(M_C)$ is closed, then by Lemma 1.65 it follows that $\alpha(M_C) = d(M_C) = \infty$, so (ii) from Theorem 1.119 holds, while if $\mathcal{R}(M_C)$ is not closed, then (iii) from Theorem 1.119 is satisfied. □

In the following four theorems we present certain necessary and sufficient conditions such that for given $A \in \mathcal{B}(\mathcal{H})$ and $B \in \mathcal{B}(\mathcal{K})$ there exists $C \in \mathcal{B}(\mathcal{K}, \mathcal{H})$ such that M_C is Fredholm consistent. The set of all such C, denoted by $S_{FC}(A, B)$, will be completely described. We consider four possible cases:

THEOREM 1.121. *Let $A \in \mathcal{B}(\mathcal{H})$ and $B \in \mathcal{B}(\mathcal{K})$ be the operators with closed ranges. Then there exists $C \in \mathcal{B}(\mathcal{K}, \mathcal{H})$ such that M_C is FC if and only if one of the following conditions is satisfied:*

(1) $A, B \in \mathcal{F}(\mathcal{H})$. *In this case*

$$S_{FC}(A, B) = \mathcal{B}(\mathcal{K}, \mathcal{H}).$$

(2) $A \in \mathcal{F}_l(\mathcal{H})$, $B \in \mathcal{F}_r(\mathcal{H})$, $d(A) = \alpha(B) = \infty$. *In this case,*

$$\begin{aligned} &S_{FC}(A, B)\\ &= \{C \in \mathcal{B}(\mathcal{K}, \mathcal{H})\colon C \text{ is given by } (1.81), C_4 \in \mathcal{F}(\mathcal{N}(B), \mathcal{R}(A)^{\perp})\}\\ &\quad \cup \{C \in \mathcal{B}(\mathcal{K}, \mathcal{H})\colon C \text{ is given by } (1.81), \mathcal{R}(C_4) \text{ is closed, } \alpha(C_4)\\ &\qquad = d(C_4) = \infty\}\\ &\quad \cup \{C \in \mathcal{B}(\mathcal{K}, \mathcal{H})\colon C \text{ is given by } (1.81), \mathcal{R}(C_4) \text{ is not closed}\}. \end{aligned}$$

(3) $\alpha(B) = d(A) = \alpha(A) = d(B) = \infty$. *In this case,*

$$S_{FC}(A, B) = \mathcal{B}(\mathcal{K}, \mathcal{H}).$$

(4) $A \in \mathcal{F}_l(\mathcal{H})$, $\alpha(B) = d(A) = d(B) = \infty$. *In this case,*

$$\begin{aligned} &S_{FC}(A, B)\\ &= \{C \in \mathcal{B}(\mathcal{K}, \mathcal{H})\colon C \text{ is given by } (1.81),\ \mathcal{R}(C_4) \text{ is closed, } \alpha(C_4) = \infty\}\\ &\quad \cup \{C \in \mathcal{B}(\mathcal{K}, \mathcal{H})\colon C \text{ is given by } (1.81),\ \mathcal{R}(C_4) \text{ is not closed}\}. \end{aligned} \tag{1.85}$$

(5) $B \in \mathcal{F}_r(\mathcal{K})$, $\alpha(A) = \alpha(B) = d(A) = \infty$. *In this case,*

$$\begin{aligned} &S_{FC}(A, B)\\ &= \{C \in \mathcal{B}(\mathcal{K}, \mathcal{H})\colon C \text{ is given by } (1.81),\ \mathcal{R}(C_4) \text{ is closed, } d(C_4) = \infty\}\\ &\quad \cup \{C \in \mathcal{B}(\mathcal{K}, \mathcal{H})\colon C \text{ is given by } (1.81),\ \mathcal{R}(C_4) \text{ is not closed}\}. \end{aligned} \tag{1.86}$$

(6) $A \in \mathcal{F}_r(\mathcal{H})$, $B \in \mathcal{F}_l(\mathcal{K})$, $\alpha(A) = d(B) = \infty$. *In this case,*

$$S_{FC}(A, B) = \mathcal{B}(\mathcal{K}, \mathcal{H}).$$

(7) $B \in \mathcal{F}_l(\mathcal{K})$, $\alpha(A) = d(A) = \infty$. *In this case,*

$$S_{FC}(A, B) = \mathcal{B}(\mathcal{K}, \mathcal{H}).$$

(8) $A \in \mathcal{F}_r(\mathcal{H})$, $\alpha(B) = d(B) = \infty$. *In this case,*

$$S_{FC}(A, B) = \mathcal{B}(\mathcal{K}, \mathcal{H}).$$

PROOF. Evidently by item (i) of Theorem 1.120, we have that if $A \in \mathcal{F}(\mathcal{H})$, $B \in \mathcal{F}(\mathcal{K})$, then M_C is FC for any $C \in \mathcal{B}(\mathcal{K}, \mathcal{H})$.

Now, we first suppose that $\alpha(B) = d(A) = \infty$ and consider four cases depending on whether the values of $\alpha(A)$ and $d(B)$ are finite or not.

Case 1(a). $\alpha(B) = d(A) = \infty$, $\alpha(A) + d(B) < \infty$. Now $A \in \mathcal{F}_l(\mathcal{H})$, $B \in \mathcal{F}_r(\mathcal{K})$, so by item (ii) of Theorem 1.120, we have that

$$\{C\colon C \text{ is given by (1.81)},\ C_4 \in \mathcal{F}(\mathcal{N}(B), \mathcal{R}(A)^{\perp})\} \subseteq S_{FC}(A,B) \tag{1.87}$$

and the existence of C such that $C_4 \in \mathcal{F}(\mathcal{N}(B), \mathcal{R}(A)^{\perp})$ is evident. Since $\mathcal{N}(B)$ and $\mathcal{R}(A)^{\perp}$ are infinite-dimensional spaces, there exists C such that $\mathcal{R}(C_4)$ is not closed and also there exists C such that $\mathcal{R}(C_4)$ is closed and $\alpha(C_4) = d(C_4) = \infty$. The condition that $\alpha(C_4) = d(C_4) = \infty$ is necessary and sufficient for $\alpha(A) + \alpha(C_4) = d(B) + d(C_4) = \infty$ to hold in item (iii)(a) of Theorem 1.120. Hence,

$$\begin{aligned} &\{C \in \mathcal{B}(\mathcal{K},\mathcal{H})\colon C \text{ is given by (1.81)},\ \mathcal{R}(C_4) \text{ is closed},\ \alpha(C_4) = d(C_4) = \infty\} \\ &\cup \{C \in \mathcal{B}(\mathcal{K},\mathcal{H})\colon C \text{ is given by (1.81)},\ \mathcal{R}(C_4) \text{ is not closed}\} \subseteq S_{FC}(A,B). \end{aligned} \tag{1.88}$$

That the reverse inclusions of those in (1.87) and (1.88) hold follows from the fact that Theorem 1.120 covers all the cases when M_C is FC.

Case 1(b). $\alpha(B) = d(A) = \alpha(A) = d(B) = \infty$. By item (iii) of Theorem 1.120, we have that M_C is FC for any $C \in \mathcal{B}(\mathcal{K},\mathcal{H})$ such that $\mathcal{R}(C_4)$ is closed. Since by item (iii) of that theorem, we have that M_C is FC for any $C \in \mathcal{B}(\mathcal{K},\mathcal{H})$ such that $\mathcal{R}(C_4)$ is not closed, we conclude that in this case M_C is FC for any $C \in \mathcal{B}(\mathcal{K},\mathcal{H})$.

Case 1(c). $\alpha(B) = d(A) = \infty$, $\alpha(A) < \infty$, $d(B) = \infty$. Then $A \in \mathcal{F}_l(\mathcal{H})$. In this case, by item (iii) of Theorem 1.120, we have that M_C is FC for any $C \in \mathcal{B}(\mathcal{K},\mathcal{H})$ such that $\mathcal{R}(C_4)$ is closed and $\alpha(A) + \alpha(C_4) = \infty$ and by item (iii), we have that M_C is FC for any $C \in \mathcal{B}(\mathcal{K},\mathcal{H})$ such that $\mathcal{R}(C_4)$ is not closed. Since in Theorem 1.120 all possible cases are listed when M_C is FC, we conclude that

$$\begin{aligned} &S_{FC}(A,B) \\ &\quad = \{C \in \mathcal{B}(\mathcal{K},\mathcal{H})\colon C \text{ is given by (1.81)},\ \mathcal{R}(C_4) \text{ is closed},\ \alpha(C_4) = \infty\} \\ &\quad\quad \cup \{C \in \mathcal{B}(\mathcal{K},\mathcal{H})\colon C \text{ is given by (1.81)},\ \mathcal{R}(C_4) \text{ is not closed}\}. \end{aligned} \tag{1.89}$$

Case 1(d). $\alpha(B) = d(A) = \infty$, $\alpha(A) = \infty$, $d(B) < \infty$. Analogously to the previous case we have that

$$\begin{aligned} &S_{FC}(A,B) \\ &\quad = \{C \in \mathcal{B}(\mathcal{K},\mathcal{H})\colon C \text{ is given by (1.81)},\ \mathcal{R}(C_4) \text{ is closed},\ d(C_4) = \infty\} \\ &\quad\quad \cup \{C \in \mathcal{B}(\mathcal{K},\mathcal{H})\colon C \text{ is given by (1.81)},\ \mathcal{R}(C_4) \text{ is not closed}\}. \end{aligned} \tag{1.90}$$

Now, we suppose that $d(A), \alpha(B) < \infty$. Then $A \in \mathcal{F}_r(\mathcal{H})$, $B \in \mathcal{F}_l(\mathcal{K})$ and for any $C \in \mathcal{B}(\mathcal{K},\mathcal{H})$ we have that $\mathcal{R}(C_4)$ is closed and $\alpha(C_4)$, $d(C_4) < \infty$. By items (iii) of Theorem 1.120, we have that M_C is FC for some $C \in \mathcal{B}(\mathcal{K},\mathcal{H})$ if and only if $\alpha(A) = d(B) = \infty$. In that case M_C is FC for any $C \in \mathcal{B}(\mathcal{K},\mathcal{H})$.

Let us suppose that $d(A) = \infty$ and $\alpha(B) < \infty$. Then $B \in \mathcal{F}_l(\mathcal{K})$ and for any $C \in \mathcal{B}(\mathcal{K},\mathcal{H})$ we have that $\mathcal{R}(C_4)$ is closed, $\alpha(C_4) < d(C_4) = \infty$. By items (iii) of Theorem 1.120, we have that M_C is FC for some $C \in \mathcal{B}(\mathcal{K},\mathcal{H})$ if and only if $\alpha(A) = \infty$. Evidently, in this case M_C is FC for any $C \in \mathcal{B}(\mathcal{K},\mathcal{H})$.

In the case when $d(A) < \infty$ and $\alpha(B) = \infty$ we can analogously conclude that M_C is FC for some $C \in \mathcal{B}(\mathcal{K}, \mathcal{H})$ if and only if $d(B) = \infty$. In that case M_C is FC for any $C \in \mathcal{B}(\mathcal{K}, \mathcal{H})$. □

THEOREM 1.122. *Let $A \in \mathcal{B}(\mathcal{H})$ have closed range and let $B \in \mathcal{B}(\mathcal{K})$ have a range that is not closed. Then there exists $C \in \mathcal{B}(\mathcal{K}, \mathcal{H})$ such that M_C is FC if and only if one of the following conditions is satisfied:*

(1) $d(A) = \alpha(B) = \infty$, $\alpha(A) < \infty$. *In this case,*

$$\begin{aligned}(1.91)\quad & S_{FC}(A,B) \\ &= \{C \in \mathcal{B}(\mathcal{K},\mathcal{H})\colon C \textit{ is given by } (1.81),\ \mathcal{R}(C_4) \textit{ is closed},\\ &\qquad d(C_4) = \infty,\ \mathcal{R}(B^*) + \mathcal{R}(C_3^* P_{\mathcal{R}(C_4)^\perp}) = \overline{\mathcal{R}(B^*)},\ \alpha(C_4) = \infty\} \\ &\quad \cup \{C \in \mathcal{B}(\mathcal{K},\mathcal{H})\colon C \textit{ is given by } (1.81),\ \mathcal{R}(C_4) \textit{ is closed},\\ &\qquad \mathcal{R}(B^*) + \mathcal{R}(C_3^* P_{\mathcal{R}(C_4)^\perp}) \neq \overline{\mathcal{R}(B^*)}\} \\ &\quad \cup \{C \in \mathcal{B}(\mathcal{K},\mathcal{H})\colon C \textit{ is given by } (1.81),\ \mathcal{R}(C_4) \textit{ is closed},\ d(C_4) < \infty\} \\ &\quad \cup \{C \in \mathcal{B}(\mathcal{K},\mathcal{H})\colon C \textit{ is given by } (1.81),\ \mathcal{R}(C_4) \textit{ is not closed}\}.\end{aligned}$$

(2) $d(A) = \infty$, $\alpha(A) + \alpha(B) < \infty$. *In this case,*

$$\begin{aligned}& S_{FC}(A,B) = \\ &\quad \{C \in \mathcal{B}(\mathcal{K},\mathcal{H})\colon C \textit{ is given by } (1.81),\ \mathcal{R}(B^*) + \mathcal{R}(C_3^* P_{\mathcal{R}(C_4)^\perp}) \neq \overline{\mathcal{R}(B^*)}\}.\end{aligned}$$

(3) $d(A) = \alpha(A) = \infty$. *In this case,* $S_{FC}(A,B) = \mathcal{B}(\mathcal{K},\mathcal{H})$.
(4) $d(A) < \infty$. *In this case,* $S_{FC}(A,B) = \mathcal{B}(\mathcal{K},\mathcal{H})$.

PROOF. According to Theorem 1.120 (iv), we distinguish the following cases.

Case (a). $d(A) = \alpha(B) = \infty$, $\alpha(A) < \infty$. Notice that for each of the cases (c)–(f) of item (iv) of Theorem 1.120, there exists C which satisfies the conditions from that case, so $S_{FC}(A,B)$ is described by (1.91).

Case (b). $d(A) = \infty$, $\alpha(A) + \alpha(B) < \infty$. In this case for any $C \in \mathcal{B}(\mathcal{K},\mathcal{H})$ we have that $\mathcal{R}(C_4)$ is closed, $d(C_4) = \infty$, and $\alpha(A) + \alpha(C_4) < \infty$, so items (c), (d), and (f) from Theorem 1.120 cannot be satisfied. Hence,

$$\begin{aligned}& S_{FC}(A,B) \\ &\quad = \{C \in \mathcal{B}(\mathcal{K},\mathcal{H})\colon C \text{ is given by } (1.81),\ \mathcal{R}(B^*) + \mathcal{R}(C_3^* P_{\mathcal{R}(C_4)^\perp}) \neq \overline{\mathcal{R}(B^*)}\}.\end{aligned}$$

Case (c). $d(A) = \alpha(A) = \infty$. In this case for any $C \in \mathcal{B}(\mathcal{K},\mathcal{H})$ we have that $\alpha(A) + \alpha(C_4) = \infty$, so by items (c)–(f) from Theorem 1.120, we get that $S_{FC}(A,B) = \mathcal{B}(\mathcal{K},\mathcal{H})$.

Case (d). $d(A) < \infty$. For any $C \in \mathcal{B}(\mathcal{K},\mathcal{H})$ we have that $\mathcal{R}(C_4)$ is closed and $d(C_4) < \infty$. Hence by item (d) of Theorem 1.120 it follows that $S_{FC}(A,B) = \mathcal{B}(\mathcal{K},\mathcal{H})$ in this case. □

THEOREM 1.123. *Let $A \in \mathcal{B}(\mathcal{H})$ have nonclosed range and let $B \in \mathcal{B}(\mathcal{K})$ have closed range. Then there exists $C \in \mathcal{B}(\mathcal{K}, \mathcal{H})$ such that M_C is FC if and only if one of the following conditions is satisfied:*

(1) $d(A) = \alpha(B) = \infty$, $d(B) < \infty$. *In this case,*

$$\begin{aligned} &S_{FC}(A, B) \\ &\quad = \{C \in \mathcal{B}(\mathcal{K}, \mathcal{H}) \colon C \text{ is given by (1.81)}, \ \mathcal{R}(C_4) \text{ is closed}, \\ &\qquad\qquad \alpha(C_4) = \infty, \mathcal{R}(A) + \mathcal{R}(C_2 P_{\mathcal{N}(C_4)}) = \overline{\mathcal{R}(A)}, \ d(C_4) = \infty\} \\ &\qquad \cup \{C \in \mathcal{B}(\mathcal{K}, \mathcal{H}) \colon C \text{ is given by (1.81)}, \ \mathcal{R}(C_4) \text{ is closed}, \\ &\qquad\qquad\qquad\qquad \mathcal{R}(A) + \mathcal{R}(C_2 P_{\mathcal{N}(C_4)}) \neq \overline{\mathcal{R}(A)}\} \\ &\qquad \cup \{C \in \mathcal{B}(\mathcal{K}, \mathcal{H}) \colon C \text{ is given by (1.81)}, \ \mathcal{R}(C_4) \text{ is closed}, \ \alpha(C_4) < \infty\} \\ &\qquad \cup \{C \in \mathcal{B}(\mathcal{K}, \mathcal{H}) \colon C \text{ is given by (1.81)}, \ \mathcal{R}(C_4) \text{ is not closed}\}. \end{aligned}$$

(2) $\alpha(B) = \infty$, $d(A) + d(B) < \infty$. *In this case,*

$$\begin{aligned} &S_{FC}(A, B) \\ &\quad = \{C \in \mathcal{B}(\mathcal{K}, \mathcal{H}) \colon C \text{ is given by (1.81)}, \ \mathcal{R}(A) + \mathcal{R}(C_2 P_{\mathcal{N}(C_4)}) \neq \overline{\mathcal{R}(A)}\}. \end{aligned}$$

(3) $\alpha(B) = d(B) = \infty$. *In this case,* $S_{FC}(A, B) = \mathcal{B}(\mathcal{K}, \mathcal{H})$.
(4) $\alpha(B) < \infty$. *In this case,* $S_{FC}(A, B) = \mathcal{B}(\mathcal{K}, \mathcal{H})$.

PROOF. The proof can be given analogously to that of Theorem 1.122. □

THEOREM 1.124. *Let $A \in \mathcal{B}(\mathcal{H})$ and $B \in \mathcal{B}(\mathcal{K})$ be the operators with nonclosed ranges. Then*

$$S_{FC}(A, B) = \mathcal{B}(\mathcal{K}, \mathcal{H}).$$

PROOF. By item (vi) of Theorem 1.120, we have that if $\mathcal{R}(A)$, $\mathcal{R}(B)$ are nonclosed, then M_C is FC for any $C \in \mathcal{B}(\mathcal{K}, \mathcal{H})$. □

As a corollary of the previous result we get the following:

THEOREM 1.125. *Let $A \in \mathcal{B}(\mathcal{H})$ and $B \in \mathcal{B}(\mathcal{K})$ be given operators. Then there exists $C \in \mathcal{B}(\mathcal{K}, \mathcal{H})$ such that M_C is FC if and only if one of the following conditions is satisfied:*

(a) $A \in \mathcal{F}(\mathcal{H})$ *and* $B \in \mathcal{F}(\mathcal{K})$.
(b) $\alpha(A) + \alpha(B) = d(A) + d(B) = \infty$.
(c) $\mathcal{R}(A)$ *is nonclosed.*
(d) $\mathcal{R}(B)$ *is nonclosed.*

Now, using Theorems 1.119 and 1.121–1.124 we can compute $\bigcap_{C \in \mathcal{B}(\mathcal{K}, \mathcal{H})} \sigma_{FC}(M_C)$:

COROLLARY 1.126. Let $A \in \mathcal{B}(\mathcal{H})$ and $B \in \mathcal{B}(\mathcal{K})$. Then

$$\begin{aligned} \bigcap_{C \in \mathcal{B}(\mathcal{K}, \mathcal{H})} \sigma_{FC}(M_C) = \Big(\rho_e(A) \cap \sigma_e(B) \cap \rho_{le}(B)\Big) \cup \Big(\rho_e(B) \cap \sigma_e(A) \cap \rho_{re}(A)\Big) \\ \cup \Big(\rho_{le}(A) \cap \sigma_e(A) \cap \rho_{le}(B)\Big) \cup \Big(\rho_{re}(A) \cap \sigma_e(B) \cap \rho_{re}(B)\Big). \end{aligned}$$

As in the case of ordinary spectrum (see [**17**]) and consistent invertibility spectrum (see [**91**]) we have the following:

COROLLARY 1.127. Let $A \in \mathcal{B}(\mathcal{H})$ and $B \in \mathcal{B}(\mathcal{K})$. Then for every $C \in \mathcal{B}(\mathcal{K}, \mathcal{H})$, we have that

$$\sigma_{FC}(M_C) \subseteq \sigma_{FC}(A) \cup \sigma_{FC}(B). \tag{1.92}$$

PROOF. Suppose that $0 \notin \sigma_{FC}(A) \cup \sigma_{FC}(A)$ (instead of 0 we can take any $\lambda \in \mathbb{C}$). Then A and B are FC operators. So by Theorem 1.119 we have 9 possible cases. We can check using Theorems 1.121–1.124 that in each of these cases we have that $S_{FC}(A, B) = \mathcal{B}(\mathcal{K}, \mathcal{H})$. Hence $0 \notin \sigma_{FC}(M_C)$ for any $C \in \mathcal{B}(\mathcal{K}, \mathcal{H})$. □

REMARK 1.128. The reverse inclusion of (1.92) does not hold in general: For $A \in \mathcal{F}_r(\mathcal{H}) \setminus \mathcal{F}(\mathcal{H})$ and $B \in \mathcal{B}(\mathcal{K})$ such that $\mathcal{R}(B)$ is closed and $\alpha(B) = d(B) = \infty$, by Theorem 1.121(8), we have that $S_{FC}(A, B) = \mathcal{B}(\mathcal{K}, \mathcal{H})$. So A is not FC and $0 \in \sigma_{FC}(A) \cup \sigma_{FC}(B)$ while $0 \notin \sigma_{FC}(M_C)$ for any $C \in \mathcal{B}(\mathcal{K}, \mathcal{H})$. Using Theorems 1.121–1.124, we can easy find necessary and sufficient conditions for the operators A and B to satisfy so that

$$\sigma_{FC}(M_C) = \sigma_{FC}(A) \cup \sigma_{FC}(B), \tag{1.93}$$

for any $C \in \mathcal{B}(\mathcal{K}, \mathcal{H})$.

Using Theorems 1.121–1.124 we can describe the set $\bigcup_{C \in \mathcal{B}(\mathcal{K},\mathcal{H})} \sigma_{FC}(M_C)$:

COROLLARY 1.129. Let $A \in \mathcal{B}(\mathcal{H})$ and $B \in \mathcal{B}(\mathcal{K})$. Then

$$\begin{aligned} \bigcup_{C \in \mathcal{B}(\mathcal{K},\mathcal{H})} \sigma_{FC}(M_C) = &\Big(\rho_{le}(A) \cap \sigma_e(A)\Big) \cup \Big(\rho_{re}(B) \cap \sigma_e(B)\Big) \\ &\cup \Big(\rho_e(A) \cap \sigma_e(B) \cap \rho_{le}(B)\Big) \cup \Big(\rho_e(B) \cap \sigma_e(A) \cap \rho_{re}(A)\Big). \end{aligned} \tag{1.94}$$

PROOF. Notice that the set $\bigcup_{C \in \mathcal{B}(\mathcal{K},\mathcal{H})} \sigma_{FC}(M_C)$ is exactly the union of two sets that are disjoint: $T = \bigcup_{C \in \mathcal{B}(\mathcal{K},\mathcal{H})} \sigma_{FC}(M_C) \setminus (\bigcap_{C \in \mathcal{B}(\mathcal{K},\mathcal{H})} \sigma_{FC}(M_C))$ and $\bigcap_{C \in \mathcal{B}(\mathcal{K},\mathcal{H})} \sigma_{FC}(M_C)$. Since the first one is described in Corollary 1.126, we describe the other one. Notice that $0 \in T$ if and only A and B satisfy conditions from one of the items given in Theorems 1.121–1.124 in which we have that $S_{FC}(A, B) \neq \mathcal{B}(\mathcal{K}, \mathcal{H})$ (instead of 0 we can consider any $\lambda \in \mathbb{C}$). Hence, by Theorems 1.121–1.124, we have that $S_{FC}(A, B) \neq \mathcal{B}(\mathcal{K}, \mathcal{H})$ if and only if one of the following conditions holds:

- $A \in \mathcal{F}_l(\mathcal{H})$, $B \in \mathcal{F}_r(\mathcal{H})$, $d(A) = \alpha(B) = \infty$,
- $A \in \mathcal{F}_l(\mathcal{H})$, $\alpha(B) = d(A) = d(B) = \infty$,
- $B \in \mathcal{F}_r(\mathcal{K})$, $\alpha(A) = \alpha(B) = d(A) = \infty$,
- $\mathcal{R}(A)$ is closed, $\mathcal{R}(B)$ is nonclosed, $d(A) = \alpha(B) = \infty$, $\alpha(A) < \infty$,
- $\mathcal{R}(A)$ is closed, $\mathcal{R}(B)$ is nonclosed, $d(A) = \infty$, $\alpha(A) + \alpha(B) < \infty$,
- $\mathcal{R}(A)$ is nonclosed, $\mathcal{R}(B)$ is closed, $d(A) = \alpha(B) = \infty$, $d(B) < \infty$,
- $\mathcal{R}(A)$ is nonclosed, $\mathcal{R}(B)$ is closed, $\alpha(B) = \infty$, $d(A) + d(B) < \infty$.

Hence,

$$\begin{aligned}T &= \{\lambda\in\mathbb{C}\colon A-\lambda\in\mathcal{F}_l(\mathcal{H})\setminus\mathcal{F}(\mathcal{H}),\ B-\lambda\in\mathcal{F}_r(\mathcal{K})\setminus\mathcal{F}(\mathcal{K})\}\\ &\cup\{\lambda\in\mathbb{C}\colon A-\lambda\in\mathcal{F}_l(\mathcal{H})\setminus\mathcal{F}(\mathcal{H}),\ \mathcal{R}(B-\lambda)\text{ is closed},\ \alpha(B-\lambda)=d(B-\lambda)=\infty\}\\ &\cup\{\lambda\in\mathbb{C}\colon B-\lambda\in\mathcal{F}_r(\mathcal{K})\setminus\mathcal{F}(\mathcal{K}),\ \mathcal{R}(A-\lambda)\text{ is closed},\ \alpha(A-\lambda)=d(A-\lambda)=\infty\}\\ &\cup\{\lambda\in\mathbb{C}\colon A-\lambda\in\mathcal{F}_l(\mathcal{H})\setminus\mathcal{F}(\mathcal{H}),\ \mathcal{R}(B-\lambda)\ \text{ is nonclosed}\}\\ &\cup\{\lambda\in\mathbb{C}\colon B-\lambda\in\mathcal{F}_r(\mathcal{H})\setminus\mathcal{F}(\mathcal{H}),\ \mathcal{R}(A-\lambda)\text{ is nonclosed}\}.\end{aligned}$$

Using Theorem 1.119 we get that the union of the second and the forth set appearing above and the third set appearing in the expression for $\bigcap_{C\in\mathcal{B}(\mathcal{K},\mathcal{H})}\sigma_{FC}(M_C)$ (Corollary 1.126) is equal to

$$\begin{aligned}&\{\lambda\in\mathbb{C}\colon A-\lambda\in\mathcal{F}_l(\mathcal{H})\setminus\mathcal{F}(\mathcal{H}),\ B-\lambda\ is\ FC\}\\ &\quad\cup\{\lambda\in\mathbb{C}\colon A-\lambda\in\mathcal{F}_l(\mathcal{H})\setminus\mathcal{F}(\mathcal{H}),\ B-\lambda\in\mathcal{F}_l(\mathcal{K})\setminus\mathcal{F}(\mathcal{K})\}.\end{aligned}$$

Also, the union of the third and the fifth set appearing in the expression for T and the forth set appearing in the expression for $\bigcap_{C\in\mathcal{B}(\mathcal{K},\mathcal{H})}\sigma_{FC}(M_C)$ (Corollary 1.126) is equal to

$$\begin{aligned}&\{\lambda\in\mathbb{C}\colon B-\lambda\in\mathcal{F}_r(\mathcal{H})\setminus\mathcal{F}(\mathcal{H}),\ A-\lambda\ is\ FC\}\\ &\quad\cup\{\lambda\in\mathbb{C}\colon B-\lambda\in\mathcal{F}_r(\mathcal{H})\setminus\mathcal{F}(\mathcal{H}),\ A-\lambda\in\mathcal{F}_r(\mathcal{K})\setminus\mathcal{F}(\mathcal{K})\}.\end{aligned}$$

Now, we have

$$\begin{aligned}\bigcup_{C\in\mathcal{B}(\mathcal{K},\mathcal{H})}\sigma_{FC}(M_C) =&\{\lambda\in\mathbb{C}\colon A-\lambda\in\mathcal{F}_l(\mathcal{H})\setminus\mathcal{F}(\mathcal{H}),\ B-\lambda\in\mathcal{F}_r(\mathcal{K})\setminus\mathcal{F}(\mathcal{K})\}\\ &\cup\{\lambda\in\mathbb{C}\colon A-\lambda\in\mathcal{F}_l(\mathcal{H})\setminus\mathcal{F}(\mathcal{H}),\ B-\lambda\ is\ FC\}\\ &\cup\{\lambda\in\mathbb{C}\colon A-\lambda\in\mathcal{F}_l(\mathcal{H})\setminus\mathcal{F}(\mathcal{H}),\ B-\lambda\in\mathcal{F}_l(\mathcal{K})\setminus\mathcal{F}(\mathcal{K})\}\\ &\cup\{\lambda\in\mathbb{C}\colon B-\lambda\in\mathcal{F}_r(\mathcal{H})\setminus\mathcal{F}(\mathcal{H}),\ A-\lambda\ is\ FC\}\\ &\cup\{\lambda\in\mathbb{C}\colon B-\lambda\in\mathcal{F}_r(\mathcal{H})\setminus\mathcal{F}(\mathcal{H}),\ A-\lambda\in\mathcal{F}_r(\mathcal{K})\setminus\mathcal{F}(\mathcal{K})\}\\ &\cup\Big(\rho_e(A)\cap\sigma_e(B)\cap\rho_{le}(B)\Big)\cup\Big(\rho_e(B)\cap\sigma_e(A)\cap\rho_{re}(A)\Big).\end{aligned}$$

Using that B is not FC if and only if $B\in\mathcal{F}_r(\mathcal{H})\setminus\mathcal{F}(\mathcal{H})$ or $B\in\mathcal{F}_l(\mathcal{H})\setminus\mathcal{F}(\mathcal{H})$, we get that the union of the first three sets on the hand-right side of the last equality is equal to

$$\{\lambda\in\mathbb{C}\colon A-\lambda\in\mathcal{F}_l(\mathcal{H})\setminus\mathcal{F}(\mathcal{H})\}$$

while the union of the first, fourth, and fifth is equal to

$$\{\lambda\in\mathbb{C}\colon B-\lambda\in\mathcal{F}_r(\mathcal{H})\setminus\mathcal{F}(\mathcal{H})\}$$

Hence (1.94) holds. □

In the following result we give an answer to Question 3:

COROLLARY 1.130. Let $A\in\mathcal{B}(\mathcal{H})$ and $B\in\mathcal{B}(\mathcal{K})$. Then

$$\bigcap_{C\in\mathcal{B}(\mathcal{K},\mathcal{H})}\sigma_{FC}(M_C)=\sigma_{FC}(M_0).$$

PROOF. Since one inclusion is trivial, we need only to show that

$$\sigma_{FC}(M_0)\subseteq\bigcap_{C\in\mathcal{B}(\mathcal{K},\mathcal{H})}\sigma_{FC}(M_C).$$

So, let $\lambda \notin \bigcap_{C\in\mathcal{B}(\mathcal{K},\mathcal{H})} \sigma_{FC}(M_C)$ be arbitrary. Without any loss of generality we can suppose that $\lambda = 0$. So, we have that there exists $C_0 \in \mathcal{B}(\mathcal{K},\mathcal{H})$ such that M_{C_0} is FC. Hence, we can conclude that for A and B the conditions from one of the items given in Theorems 1.121–1.124 must hold. It can be checked that in all of these cases, we have $0 \in S_{FC}(A,B)$; i.e., $0 \notin \sigma_{FC}(M_0)$. $\square$

CHAPTER 2

Invertibility of an operator $A + CX$

As a particular instance of completion problems, the following question is of interest: If $A \in \mathcal{B}(\mathcal{H})$, $B \in \mathcal{B}(\mathcal{K})$, and $C \in \mathcal{B}(\mathcal{K}, \mathcal{H})$ are fixed, does there exist an operator $X \in \mathcal{B}(\mathcal{H}, \mathcal{K})$ such that the operator $M_X \in \mathcal{B}(\mathcal{H} \oplus \mathcal{K})$ given by

$$M_X = \begin{bmatrix} A & C \\ X & B \end{bmatrix} : \begin{bmatrix} \mathcal{H} \\ \mathcal{K} \end{bmatrix} \to \begin{bmatrix} \mathcal{H} \\ \mathcal{K} \end{bmatrix}$$

is of a prescribed type?

The first to ever address any questions dealing with completions of the operator matrices of the form M_X (for separable Hilbert spaces not necessarily of finite dimension) was K. Takahashi. More specifically, in the paper [**160**] he gave necessary and sufficient conditions for the existence of $X \in \mathcal{B}(\mathcal{H})$ such that M_X is invertible. The key result obtained in [**160**] that allowed the author to solve the problem of completion of M_X to invertibility completely proved to also be the one that characterizes the pairs of operators A, C such that $A + CX$ is invertible for some $X \in \mathcal{B}(\mathcal{H}, \mathcal{K})$. This result is of interest in connection with the spectrum assignment problem in systems theory.

The main link between the problem of completion of the operator matrix of the form M_X to an invertible operator and that of the existence of $X \in \mathcal{B}(\mathcal{H}, \mathcal{K})$ such that $A + CX$ is invertible is established directly by the equality

$$\begin{bmatrix} I & X \\ C & A \end{bmatrix} = \begin{bmatrix} I & 0 \\ C & I \end{bmatrix} \begin{bmatrix} I & X \\ 0 & A - CX \end{bmatrix}.$$

Motivated by the connection between these two problems, in this chapter we investigate further problems along the direction pointed out by Takahashi of characterizing all pairs of operators A, C for which $A + CX$ is of a certain type for some X. The following set of questions seems to be a natural continuation of K. Takahashi research: For which operators $A \in \mathcal{B}(\mathcal{H}, \mathcal{K})$ and $C \in \mathcal{B}(\mathcal{L}, \mathcal{K})$ does there exist $X \in \mathcal{B}(\mathcal{H}, \mathcal{L})$ such that $A + CX$ is injective, surjective, with dense range, with closed range, or Fredholm? Later, we completely answer some of these questions, and provide partial answers to the others.

2.1. Auxiliary results

All Hilbert spaces under consideration in this chapter are assumed to be separable. The following lemma will be needed in the sequel and is given without proof since it is rather straightforward.

LEMMA 2.1. (i) *If $K \subseteq \mathcal{H}$ is an operator range and $M \subseteq K$ is a subspace such that* $\operatorname{codim}_K M = \infty$, *then there exists an operator range $L \subseteq K$ such that $M \cap L = \{0\}$.*

(ii) *If $L \subseteq \mathcal{K}$ is an infinite-dimensional operator range and $\mathcal{H}$ is a separable Hilbert space, then there exists an injective $T \in \mathcal{B}(\mathcal{H}, \mathcal{K})$ such that $\mathcal{R}(T) \subseteq L$.*

In this chapter by an *operator range* we shall mean a subspace $\mathcal{K} \subseteq \mathcal{H}$ of a separable Hilbert space $\mathcal{H}$ such that $\mathcal{R}(A) = \mathcal{K}$ for some separable Hilbert space $\mathcal{H}_0$ and some $A \in \mathcal{B}(\mathcal{H}_0, \mathcal{H})$.

Two easy facts about operator ranges that we need are stated as follows.

LEMMA 2.2. (1) *If $M \subseteq \mathcal{K}$ is an operator range and $\mathcal{H}$ is an infinite-dimensional Hilbert space, then $M = \mathcal{R}(A)$ for some $A \in \mathcal{B}(\mathcal{H}, \mathcal{K})$.*

(2) *If $M \subseteq \mathcal{H}$ is a closed subspace and $K \subseteq \mathcal{H}$ is an operator range, then there is an operator range $S \subseteq K$ such that $K = (M \cap K) \oplus S$.*

Below we list some facts that will be used throughout the chapter.

THEOREM 2.3 ([**148**]). *If $\mathcal{H}$ and $\mathcal{K}$ are infinite-dimensional Banach spaces and $A \in \mathcal{B}(\mathcal{H}, \mathcal{K})$ is a semi-Fredholm operator, then $A + K$ is semi-Fredholm and $\operatorname{ind}(A + K) = \operatorname{ind}(K)$ for every compact operator $K \in \mathcal{B}(\mathcal{H}, \mathcal{K})$.*

THEOREM 2.4 ([**148**]). *If $A \in \mathcal{F}_+(\mathcal{H}, \mathcal{K})$ and $M \subseteq \mathcal{H}$ is a closed subspace, then the subspace $A[M]$ is also closed.*

The proof of next lemma can be found in [**111**] but here we give a more detail proof.

LEMMA 2.5. *If the operator range M of a separable Hilbert space $\mathcal{H}$ is not closed, then there exists a closed infinite-dimensional subspace F such that $F \cap M = \{0\}$.*

PROOF. Without loss of generality, assume that M is an operator range of positive operator $A \in \mathcal{B}(\mathcal{H})$ since $\mathcal{R}(A) = \mathcal{R}(\sqrt{AA^*})$. Also, we can suppose that $\mathcal{R}(A)$ is dense in $\mathcal{H}$ since if it is not the case, instead of $\mathcal{H}$ we can consider $\overline{\mathcal{R}(A)}$. Let E be a spectral measure of A. Since $\mathcal{N}(A) = \{0\}$ it follows that $E(\{0\}) = 0$ which implies that 0 is an accumulation point of the $\sigma(A)$. Indeed, if 0 is not an accumulation point of the $\sigma(A)$, then $\{0\}$ is a relatively open in $\sigma(A)$, and by the spectral theorem (see Theorem 3.15 in [**125**]) we have that $E(\{0\}) \neq 0$. So, we can choose the sequence $(t_n)_{n\in\mathbb{N}_0}$ of positive number decreasing to zero such that $t_0 = \|A\|$. Furthermore, we can choose $(t_n)_{n\in\mathbb{N}_0}$ such that $E_k = E((t_k, t_{k-1}]) \neq 0$ using the fact that if S is relatively open set in $\sigma(A)$, then $E(\{S\}) \neq 0$. Let $\mathcal{H}_n = \bigvee\{\mathcal{R}(E_{(2k-1)2^{n-1}} \colon k \in \mathbb{N}\}$. Then $\{\mathcal{H}_n\}_{n\in\mathbb{N}}$ is a pairwise orthogonal family of subspaces and $\mathcal{H} = \bigoplus \mathcal{H}_n$. Let P_n be the projector onto $\mathcal{H}_n$; i.e., $P_n = \sum_{n\in\mathbb{N}} E_{(2k-1)2^{n-1}}$. We have that $j(A|\mathcal{H}_n) \leq j(A|E_{(2k+1)2^{n-1}})$ and since $\sigma(A|E_{(2k+1)2^{n-1}}) \subseteq [t_{(2k+1)2^{n-1}}, t_{(2k-1)2^{n-1}}]$ and A is a positive operator, we have that $j(A|\mathcal{H}_n) \leq j(A|E_{(2k+1)2^{n-1}}) \leq t_{(2k-1)2^{n-1}}$. Since $(t_n)_{n\in\mathbb{N}_0}$ is decreasing to zero, we get that $j(A|\mathcal{H}_n) = 0$ which implies using the fact that A is injective that $\mathcal{R}(A)$ is not closed. Hence we can take $x_n \in \mathcal{R}(A|_{\mathcal{H}_n})$, for each $n \in \mathbb{N}$. Let $\mathcal{H}_0 = \bigvee\{x_n \colon n \in \mathbb{N}\}$. Evidently, $\dim \mathcal{H}_0 = \infty$. If we suppose that $x \in \mathcal{H}_0 \cap \mathcal{R}(A)$, then $x = Ay = \sum_{n\in\mathbb{N}} \alpha_n x_n$. We have that $P_n x = \alpha_n x_n = P_n Ay = P_n A P_n y$, so $\alpha_n = 0$. Hence $\mathcal{H}_0 \cap \mathcal{R}(A) = \{0\}$. □

LEMMA 2.6 ([**77**]). *Let $\mathcal{H}, \mathcal{K}$ be separable Hilbert spaces, let $M, L \subseteq \mathcal{K}$ be operator ranges, and let $T \in \mathcal{B}(\mathcal{H}, \mathcal{K})$ be such that $\mathcal{R}(T) \subseteq M + L$. Then $T = T_1 + T_2$ for some $T_1, T_2 \in \mathcal{B}(\mathcal{H}, \mathcal{K})$ with $\mathcal{R}(T_1) \subseteq M$, $\mathcal{R}(T_2) \subseteq L$.*

The proof of the next theorem can be found in [**77**] but we give it for the completeness:

THEOREM 2.7. *Let $\mathcal{R}_1$ and $\mathcal{R}_2$ be operator ranges. Then $\mathcal{R}_1 + \mathcal{R}_2 = \mathcal{H}$ if and only if there are disjoint closed subspaces $\mathcal{M}_1 \subset \mathcal{R}_1$ and $\mathcal{M}_2 \subset \mathcal{R}_2$ such that $\mathcal{M}_1 + \mathcal{M}_2 = \mathcal{H}$.*

PROOF. We can suppose that A and B are positive operators with $\mathcal{R}(A) = \mathcal{R}_1$, $\mathcal{R}(B) = \mathcal{R}_2$ (if not, take $\sqrt{AA^*}$ and $\sqrt{BB^*}$ instead of A and B). Then $\mathcal{R}((A^2 + B^2)^{\frac{1}{2}}) = \mathcal{R}(A) + \mathcal{R}(B) = \mathcal{H}$. Since $(A^2 + B^2)^{\frac{1}{2}}$ is a positive operator it follows that it is invertible. Hence, there is a positive number δ such that

$$\delta^2 \|x\|^2 \leq ((A^2 + B^2)x, x) = \|Ax\|^2 + \|Bx\|^2$$

for all $x \in \mathcal{H}$. Let E be the spectral measure of A, and choose a positive number $\varepsilon < \delta$. If $x \in \mathcal{H}$, then

$$\begin{aligned}(B^2 E[0,\varepsilon]x, E[0,\varepsilon]x) = \|BE[0,\varepsilon]x\|^2 &\geq \delta^2 \|E[0,\varepsilon]x\|^2 - \|AE[0,\varepsilon]x\|^2 \\ &\geq (\delta^2 - \varepsilon^2)\|E[0,\varepsilon]x\|^2.\end{aligned}$$

Since

$$\|E[0,\varepsilon]B^2 E[0,\varepsilon]x\| \|E[0,\varepsilon]x\| \geq (B^2 E[0,\varepsilon]x, E[0,\varepsilon]x)$$

and

$$\|B^2 E[0,\varepsilon]x\| \|E[0,\varepsilon]x\| \geq (B^2 E[0,\varepsilon]x, E[0,\varepsilon]x)$$

we get

$$\begin{aligned}\|B^2 E[0,\varepsilon]x\| &\geq (\delta^2 - \varepsilon^2)\|E[0,\varepsilon]x\|, \\ \|E[0,\varepsilon]B^2 E[0,\varepsilon]x\| &\geq (\delta^2 - \varepsilon^2)\|E[0,\varepsilon]x\|.\end{aligned}$$

The first of these inequalites show that $\mathcal{M}_2 = B^2\mathcal{R}(E[0,\varepsilon])$ is closed. Also, if $\mathcal{M}_1$ is the range of $E[0,\varepsilon]^{\perp}$, then

$$\mathcal{M}_1 + \mathcal{M}_2 = \mathcal{M}_1 \oplus E[0,\varepsilon]\mathcal{M}_2 = \mathcal{M}_1 \oplus \mathcal{R}(E[0,\varepsilon]B^2 E[0,\varepsilon])$$

is closed. Finally, we can check $\mathcal{M}_1^{\perp} \cap \mathcal{M}_2^{\perp} = \{0\}$ and $\mathcal{M}_1 \cap \mathcal{M}_2 = \{0\}$. □

The proof of the following result can be found in [**61**] (see Theorem 4.1 there):

THEOREM 2.8. *Let $A, B \in \mathcal{B}(\mathcal{H})$ be closed range operators. Then AB has a closed range if and only if $\mathcal{N}(A) + \mathcal{R}(B)$ is closed.*

2.2. Invertibility of an operator $A + CX$

In an attempt to solve the problem of the completions of the operator matrix

$$M_X = \begin{bmatrix} A & C \\ X & B \end{bmatrix} : \begin{bmatrix} \mathcal{H} \\ \mathcal{K} \end{bmatrix} \to \begin{bmatrix} \mathcal{H} \\ \mathcal{K} \end{bmatrix}$$

to the invertibility, K. Takahashi in his paper [**160**] completely answered a question of the existence of $X \in \mathcal{B}(\mathcal{H}, \mathcal{K})$ such that $S + RX$ is invertible, where $S \in \mathcal{B}(\mathcal{H})$ and $R \in \mathcal{B}(\mathcal{K}, \mathcal{H})$ are given operators.

For operators $S \in \mathcal{B}(\mathcal{H})$ and $R \in \mathcal{B}(\mathcal{K}, \mathcal{H})$, let

$$\mathcal{N}(S|R) = \{G \in \mathcal{B}(\mathcal{K}, \mathcal{H}) \colon \mathcal{R}(SG) \subseteq \mathcal{R}(R)\}. \tag{2.1}$$

By the well-known R. G. Douglas result [**68**], we know that $G \in \mathcal{N}(S|R)$ if and only if there exist $H \in \mathcal{B}(\mathcal{K})$ such that $SG = -RH$ which is equivalent with the fact that $\mathcal{R}\left(\begin{bmatrix} G \\ H \end{bmatrix}\right) \subseteq \mathcal{N}(\begin{bmatrix} S & R \end{bmatrix})$.

First, we prove the following auxiliary result:

LEMMA 2.9. *Let $S \in \mathcal{B}(\mathcal{H})$ and $R \in \mathcal{B}(\mathcal{K}, \mathcal{H})$ be such that $\begin{bmatrix} S & R \end{bmatrix} : \mathcal{H} \oplus \mathcal{K} \to \mathcal{H}$ is right invertible. Then the following are equivalent:*

(i) *$S + RX$ is invertible for some $X \in \mathcal{B}(\mathcal{H}, \mathcal{K})$.*

(ii) $\mathcal{R}\left(\begin{bmatrix} I \\ X \end{bmatrix}\right) \oplus \mathcal{N}(\begin{bmatrix} S & R \end{bmatrix}) = \mathcal{H} \oplus \mathcal{K}$.

(iii) $\dim(\mathcal{N}(\begin{bmatrix} S & R \end{bmatrix})) = \dim \mathcal{K}$ *and operator* $\begin{bmatrix} I & G \\ X & H \end{bmatrix}$ *is invertible, where* $\begin{bmatrix} G \\ H \end{bmatrix} : \mathcal{K} \to \mathcal{H} \oplus \mathcal{K}$ *is a left invertible operator whose range is equal to* $\dim(\mathcal{N}(\begin{bmatrix} S & R \end{bmatrix}))$.

PROOF. (i)$\Rightarrow$(ii): Since $S + RX = \begin{bmatrix} S & R \end{bmatrix} \begin{bmatrix} I \\ X \end{bmatrix}$ is surjective, it follows that $\mathcal{N}(\begin{bmatrix} S & R \end{bmatrix})^{\perp} \subseteq \mathcal{R}\left(\begin{bmatrix} I \\ X \end{bmatrix}\right) + \mathcal{N}(\begin{bmatrix} S & R \end{bmatrix})$, so

$$\mathcal{R}\left(\begin{bmatrix} I \\ X \end{bmatrix}\right) + \mathcal{N}(\begin{bmatrix} S & R \end{bmatrix}) = \mathcal{H} + \mathcal{K}.$$

From the injectivity of $S + RX$ it follows that $\mathcal{R}\left(\begin{bmatrix} I \\ X \end{bmatrix}\right) \cap \mathcal{N}(\begin{bmatrix} S & R \end{bmatrix}) = \{0\}$. Hence, (ii) holds.

(ii)$\Rightarrow$(i): Since

$$\mathcal{H} + \mathcal{K} = \begin{bmatrix} S & R \end{bmatrix} (\mathcal{H} + \mathcal{K}) = \begin{bmatrix} S & R \end{bmatrix} \mathcal{R}\left(\begin{bmatrix} I \\ X \end{bmatrix}\right) = \mathcal{R}(S + RX),$$

it follows that $S + RX$ is surjective. If $(S + RX)x = 0$,then

$$\begin{bmatrix} S & R \end{bmatrix} \begin{bmatrix} I \\ X \end{bmatrix} x = 0$$

which implies that $\begin{bmatrix} I \\ X \end{bmatrix} x \in \mathcal{R}\left(\begin{bmatrix} I \\ X \end{bmatrix}\right) \cap \mathcal{N}(\begin{bmatrix} S & R \end{bmatrix})$; i.e., $x = 0$.

(ii)$\Rightarrow$(iii): We will show that $\dim(\mathcal{N}(\begin{bmatrix} S & R \end{bmatrix})) = \dim \mathcal{K}$ and the rest of the proof will follow immediately. Let $P \in \mathcal{B}(0 \oplus \mathcal{K}, \mathcal{N}(\begin{bmatrix} S & R \end{bmatrix}))$ be defined by $P\begin{bmatrix} 0 \\ k \end{bmatrix} = P_{\mathcal{N}(\begin{bmatrix} S & R \end{bmatrix})} \begin{bmatrix} 0 \\ k \end{bmatrix}$. We prove that P is invertible. Indeed if $P\begin{bmatrix} 0 \\ k \end{bmatrix} = 0$, then $\begin{bmatrix} 0 \\ k \end{bmatrix} \in \mathcal{R}\left(\begin{bmatrix} I \\ X \end{bmatrix}\right)$ which implies that $k = 0$. Hence P is injective. To show that P is surjective, let's take arbitrary $\begin{bmatrix} x \\ y \end{bmatrix} \in \mathcal{N}(\begin{bmatrix} S & R \end{bmatrix})$. It can be checked that $P\begin{bmatrix} 0 \\ y - Xx \end{bmatrix} = \begin{bmatrix} x \\ y \end{bmatrix}$.

(iii)$\Rightarrow$(ii): It is evident. □

The answer to the question of the existence of $X \in \mathcal{B}(\mathcal{H}, K)$ such that $S+RX$ is invertible is given in the following theorem whose proof will follow as a consequence of the series of the result:

THEOREM 2.10. *Let $S \in \mathcal{B}(\mathcal{H})$ and $R \in \mathcal{B}(\mathcal{K}, \mathcal{H})$ be such that $\begin{bmatrix} S & R \end{bmatrix} : \mathcal{H} \oplus \mathcal{K} \to \mathcal{H}$ is right invertible. The following hold:*

(1) *If R is compact, there exists $X \in \mathcal{B}(\mathcal{H}, \mathcal{K})$ such that $S+RX$ is invertible if and only if S is a Weyl operator.*
(2) *If R is noncompact, there exists $X \in \mathcal{B}(\mathcal{H}, \mathcal{K})$ such that $S+RX$ is invertible if and only if $\mathcal{N}(S|R)$ contains a noncompact operator.*

One direction of part (1) of Theorem 2.10 follows from the next lemma while the opossite one follows from the fact that if $A \in \mathcal{B}(\mathcal{H})$ is an invertible operator, then A is a Weyl operator and for any compact operator $K \in \mathcal{B}(\mathcal{H})$ we have that $A+K$ is a Weyl operator too.

LEMMA 2.11. *Let $S \in \mathcal{B}(\mathcal{H})$ and $R \in \mathcal{B}(\mathcal{K}, \mathcal{H})$ be such that $\begin{bmatrix} S & R \end{bmatrix} : \mathcal{H} \oplus \mathcal{K} \to \mathcal{H}$ is right invertible. If S is Weyl, then there exists $X \in \mathcal{B}(\mathcal{H}, \mathcal{K})$ such that $S+RX$ is an invertible operator.*

PROOF. Since $\mathcal{R}(S) + \mathcal{R}(R) = \mathcal{H}$ it follows that $\mathcal{R}(P_{\mathcal{R}(S)^\perp} R) = \mathcal{R}(S)^\perp$. We have that $\mathcal{R}(P_{\mathcal{R}(S)^\perp} R) = \mathcal{R}(S)^\perp$. Now using the fact that for arbitrary operator A, $\dim(\mathcal{R}(A)) = \dim(\mathcal{R}(A^*))$ and that S is Weyl, we have the following:

$$\dim(\mathcal{N}(P_{\mathcal{R}(S)^\perp} R)^\perp) = \dim(\mathcal{R}(S)^\perp) = \dim(\mathcal{N}(S)). \tag{2.2}$$

Hence there exists $X \in \mathcal{B}(\mathcal{H}, \mathcal{K})$ such that

$$X = \begin{bmatrix} 0 & X_1 \\ 0 & 0 \end{bmatrix} : \begin{bmatrix} \mathcal{N}(S)^\perp \\ \mathcal{N}(S) \end{bmatrix} \to \begin{bmatrix} \mathcal{N}(P_{\mathcal{R}(S)^\perp} R)^\perp \\ \mathcal{N}(P_{\mathcal{R}(S)^\perp} R) \end{bmatrix},$$

where X_1 is invertible. Since

$$S = \begin{bmatrix} S_1 & 0 \\ 0 & 0 \end{bmatrix} : \begin{bmatrix} \mathcal{N}(S)^\perp \\ \mathcal{N}(S) \end{bmatrix} \to \begin{bmatrix} \mathcal{R}(S) \\ \mathcal{R}(S)^\perp \end{bmatrix}$$

and

$$R = \begin{bmatrix} R_1 & R_2 \\ R_3 & 0 \end{bmatrix} : \begin{bmatrix} \mathcal{N}(P_{\mathcal{R}(S)^\perp} R)^\perp \\ \mathcal{N}(P_{\mathcal{R}(S)^\perp} R) \end{bmatrix} \to \begin{bmatrix} \mathcal{R}(S) \\ \mathcal{R}(S)^\perp \end{bmatrix}$$

we have that

$$S+RX = \begin{bmatrix} S_1 & R_1 X_1 \\ 0 & R_3 X_1 \end{bmatrix} : \begin{bmatrix} \mathcal{N}(S)^\perp \\ \mathcal{N}(S) \end{bmatrix} \to \begin{bmatrix} \mathcal{R}(S) \\ \mathcal{R}(S)^\perp \end{bmatrix}.$$

Now the proof follows from the fact that $R_3 X_1$; i.e., R_3 is invertible. Indeed if $x \in \mathcal{N}(R_3)$, then $Rx \in \mathcal{R}(S)$ which implies that $x \in \mathcal{N}(P_{\mathcal{R}(S)^\perp} R)$, so $x = 0$. Also, from $\mathcal{R}(P_{\mathcal{R}(S)^\perp} R) = \mathcal{R}(S)^\perp$ we have that R_3 is surjective. □

To prove the part (2) of Theorem 2.10 we need the following two results:

LEMMA 2.12. *Let $S \in \mathcal{B}(\mathcal{H})$ and $R \in \mathcal{B}(\mathcal{K}, \mathcal{H})$ be such that $\begin{bmatrix} S & R \end{bmatrix} : \mathcal{H} \oplus \mathcal{K} \to \mathcal{H}$ is right invertible. If $\dim(\mathcal{N}(\begin{bmatrix} S & R \end{bmatrix})) = \dim \mathcal{K}$ and R is left invertible, then there exists $X \in \mathcal{B}(\mathcal{H}, \mathcal{K})$ such that $S+RX$ is an invertible operator.*

PROOF. Since $\dim(\mathcal{N}([\begin{array}{cc} S & R \end{array}])) = \dim \mathcal{K}$, there exists a left invertible operator

$$\left[\begin{array}{c} G \\ H \end{array}\right] : \mathcal{K} \to \mathcal{H} \oplus \mathcal{K}$$

whose range is equal to $\mathcal{N}([\begin{array}{cc} S & R \end{array}])$. Now $SG = -RH$ and R is left invertible, so $H = -R_l^{-1}SG$ which implies that $\mathcal{N}(G) \subseteq \mathcal{N}(H)$ which together with left invertibility of $\left[\begin{array}{c} G \\ H \end{array}\right]$ implies that G is injective. Also, it follows that $\mathcal{R}(H^*) \subseteq \mathcal{R}(G^*)$. Since $\mathcal{R}(G^*) + \mathcal{R}(H^*)$ is closed we have that $\mathcal{R}(G^*)$ is closed. Hence G is left invertible. Now, there exists an operator $F \in \mathcal{B}(\mathcal{H}, \mathcal{K})$ such that $H - FG$ is invertible (for example let $F = G_l^{-1} - R_l^{-1}S$). Now, by

$$\left[\begin{array}{cc} I & G \\ F & H \end{array}\right] = \left[\begin{array}{cc} I & 0 \\ F & I \end{array}\right] \left[\begin{array}{cc} I & G \\ 0 & H - FG \end{array}\right],$$

we have that $\left[\begin{array}{cc} I & G \\ F & H \end{array}\right]$ is invertible and the proof follows directly by Lemma 2.9. □

COROLLARY 2.13. Let $S \in \mathcal{B}(\mathcal{H})$ and $R \in \mathcal{B}(\mathcal{K}, \mathcal{H})$ be such that $\left[\begin{array}{cc} S & R \end{array}\right] : \mathcal{H} \oplus \mathcal{K} \to \mathcal{H}$ is right invertible and let R be a noncompact operator. If S is not left semi-Fredholm, then there exists $F \in \mathcal{B}(\mathcal{H}, \mathcal{K})$ such that $S + RF$ is invertible.

PROOF. Let E be a spectral measure of R^*R. First we prove that there exists $\epsilon > 0$ such that $\mathcal{K}_\epsilon = \mathcal{R}(E(\epsilon, \infty))$ is infinite dimensional. Let $R_\epsilon = RE(\epsilon, \infty)$. Then

$$\|(R - R_\epsilon)x\|^2 = \int \lambda \chi_{\{\epsilon\}} dE_\lambda \leq \epsilon \|x\|^2.$$

If for each $\epsilon > 0$, $\mathcal{K}_\epsilon$ is finite dimensional, then it follows that R is a compact as a limit of finite-dimensional operators R_ϵ. Hence there exists $\epsilon_0 > 0$ such that $\mathcal{K}_{\epsilon_0} = \mathcal{R}(E(\epsilon_0, \infty))$ is infinite dimensional. Since the set of right invertible operators is open in the set of bounded linear operators, we get that for small enough $\epsilon > 0$ such that $\epsilon < \epsilon_0$, the operator $\left[\begin{array}{cc} S & R_1 \end{array}\right]$ is right invertible; i.e., $\left[\begin{array}{cc} S & R|_{\mathcal{K}_\epsilon} \end{array}\right]$ is right invertible, where $\mathcal{K}_\epsilon$ is an infinite-dimensional closed subspace. Also, we have that

$$\|R_\epsilon x\|^2 = (R^*RE(\epsilon, \infty)x, x) = \int \lambda \chi_{\mathcal{R}(E(\epsilon,\infty))} dE_\lambda \leq \epsilon \|E((\epsilon, \infty))x\|^2,$$

which implies that $R|_{\mathcal{K}_1}$ is left invertible. Furthermore, we have

$$\dim(\mathcal{N}([\begin{array}{cc} S & R|_{\mathcal{K}_1} \end{array}])) = \infty.$$

In fact, if $\dim(\mathcal{N}([\begin{array}{cc} S & R|_{\mathcal{K}_1} \end{array}])) < \infty$, then $\left[\begin{array}{cc} S & R|_{\mathcal{K}_1} \end{array}\right]$ is left semi-Fredholm, which contradicts the assumption that S is not left semi-Fredholm. Now, by Lemma 2.12 there exists $F_1 \in \mathcal{B}(\mathcal{H}, \mathcal{K}_1)$ such that $S+(R|_{\mathcal{K}_1})F_1$ is invertible. Define $F = JF_1$ where $J : \mathcal{K}_1 \to \mathcal{K}$ is the inclusion mapping. Then $S + RF$ is invertible. □

Now, we are ready to give the proof of Thorem 2.10(2):
Suppose that there exists $X \in \mathcal{B}(\mathcal{H}, \mathcal{K})$ such that $S + RX$ is invertible. Then

$$R = S(S + RX)^{-1}R + RX(S + RX)^{-1}R$$

implies that $\mathcal{R}(S(S+RX)^{-1}R) \subseteq \mathcal{R}(R)$; i.e., noncompact operator $(S+RX)^{-1}R \in \mathcal{N}(S|R)$.

In the other direction, suppose that $\mathcal{N}(S|R)$ contains a noncompact operator D. By Corollary 2.13, we have only to consider the case when S is left semi-Fredholm and it suffices to show that there exists a subspace $\mathcal{K}_1$ of $\mathcal{K}$ such that $R_1 = R|_{\mathcal{K}_1}$ is left invertible and $\begin{bmatrix} S & R_1 \end{bmatrix}$ is a right invertible operator with $\dim(\mathcal{N}(\begin{bmatrix} S & R_1 \end{bmatrix})) = \dim \mathcal{K}_1 = \infty$. Thus we assume that S is left semi-Fredholm. Since D is noncompact, the condition $D \in \mathcal{N}(S|R)$ implies $\dim(\mathcal{N}(\begin{bmatrix} S & R \end{bmatrix})) = \infty$, so there exists a left invertible $\begin{bmatrix} G \\ H \end{bmatrix} : \mathcal{K} \to \mathcal{H} \oplus \mathcal{K}$ whose range is equal to $\mathcal{N}(\begin{bmatrix} S & R \end{bmatrix})$. Then $\mathcal{R}(D) \subseteq \mathcal{R}(G)$ and G is noncompact. Also, from $SG+RH=0$ it follows that RH is noncompact. Now similarly as in Corollary 2.13 using the spectral measure of $(RH)^*(RH)$, there exists an infinite-dimensional subspace $\mathcal{K}_0$ of $\mathcal{K}$ such that $RH|_{\mathcal{K}_0}$ is left invertible. Let $\mathcal{K}_1 = \mathcal{R}(H|_{\mathcal{K}_0}) \oplus \mathcal{R}(H)^\perp$. Notice that $\mathcal{R}(H|_{\mathcal{K}_0})$ is a closed subspace since $\mathcal{R}(RH|_{\mathcal{K}_0})$ is closed so $R^{-1}[RH|_{\mathcal{K}_0}] = \mathcal{R}(H|_{\mathcal{K}_0}) \oplus \mathcal{N}(R)$ is closed as an inverse image of a closed subspace. Let us show that $\mathcal{K}_1$ satisfies the required conditions. Since $\begin{bmatrix} S & R \end{bmatrix}$ is right invertible and $\mathcal{R}(S)$ is closed, it follows from $\mathcal{N}(\begin{bmatrix} S & R \end{bmatrix}) = \mathcal{R}\left(\begin{bmatrix} G \\ H \end{bmatrix}\right)$ that $P^{\mathtt{rst}}_{\mathcal{R}(S)^\perp} R|_{\mathcal{R}(H)^\perp} : \mathcal{R}(H)^\perp \to \mathcal{R}(S)^\perp$ is invertible and $\begin{bmatrix} S & R_1 \end{bmatrix}$ is right invertible, where $R_1 = R|_{\mathcal{K}_1}$. The left invertibility of $RH|_{\mathcal{K}_0}$ implies that of $R|_{\mathcal{R}(H|_{\mathcal{K}_0})} : \mathcal{R}(H|_{\mathcal{K}_0}) \to \mathcal{H}$, which together with the invertibility of $P^{\mathtt{rst}}_{\mathcal{R}(S)^\perp} R|_{\mathcal{R}(H)^\perp}$ shows that R_1 is left invertible. Since

$$\mathcal{R}\left(\begin{bmatrix} G \\ H \end{bmatrix}(\mathcal{K}_0)\right) \subseteq \mathcal{N}(\begin{bmatrix} S & R_1 \end{bmatrix}),$$

we have that $\dim(\mathcal{N}(\begin{bmatrix} S & R_1 \end{bmatrix})) \geq \dim \mathcal{K}_0$. Hence

$$\dim(\mathcal{N}(\begin{bmatrix} S & R_1 \end{bmatrix})) = \dim \mathcal{K}_1 = \infty.$$

2.3. Injectivity of an operator $A+CX$

There are several papers [**35, 90, 92, 133, 170**], which deal with various completions of the operator matrix of the form M_X but the only one of them that addresses completion to injectivity is [**35**], and it does it under the special assumption that the operator C is with closed range. Recently, the problem of completing the operator matrix M_X to injectivity is solved and the solution is presented in [**55**].

In this section we completely answer the following question: For given $A \in \mathcal{B}(\mathcal{H},\mathcal{K})$, $C \in \mathcal{B}(\mathcal{L},\mathcal{K})$ does there exist $X \in \mathcal{B}(\mathcal{H},\mathcal{L})$ such that $A+CX$ is injective? Instead of looking for such an operator $X \in \mathcal{B}(\mathcal{H},\mathcal{L})$, we equivalently ask the question of the existence of an operator $T \in \mathcal{B}(\mathcal{H},\mathcal{K})$ such that $\mathcal{R}(T) \subseteq \mathcal{R}(C)$ and $A+T$ is injective. As an application of the results obtained, for given operators $A \in \mathcal{B}(\mathcal{H})$, $B \in \mathcal{B}(\mathcal{K})$, and $C \in \mathcal{B}(\mathcal{K},\mathcal{H})$ some conditions for the existence of an operator X such that the 2×2 operator matrix $M_X = \begin{bmatrix} A & C \\ X & B \end{bmatrix}$ is injective are presented.

First we address some cases when for given $A \in \mathcal{B}(\mathcal{H},\mathcal{K})$, $C \in \mathcal{B}(\mathcal{L},\mathcal{K})$ there is no $T \in \mathcal{B}(\mathcal{H},\mathcal{K})$ such that $\mathcal{R}(T) \subseteq \mathcal{R}(C)$ and $A+T$ is injective.

THEOREM 2.14. *Let $A \in \mathcal{B}(\mathcal{H},\mathcal{K})$, $C \in \mathcal{B}(\mathcal{L},\mathcal{K})$ and assume that the following two conditions hold:*

(i) $\mathcal{N}(A)^\perp \cap A^{-1}[\mathcal{R}(C)]$ *does not contain a closed infinite-dimensional subspace.*
(ii) $\operatorname{codim}_{\mathcal{R}(C)}(\mathcal{R}(A) \cap \mathcal{R}(C)) < \dim(\mathcal{N}(A)) < \infty$.

Then there is no $T \in \mathcal{B}(\mathcal{H},\mathcal{K})$ such that $\mathcal{R}(T) \subseteq \mathcal{R}(C)$ and $A+T$ is injective.

PROOF. Suppose, to the contrary, that there is such an operator T. Set $U = \mathcal{R}(A) \cap \mathcal{R}(C)$ and let F be any subspace of $\mathcal{K}$ such that $U \oplus F = \mathcal{R}(C)$. By assumption, F is of finite dimension and thus both U and F are operator ranges. By Corollary 1 of [**77**] we can write $T = T_1 + T_2$, where $T_1, T_2 \in \mathcal{B}(\mathcal{H},\mathcal{K})$ are such that $\mathcal{R}(T_1) \subseteq U$ and $\mathcal{R}(T_2) \subseteq F$. From $\mathcal{N}(A+T_1) \cap \mathcal{N}(T_2) = \{0\}$, it follows that

$$\dim(\mathcal{N}(A+T_1)) \leq \operatorname{codim}(\mathcal{N}(T_2)) \leq \dim F < \dim(\mathcal{N}(A)),$$

by (ii). Since $\mathcal{R}(T_1) \subseteq \mathcal{R}(A)$ we have $T_1 = AW$ for some $W \in \mathcal{B}(\mathcal{H})$. Clearly $\mathcal{R}(W) \subseteq A^{-1}[U]$. Now we show that W must be compact.

Towards a contradiction suppose that W is not a compact operator. Then there exists a closed subspace $M \subseteq \mathcal{R}(W)$ of infinite dimension. Let $M_0 = M \cap \mathcal{N}(A)^\perp$. Since $\dim(\mathcal{N}(A)) < \infty$, it is easy to check that M_0 is a closed infinite-dimensional subspace. Also, $M_0 \subseteq \mathcal{N}(A)^\perp \cap A^{-1}[U]$, contradicting our assumption.

We have thus shown that W is a compact operator. We now set $K = \mathcal{R}(I+W) \cap \mathcal{N}(A)$ and let $S \subseteq \mathcal{N}(A)$ be any subspace such that $K \oplus S = \mathcal{N}(A)$. For $P_1 = (I+W)^{-1}[K] \cap \mathcal{N}(I+W)^\perp$ we have

$$\mathcal{N}(A+T_1) = (I+W)^{-1}[\mathcal{N}(A)] = P_1 \oplus \mathcal{N}(I+W)$$

so

$$\dim P_1 + \dim(\mathcal{N}(I+W)) = \dim(\mathcal{N}(A+T_1)) < \dim(\mathcal{N}(A)) = \dim K + \dim S;$$

i.e., $\dim(\mathcal{N}(I+W)) < \dim S$, since clearly $\dim P_1 = \dim K$. On the other hand

$$\dim S \leq \dim(\mathcal{R}(I+W)^\perp) = \dim(\mathcal{N}(I+W))$$

(since W is compact) and we have finally reached a contradiction. $\square$

REMARK 2.15. From the previous theorem, we have that if there exists $T \in \mathcal{B}(\mathcal{H},\mathcal{K})$ such that $\mathcal{R}(T) \subseteq \mathcal{R}(C)$ and $A+T$ is injective, then one of the following conditions must hold:

(i$*$) $\mathcal{N}(A)^\perp \cap A^{-1}[\mathcal{R}(C)]$ contains a closed infinite-dimensional subspace.
(ii$*$) $\dim(\mathcal{N}(A)) = \infty$.
(iii$*$) $\operatorname{codim}_{\mathcal{R}(C)}(\mathcal{R}(A) \cap \mathcal{R}(C)) \geq \dim(\mathcal{N}(A))$, $\dim(\mathcal{N}(A)) < \infty$.

In the following theorem we prove that the condition (i$*$) in the above remark is also sufficient for the existence of an operator T with the required properties:

THEOREM 2.16. *Let $A \in \mathcal{B}(\mathcal{H},\mathcal{K})$, $C \in \mathcal{B}(\mathcal{L},\mathcal{K})$. If $\mathcal{N}(A)^\perp \cap A^{-1}[\mathcal{R}(C)]$ contains a closed infinite-dimensional subspace, then there exists $T \in \mathcal{B}(\mathcal{H},\mathcal{K})$ such that $\mathcal{R}(T) \subseteq \mathcal{R}(C)$ and $A+T$ is injective.*

PROOF. Let $F \subseteq \mathcal{N}(A)^\perp \cap A^{-1}[\mathcal{R}(C)]$ be a closed infinite-dimensional subspace and let $P \subseteq F$ be a closed infinite-dimensional subspace whose orthogonal complement in F, which we denote by Q, is also of infinite dimension. Fix an injective $L \in \mathcal{B}(F, P)$ and let

$$L = \begin{bmatrix} L_1 & L_2 \end{bmatrix} : \begin{bmatrix} P \\ Q \end{bmatrix} \to P.$$

Define $W_{2,0} \in \mathcal{B}(F, \mathcal{H})$ by

$$W_{2,0} = \begin{bmatrix} L_1 - 1_P & L_2 \\ 0 & 0 \\ 0 & 0 \end{bmatrix} : \begin{bmatrix} P \\ Q \end{bmatrix} \to \begin{bmatrix} P \\ Q \\ F^\perp \end{bmatrix}.$$

If $\pi_F \in \mathcal{B}\left(\mathcal{N}(A)^\perp, F\right)$ is the orthogonal projector onto F, set $W_2 = W_{2,0} \cdot \pi_F \in \mathcal{B}\left(\mathcal{N}(A)^\perp, \mathcal{H}\right)$. Finally pick any injective $W_1 \in \mathcal{B}(\mathcal{N}(A), \mathcal{H})$ such that $\mathcal{R}(W_1) \subseteq Q$. It is immediate that for $W = \begin{bmatrix} W_1 & W_2 \end{bmatrix} : \begin{bmatrix} \mathcal{N}(A) \\ \mathcal{N}(A)^\perp \end{bmatrix} \to \mathcal{H}$ we have $\mathcal{R}(W) \subseteq F$ and so $\mathcal{R}(AW) \subseteq R(C)$. We claim that $\mathcal{N}(A(I + W)) = \{0\}$.

Indeed, suppose $u \in \mathcal{N}(A), v \in \mathcal{N}(A)^\perp$ are such that $(I + W)(u + v) \in \mathcal{N}(A)$. From $(I + W)(u + v) = u + (v + W_1 u + W_2 v)$ and $v + W_1 u + W_2 v \in \mathcal{N}(A)^\perp$, it follows that $v + W_1 u + W_2 v = 0$, whence $v \in F$ and consequently $W_2 v = W_{2,0} v$. We now have $v = -W_1 u - W_{2,0} v$. Since $\mathcal{R}(W_1) \subseteq Q$ and $\mathcal{R}(W_{2,0}) \subseteq P$ this further gives $-W_{2,0} v = \pi_P(v)$, where $\pi_P \in \mathcal{B}(F, \mathcal{H})$ is the orthogonal projector onto P, viewed as an operator having $\mathcal{H}$ as codomain. But from

$$W_{2,0} + \pi_P = \begin{bmatrix} L_1 & L_2 \\ 0 & 0 \\ 0 & 0 \end{bmatrix} : \begin{bmatrix} P \\ Q \end{bmatrix} \to \begin{bmatrix} P \\ Q \\ F^\perp \end{bmatrix},$$

we see that $\mathcal{N}(W_{2,0} + \pi_P) = \{0\}$. So $v = 0$ and thus $W_1 u = 0$ leading to $u = 0$, by injectivity of W_1. The above discussion shows that the operator $T = AW$ is as required. □

The following theorem gives a special case when the condition (i∗) from the earlier remark must itself be necessary for the existence of an operator T with the required properties:

THEOREM 2.17. *Let $A \in \mathcal{B}(\mathcal{H}, \mathcal{K})$ and $C \in \mathcal{B}(\mathcal{L}, \mathcal{K})$. If* $\operatorname{codim}_{\mathcal{R}(C)} (\mathcal{R}(A) \cap \mathcal{R}(C)) < \infty$, $\dim(\mathcal{N}(A)) = \infty$, *and there exists $T \in \mathcal{B}(\mathcal{H}, \mathcal{K})$ such that $\mathcal{R}(T) \subseteq \mathcal{R}(C)$ and $A + T$ is injective, then $\mathcal{N}(A)^\perp \cap A^{-1}[\mathcal{R}(C)]$ contains a closed infinite-dimensional subspace.*

PROOF. Let R_0 be a subspace of $\mathcal{N}(A)$ of dimension $\operatorname{codim}_{\mathcal{R}(C)} \mathcal{R}(A) \cap \mathcal{R}(C) + 1$. Then $\mathcal{H}_0 = R_0 \oplus \mathcal{N}(A)^\perp$ is closed. Put $A_0 = A|_{\mathcal{H}_0} \in \mathcal{B}(\mathcal{H}_0, \mathcal{H})$ and $T_0 = T|_{\mathcal{H}_0} \in \mathcal{B}(\mathcal{H}_0, \mathcal{H})$. Obviously, $\mathcal{R}(A_0) = \mathcal{R}(A)$ and $\mathcal{N}(A_0) = R_0$. Since $A_0 + T_0$ is an injective operator, by Theorem 2.14 it follows that there exists an infinite-dimensional closed subspace M contained in the orthogonal complement of $\mathcal{N}(A_0)$ in $\mathcal{H}_0$ (and this is precisely $\mathcal{N}(A)^\perp$) for which $A(M) = A_0(M) \subseteq \mathcal{R}(C)$. □

Finally, collecting all the previously obtained results, in the following theorem, for given $A \in \mathcal{B}(\mathcal{H}, \mathcal{K})$ and $C \in \mathcal{B}(\mathcal{L}, \mathcal{K})$, we present necessary and sufficient conditions for the existence of $X \in \mathcal{B}(\mathcal{H}, \mathcal{L})$ such that $A + CX$ is injective.

THEOREM 2.18. *Let $A \in \mathcal{B}(\mathcal{H},\mathcal{K})$ and $C \in \mathcal{B}(\mathcal{L},\mathcal{K})$. There exists $X \in \mathcal{B}(\mathcal{H},\mathcal{L})$ such that $A + CX$ is injective if and only if one of the following conditions is satisfied:*

(1) *$\mathcal{N}(A)^{\perp} \cap A^{-1}(\mathcal{R}(C))$ contains a closed infinite-dimensional subspace.*
(2) *$\operatorname{codim}_{\mathcal{R}(C)}(\mathcal{R}(A) \cap \mathcal{R}(C)) \geq \dim(\mathcal{N}(A))$.*

PROOF. As before, we note that the existence of $X \in \mathcal{B}(\mathcal{H},\mathcal{L})$ such that $A+CX$ is injective is equivalent with the existence of $T \in \mathcal{B}(\mathcal{H},\mathcal{K})$ such that $\mathcal{R}(T) \subseteq \mathcal{R}(C)$ and $A + T$ is injective.

If such T exists, then by Theorem 2.14 we have that one of the following conditions must hold:

(i) $\mathcal{N}(A)^{\perp} \cap A^{-1}(\mathcal{R}(C))$ contains a closed infinite-dimensional subspace.
(ii) $\dim(\mathcal{N}(A)) = \infty$.
(iii) $\operatorname{codim}_{\mathcal{R}(C)}(\mathcal{R}(A) \cap \mathcal{R}(C)) \geq \dim(\mathcal{N}(A))$.

So, suppose that (ii) holds. If $\operatorname{codim}_{\mathcal{R}(C)}(\mathcal{R}(A) \cap \mathcal{R}(C)) = \infty$, then we have that (2) holds. So, we suppose that $\operatorname{codim}_{\mathcal{R}(C)}(\mathcal{R}(A) \cap \mathcal{R}(C)) < \infty$. By Theorem 2.17 we get that then (1) must hold.

On the contrary, if (1) is satisfied, then Theorem 2.16 guarantees the existence of $T \in \mathcal{B}(\mathcal{H},\mathcal{K})$ such that $\mathcal{R}(T) \subseteq \mathcal{R}(C)$ and $A + T$ is injective. Suppose that (2) holds. There exists an operator range $F \subseteq \mathcal{R}(C)$ such that $F \cap \mathcal{R}(A) = \{0\}$ and $\dim F \geq \dim(\mathcal{N}(A))$: If $\operatorname{codim}_{\mathcal{R}(C)}(\mathcal{R}(A) \cap \mathcal{R}(C)) = \infty$, use (i) of Lemma 2.1; otherwise this is trivial. Let $T_0 \in \mathcal{B}(\mathcal{N}(A),\mathcal{K})$ be an injective operator such that $\mathcal{R}(T_0) \subseteq F$ and let

$$T = \begin{bmatrix} 0 & T_0 \end{bmatrix} : \begin{bmatrix} \mathcal{N}(A)^{\perp} \\ \mathcal{N}(A) \end{bmatrix} \to \mathcal{K}.$$

One readily checks that $\mathcal{R}(T) \subseteq \mathcal{R}(C)$ and that $A + T$ is injective. □

2.4. Surjectivity of an operator $A + CX$

In this section, we address the question of for which operators $A \in \mathcal{B}(\mathcal{H},\mathcal{K})$ and $C \in \mathcal{B}(\mathcal{L},\mathcal{K})$ there exists an operator $X \in \mathcal{B}(\mathcal{H},\mathcal{L})$ such that the operator $A+CX$ is surjective, and we completely answer it. The following theorem presents the main result of this section.

THEOREM 2.19. *Let $A \in \mathcal{B}(\mathcal{H},\mathcal{K})$ and $C \in \mathcal{B}(\mathcal{L},\mathcal{K})$. Then there exists an operator $X \in \mathcal{B}(\mathcal{H},\mathcal{L})$ such that $A+CX$ is surjective if and only if $\mathcal{R}(A)+\mathcal{R}(C) = \mathcal{K}$ and one of the following two conditions holds:*

(1) *$\mathcal{R}(A)$ is not closed.*
(2) *$\mathcal{R}(A)$ is closed and either $d(A) \leq \alpha(A)$ or $A^{-1}[\mathcal{R}(C)]$ contains a closed infinite-dimensional subspace.*

The proof of Theorem 2.19 will be given in a form of a series of results to be presented below. Using a result of R. G. Douglas from [**68**] we make the following observation that will be used in the sequel without any explicit mention: If $A \in \mathcal{B}(\mathcal{H},\mathcal{K})$ and $C \in \mathcal{B}(\mathcal{L},\mathcal{K})$, then there exists $X \in \mathcal{B}(\mathcal{H},\mathcal{L})$ such that $A + CX$ is surjective if and only if $A+T$ is surjective for some $T \in \mathcal{B}(\mathcal{H},\mathcal{K})$ with $\mathcal{R}(T) \subseteq \mathcal{R}(C)$. Clearly a necessary condition for this to hold is that $\mathcal{R}(A) + \mathcal{R}(C) = \mathcal{K}$.

THEOREM 2.20. *Let $A \in \mathcal{B}(\mathcal{H},\mathcal{K})$, $C \in \mathcal{B}(\mathcal{L},\mathcal{K})$ and assume that $A^{-1}[\mathcal{R}(C)]$ contains a closed infinite-dimensional subspace M. Then there exists an operator $T \in \mathcal{B}(\mathcal{H},\mathcal{K})$ such that $\mathcal{R}(T) \subseteq \mathcal{R}(C)$ and $\mathcal{R}(A+T) = \mathcal{R}(A) + \mathcal{R}(C)$.*

PROOF. Let $M \subseteq A^{-1}[\mathcal{R}(C)]$ be a closed infinite-dimensional subspace. Then there exists $S \in \mathcal{B}(M, \mathcal{K})$ such that $\mathcal{R}(S) = \mathcal{R}(C)$. If

$$A = \begin{bmatrix} A_1 & A_2 \end{bmatrix} : \begin{bmatrix} M \\ M^\perp \end{bmatrix} \to \mathcal{K},$$

define $T \in \mathcal{B}(\mathcal{H}, \mathcal{K})$ by

$$T = \begin{bmatrix} S - A_1 & 0 \end{bmatrix} : \begin{bmatrix} M \\ M^\perp \end{bmatrix} \to \mathcal{K}.$$

Since $\mathcal{R}(A_1) = A[M] \subseteq \mathcal{R}(C)$, it is straightforward that $\mathcal{R}(T) \subseteq \mathcal{R}(C)$, as well as that $\mathcal{R}(A + T) = \mathcal{R}(A) + \mathcal{R}(C)$. □

As an immediate corollary of Theorem 2.20 we have the following:

THEOREM 2.21. *Let $A \in \mathcal{B}(\mathcal{H}, \mathcal{K})$, $C \in \mathcal{B}(\mathcal{L}, \mathcal{K})$ be such that $\mathcal{R}(A) + \mathcal{R}(C) = \mathcal{K}$ and assume that $A^{-1}[\mathcal{R}(C)]$ contains a closed infinite-dimensional subspace. Then there exists an operator $T \in \mathcal{B}(\mathcal{H}, \mathcal{K})$ such that $\mathcal{R}(T) \subseteq \mathcal{R}(C)$ and $A + T$ is surjective.*

Theorem 2.20 directly solves our problem also in the case when $\mathcal{R}(A)$ is not closed. More precisely, we have the following theorem.

THEOREM 2.22. *Let $A \in \mathcal{B}(\mathcal{H}, \mathcal{K})$, $C \in \mathcal{B}(\mathcal{L}, \mathcal{K})$ be such that $\mathcal{R}(A) + \mathcal{R}(C) = \mathcal{K}$. If $\mathcal{R}(A)$ is not closed, then $A^{-1}[\mathcal{R}(C)]$ contains a closed infinite-dimensional subspace M and consequently there exists an operator $T \in \mathcal{B}(\mathcal{H}, \mathcal{K})$ such that $\mathcal{R}(T) \subseteq \mathcal{R}(C)$ and $A + T$ is surjective.*

PROOF. By Theorem 2.7, we can write $\mathcal{K} = F_A \oplus F_C$, where F_A, F_C are closed subspaces and $F_A \subseteq \mathcal{R}(A)$, $F_C \subseteq \mathcal{R}(C)$. For $M = \mathcal{R}(A) \cap F_C$, we have $\mathcal{R}(A) = F_A \oplus M$. The subspace $M_0 = A^{-1}[M] = A^{-1}[F_C]$ is closed. M is infinite dimensional because otherwise $\mathcal{R}(A)$ would be closed; thus so is M_0. Clearly $A[M_0] \subseteq \mathcal{R}(C)$. The rest follows by Theorem 2.21. □

The next theorem is a partial converse of Theorem 2.20.

THEOREM 2.23. *Let $A \in \mathcal{B}(\mathcal{H}, \mathcal{K})$, $C \in \mathcal{B}(\mathcal{L}, \mathcal{K})$ be such that $\mathcal{R}(A) + \mathcal{R}(C) = \mathcal{K}$, $\mathcal{R}(A)$ is closed, and $\alpha(A) < d(A)$. If there exists an operator $T \in \mathcal{B}(\mathcal{H}, \mathcal{K})$ such that $\mathcal{R}(T) \subseteq \mathcal{R}(C)$ and $A + T$ is surjective, then $A^{-1}[\mathcal{R}(C)]$ contains a closed infinite-dimensional subspace M.*

PROOF. Using Lemma 2.2, we can write $\mathcal{K} = \mathcal{R}(A) \oplus F$, for some operator range $F \subseteq \mathcal{R}(C)$. Also, by Lemma 2.6 we have $T = T_A + T_F$ for some $T_A, T_F \in \mathcal{B}(\mathcal{H}, \mathcal{K})$ such that $\mathcal{R}(T_A) \subseteq \mathcal{R}(A)$ and $\mathcal{R}(T_F) \subseteq F$. From $\mathcal{R}(T) \subseteq \mathcal{R}(C)$, it follows that $\mathcal{R}(T_A) \subseteq \mathcal{R}(A) \cap \mathcal{R}(C)$. We have $T_A = AW$ for some $W \in \mathcal{B}(\mathcal{H})$; clearly $\mathcal{R}(W) \subseteq A^{-1}[\mathcal{R}(C)]$.

There is equality $\mathcal{R}(A(I + W)) = \mathcal{R}(A)$. Indeed, for $z \in \mathcal{R}(A)$ we have $\mathcal{R}(A) \ni z = (A + T)x = A(I + W)x + T_F x$, for some $x \in \mathcal{H}$, which gives $z = A(I+W)x$, since $A(I+W)x \in \mathcal{R}(A)$ and $T_F x \in F$. From this it directly follows that $\mathcal{R}(I + W) + \mathcal{N}(A) = \mathcal{H}$.

We will show that W cannot be a compact operator, from which the existence will follow of a closed subspace $M \subseteq \mathcal{R}(W) \subseteq A^{-1}[\mathcal{R}(C)]$ of infinite dimension, so the proof of the theorem will be complete. To prove this, we first note that

$T_F\left((I+W)^{-1}[\mathcal{N}(A)]\right)=F$. Indeed, let $z\in F$. Then $F\ni z=A(I+W)x+T_Fx$, for some $x\in\mathcal{H}$, whence $x\in(I+W)^{-1}[\mathcal{N}(A)]$ and $T_Fx=z$.

Now if W were compact, then, letting $P=\mathcal{R}(I+W)\cap\mathcal{N}(A)$, $P_0=(I+W)^{-1}[P]\cap\mathcal{N}(I+W)^{\perp}$ and keeping in mind that $\mathcal{R}(I+W)+\mathcal{N}(A)=\mathcal{H}$, we would have

$$\begin{aligned}\dim((I+W)^{-1}[\mathcal{N}(A)])&=\dim P_0+\dim(\mathcal{N}(I+W))=\dim P+\dim(\mathcal{R}(I+W)^{\perp})\\&=\dim(\mathcal{N}(A))<\dim F,\end{aligned}$$

by one of the assumptions of the theorem, contradicting $T_F\left((I+W)^{-1}[\mathcal{N}(A)]\right)=F$. □

The analysis below completes the proof of Theorem 2.19.

THEOREM 2.24. *Let $A\in\mathcal{B}(\mathcal{H},\mathcal{K})$, $C\in\mathcal{B}(\mathcal{L},\mathcal{K})$ be such that $\mathcal{R}(A)$ is closed and $\mathcal{R}(A)+\mathcal{R}(C)=\mathcal{K}$. If $d(A)\leq\alpha(A)$, then there exists an operator $T\in\mathcal{B}(\mathcal{H},\mathcal{K})$ such that $\mathcal{R}(T)\subseteq\mathcal{R}(C)$ and $A+T$ is surjective.*

PROOF. As before we can write $\mathcal{K}=\mathcal{R}(A)\oplus F$, for some operator range $F\subseteq\mathcal{R}(C)$. Since $d(A)\leq\alpha(A)$, we can fix an operator $T_0\in\mathcal{B}(\mathcal{N}(A),\mathcal{K})$ such that $\mathcal{R}(T_0)=F$. For $T\in\mathcal{B}(\mathcal{H},\mathcal{K})$ defined by

$$T=\begin{bmatrix}0 & T_0\end{bmatrix}:\begin{bmatrix}\mathcal{N}(A)^{\perp}\\ \mathcal{N}(A)\end{bmatrix}\to\mathcal{K},$$

it is easy to verify that T meets all the requirements of the theorem. □

2.5. Right and left invertibility of an operator $A+CX$

In this section, we characterize all operators $S\in\mathcal{B}(\mathcal{H},\mathcal{K})$ and $R\in\mathcal{B}(\mathcal{L},\mathcal{K})$ for which there exists an operator $X\in\mathcal{B}(\mathcal{H},\mathcal{L})$ such that $S+RX$ is a right invertible operator as well as when it is left invertible.

First we answer the question of the existence of $X\in\mathcal{B}(\mathcal{H},\mathcal{K})$ such that $S+RX$ is right invertible. Notice that the right invertibility of $\begin{bmatrix}S & R\end{bmatrix}:\mathcal{H}\oplus\mathcal{K}\to\mathcal{H}$ is clearly a necessary condition for the existence of X such that $S+RX$ is right invertible.

THEOREM 2.25. *Let $S\in\mathcal{B}(\mathcal{H})$ and $R\in\mathcal{B}(\mathcal{K},\mathcal{H})$ be such that $\begin{bmatrix}S & R\end{bmatrix}:\mathcal{H}\oplus\mathcal{K}\to\mathcal{H}$ is right invertible. The following hold:*

1. *If R is compact, there exists $X\in\mathcal{B}(\mathcal{H},\mathcal{K})$ such that $S+RX$ is right invertible if and only if S is a right semi-Fredholm operator and $\operatorname{ind}(S)\geq 0$.*
2. *If R is noncompact, there exists $X\in\mathcal{B}(\mathcal{H},\mathcal{K})$ such that $S+RX$ is right invertible if and only if $\mathcal{N}(S|R)$ contains a noncompact operator.*

PROOF. (1) If R is compact and for some $X\in\mathcal{B}(\mathcal{H},\mathcal{K})$, $S+RX$ is right invertible, then we have that S is a right semi-Fredholm operator and $\operatorname{ind}(S)=\operatorname{ind}(S+RX)=\alpha(S+RX)\geq 0$. To prove the opposite direction, let us assume that S is a right semi-Fredholm operator and $\operatorname{ind}(S)\geq 0$. Since $\begin{bmatrix}S & R\end{bmatrix}$ is right invertible, we have that $\mathcal{R}(P_{\mathcal{R}(S)^{\perp}}R)=\mathcal{R}(S)^{\perp}$. Since

$$\dim(\mathcal{N}(P_{\mathcal{R}(S)^{\perp}}R)^{\perp})=\dim(\mathcal{R}(P_{\mathcal{R}(S)^{\perp}}R))=\dim(\mathcal{R}(S)^{\perp})\leq\dim(\mathcal{N}(S)),\tag{2.3}$$

we have that there exists a right invertible $X_1 \in \mathcal{B}(\mathcal{N}(S), \mathcal{N}(P_{\mathcal{R}(S)^\perp} R)^\perp)$. Let

$$X = \begin{bmatrix} 0 & X_1 \\ 0 & 0 \end{bmatrix} : \begin{bmatrix} \mathcal{N}(S)^\perp \\ \mathcal{N}(S) \end{bmatrix} \to \begin{bmatrix} \mathcal{N}(P_{\mathcal{R}(S)^\perp} R)^\perp \\ \mathcal{N}(P_{\mathcal{R}(S)^\perp} R) \end{bmatrix}.$$

Now, it is easy to verify that $\mathcal{R}(S + RX) = \mathcal{H}$; i.e., $S + RX$ is right invertible.

(2) If R is noncompact and for some $X \in \mathcal{B}(\mathcal{H}, K)$, $S + RX$ is right invertible, then we have that $(S + RX)_r^{-1} R \in \mathcal{N}(S|R)$. Indeed

$$S(S + RX)_r^{-1} R = R - RX(S + RX)_r^{-1} R,$$

and $(S+RF)_r^{-1} R$ is a noncompact operator. The reverse implication follows directly from Theorem 2.10 since in the case when R is noncompact and $\mathcal{N}(S|R)$ contains a noncompact operator we have that there exists $X \in \mathcal{B}(\mathcal{H}, K)$ such that $S + RX$ is invertible. □

By Theorem 2.25 and Theorem 2.10 we get the following corollary:

COROLLARY 2.26. Let $S \in \mathcal{B}(\mathcal{H})$ and $R \in \mathcal{B}(\mathcal{K}, \mathcal{H})$ be noncompact operators. Then the following are equivalent:

(1) There exists $X \in \mathcal{B}(\mathcal{H}, \mathcal{K})$ such that $S + RX$ is right invertible,
(2) There exists $X \in \mathcal{B}(\mathcal{H}, \mathcal{K})$ such that $S + RX$ is invertible.

In the following theorem we consider the more complicated problem of the existence of $X \in \mathcal{B}(\mathcal{H}, \mathcal{K})$ for which the operator $S + RX$ is left invertible in the case when $S \in \mathcal{B}(\mathcal{H})$ and $R \in \mathcal{B}(\mathcal{K}, \mathcal{H})$ are given operators:

THEOREM 2.27. *Let $S \in \mathcal{B}(\mathcal{H})$ and $R \in \mathcal{B}(\mathcal{K}, \mathcal{H})$. The following hold:*

(1) *If R is compact, then there exists $X \in \mathcal{B}(\mathcal{H}, \mathcal{K})$ such that $S + RX$ is left invertible if and only if S is left semi-Fredholm with* $\mathrm{ind}(S) \leq 0$ *and* $\mathrm{codim}_{\mathcal{R}(R)}(\mathcal{R}(S) \cap \mathcal{R}(R)) \geq \dim(\mathcal{N}(S))$.
(2) *If R is noncompact, then there exists $X \in \mathcal{B}(\mathcal{H}, \mathcal{K})$ such that $S + RX$ is left invertible.*

PROOF. (1) If R is compact and there exists $X \in \mathcal{B}(\mathcal{H}, \mathcal{K})$ such that $S + RX$ is left invertible, then S is left semi-Fredholm and $\mathrm{ind}(S) = \mathrm{ind}(S + RX) \leq 0$. Also, by Theorem 2.18, we have that at least one of the following conditions must be satisfied:

(1) $\mathcal{N}(S)^\perp \cap S^{-1}(\mathcal{R}(R))$ contains a closed infinite-dimensional subspace.
(2) $\mathrm{codim}_{\mathcal{R}(R)}(\mathcal{R}(S) \cap \mathcal{R}(R)) \geq \dim(\mathcal{N}(S))$.

If (1) is satisfied, then there exists a closed infinite-dimensional subspace $\mathcal{M}$ such that $S[\mathcal{M}] \subseteq \mathcal{R}(R)$. Since S is left semi-Fredholm, it follows that $S[\mathcal{M}]$ is closed and infinite dimensional, so R is not compact. Thus, $\mathrm{codim}_{\mathcal{R}(R)}(\mathcal{R}(S) \cap \mathcal{R}(R)) \geq \dim(\mathcal{N}(S))$.

For the opposite direction suppose that S is left semi-Fredholm and suppose that $\mathrm{codim}_{\mathcal{R}(R)}(\mathcal{R}(S) \cap \mathcal{R}(R)) \geq \dim(\mathcal{N}(S))$. Then there exists a subspace T of $\mathcal{R}(R)$ such that $\mathcal{R}(R) = \mathcal{R}(S) \cap \mathcal{R}(R) + T$, $\mathcal{R}(S) \cap \mathcal{R}(R) \cap T = \{0\}$, and $\dim T \geq \dim(\mathcal{N}(S))$. Let $T' = R^{-1}[T] \cap \mathcal{N}(R)^\perp$. Let $F_0 : \mathcal{N}(S) \to \mathcal{H}$ be injective with range in T'. Now, let us define an operator X by $X|_{\mathcal{N}(S)^\perp} = 0$ and $X|_{\mathcal{N}(S)} = F_0$. Evidently, $S + RX$ is left invertible.

(2) First suppose that $\mathcal{R}(R)$ is closed. Then $\mathcal{M}=S^{-1}[\mathcal{R}(R)]$ is closed. In that case operators S and R have the following representations:

$$S=\begin{bmatrix} S_1 & S_2 \\ 0 & S_3 \end{bmatrix}:\begin{bmatrix} \mathcal{M} \\ \mathcal{M}^\perp \end{bmatrix}\to\begin{bmatrix} \mathcal{R}(R) \\ \mathcal{R}(R)^\perp \end{bmatrix},$$

$$R=\begin{bmatrix} R_1 & 0 \\ 0 & 0 \end{bmatrix}:\begin{bmatrix} \mathcal{N}(R)^\perp \\ \mathcal{N}(R) \end{bmatrix}\to\begin{bmatrix} \mathcal{R}(R) \\ \mathcal{R}(R)^\perp \end{bmatrix},$$

where S_3 is injective and R_1 is invertible. Since $\mathcal{R}(R)$ is infinite dimensional, there exists a left invertible $T:\mathcal{M}\to\mathcal{R}(R)$.

In the case when $\mathcal{R}(S_3)$ is closed, since S_3 is injective for

$$X=\begin{bmatrix} R_1^{-1}(T-S_1) & -R_1^{-1}S_2 \\ 0 & 0 \end{bmatrix}:\begin{bmatrix} \mathcal{M} \\ \mathcal{M}^\perp \end{bmatrix}\to\begin{bmatrix} \mathcal{N}(R)^\perp \\ \mathcal{N}(R) \end{bmatrix},$$

we have that

$$S+RX=\begin{bmatrix} T & 0 \\ 0 & S_3 \end{bmatrix}:\begin{bmatrix} \mathcal{M} \\ \mathcal{M}^\perp \end{bmatrix}\to\begin{bmatrix} \mathcal{R}(R) \\ \mathcal{R}(R)^\perp \end{bmatrix};$$

i.e., $S+RX$ is left invertible. In the case when $\mathcal{R}(S_3)$ is nonclosed, take $T_0:\mathcal{M}\to\mathcal{R}(R)$ to be a left invertible operator such that $d(T_0)=\infty$. By Theorem 1.18 we know that there exists an operator $Y\in\mathcal{B}(\mathcal{M}^\perp,\mathcal{R}(R))$ such that

$$\begin{bmatrix} T_0 & Y \\ 0 & S_3 \end{bmatrix}:\begin{bmatrix} \mathcal{M} \\ \mathcal{M}^\perp \end{bmatrix}\to\begin{bmatrix} \mathcal{R}(R) \\ \mathcal{R}(R)^\perp \end{bmatrix}$$

has closed range. Let

$$X=\begin{bmatrix} R_1^{-1}(T_0-S_1) & R_1^{-1}(Y-S_2) \\ 0 & 0 \end{bmatrix}:\begin{bmatrix} \mathcal{M} \\ \mathcal{M}^\perp \end{bmatrix}\to\begin{bmatrix} \mathcal{N}(R)^\perp \\ \mathcal{N}(R) \end{bmatrix}.$$

It can be checked that $S+RX=\begin{bmatrix} T_0 & Y \\ 0 & S_3 \end{bmatrix}$ and that this is a left invertible operator.

If $\mathcal{R}(R)$ is nonclosed, then there exists a closed infinite-dimensional subspace $\mathcal{M}$ of $\mathcal{H}$ such that $\mathcal{R}(R|_{\mathcal{M}})$ is closed and infinite dimensional. Let $T\in\mathcal{B}(\mathcal{H})$ be such that $\mathcal{R}(T)=\mathcal{M}$. Then $\mathcal{R}(RT)$ is closed and RT is a noncompact operator, so we can apply the first part of the proof. □

If we compare Theorems 2.10 and 2.27 in the case when R is a compact operator, it is not so evident that if $\begin{bmatrix} S & R \end{bmatrix}:\mathcal{H}\oplus\mathcal{K}\to\mathcal{H}$ is right invertible and S is Weyl, the condition $\operatorname{codim}_{\mathcal{R}(R)}(\mathcal{R}(S)\cap\mathcal{R}(R))\geq\dim(\mathcal{N}(S))$ from Theorem 2.27(1) is satisfied. Indeed, suppose that T is a subspace such that $\mathcal{R}(R)=\mathcal{R}(R)\cap\mathcal{R}(S)\oplus T$. Then $\operatorname{codim}_{\mathcal{R}(R)}(\mathcal{R}(S)\cap\mathcal{R}(R))=\dim T$ and since $\mathcal{R}(S)+\mathcal{R}(R)=\mathcal{H}$ we have that $\mathcal{R}(S)+T=\mathcal{H}$. Hence $\dim T\geq\dim(\mathcal{R}(S)^\perp)$. Since S is Weyl it follows that $\dim T\geq\dim(\mathcal{N}(S))$.

2.6. Fredholmness of an operator $A+CX$

We start this section with a characterization of all pairs of operators (S,R) for which the operator $S+RX$ is left semi-Fredholm for some $X\in\mathcal{B}(\mathcal{H},\mathcal{K})$.

THEOREM 2.28. *Let $S\in\mathcal{B}(\mathcal{H})$ and $R\in\mathcal{B}(\mathcal{K},\mathcal{H})$. The following hold:*

(1) *If R is compact, then there exists $X\in\mathcal{B}(\mathcal{H},\mathcal{K})$ such that $S+RX$ is left semi-Fredholm if and only if S is left semi-Fredholm.*

(2) *If R is noncompact, then there exists $X \in \mathcal{B}(\mathcal{H},\mathcal{K})$ such that $S+RX$ is left semi-Fredholm.*

PROOF. (1) If R is compact and there exists $X \in \mathcal{B}(\mathcal{H},\mathcal{K})$ such that $S+RX$ is left semi-Fredholm, then S is left semi-Fredholm. If S is left semi-Fredholm, take $X=0$.

(2) If R is noncompact, then by Theorem 2.27, there exists $X \in \mathcal{B}(\mathcal{H},K)$ such that $S+RX$ is left invertible; i.e., it is left semi-Fredholm. □

Now, we consider the case of right semi-Fredholmness which is a very complicated one. Notice that the right semi-Fredholmness of $\begin{bmatrix} S & R \end{bmatrix} : \mathcal{H}\oplus\mathcal{K} \to \mathcal{H}$ is clearly a necessary condition for the existence of X such that $S+RX$ is right semi-Fredholm.

THEOREM 2.29. *Let $S \in \mathcal{B}(\mathcal{H})$ and $R \in \mathcal{B}(\mathcal{K},\mathcal{H})$. The following hold:*

(1) *If R is compact, then there exists $X \in \mathcal{B}(\mathcal{H},\mathcal{K})$ such that $S+RX$ is right semi-Fredholm if and only if S is right semi-Fredholm.*
(2) *If R is noncompact, then there exists $X \in \mathcal{B}(\mathcal{H},\mathcal{K})$ such that $S+RX$ is right semi-Fredholm if and only if $\begin{bmatrix} S & R \end{bmatrix} : \mathcal{H}\oplus\mathcal{K} \to \mathcal{H}$ is right semi-Fredholm and $\mathcal{N}(S|R)$ contains a noncompact operator.*

PROOF. (1) If R is compact and there exists $X \in \mathcal{B}(\mathcal{H},\mathcal{K})$ such that $S+RX$ is right semi-Fredholm, then S is right semi-Fredholm. If S is right semi-Fredholm, take $X=0$.

(2) Suppose that R is noncompact and that there exists $X \in \mathcal{B}(\mathcal{H},\mathcal{K})$ such that $S+RX$ is right semi-Fredholm. Since

$$S+RX = \begin{bmatrix} S & R \end{bmatrix}\begin{bmatrix} I \\ X \end{bmatrix},$$

it follows that $\begin{bmatrix} S & R \end{bmatrix}$ is a right semi-Fredholm operator and that $\mathcal{R}\left(\begin{bmatrix} I \\ X \end{bmatrix}\right)+\mathcal{N}\left(\begin{bmatrix} S & R \end{bmatrix}\right)$ is a subspace of finite codimension in $\mathcal{H}\oplus\mathcal{K}$. Now, let $\begin{bmatrix} G \\ H \end{bmatrix} : \mathcal{K} \to \mathcal{H}\oplus\mathcal{K}$ be such that

$$\mathcal{R}\left(\begin{bmatrix} G \\ H \end{bmatrix}\right) = \mathcal{N}\left(\begin{bmatrix} S & R \end{bmatrix}\right). \tag{2.4}$$

Now, we have that $W = \begin{bmatrix} I & G \\ X & H \end{bmatrix}$ is right semi-Fredholm. If $\mathcal{N}(S|R)$ contains only compact operators, then G is a compact operator, so $W = \begin{bmatrix} I & 0 \\ X & H \end{bmatrix}$ is right semi-Fredholm. This implies that H is right semi-Fredholm, so there exists $T \in \mathcal{B}(\mathcal{H})$ such that $HT = I+K$, for some compact operator $K \in \mathcal{B}(\mathcal{H})$. Since $SG+RH=0$, we have that $SGT+R+RK=0$; i.e., R is a compact operator. Thus we get a contradiction, so $\mathcal{N}(S|R)$ contains a noncompact operator.

For the converse, if $\begin{bmatrix} S & R \end{bmatrix}$ is right semi-Fredholm and $\mathcal{N}(S|R)$ contains a noncompact operator, the proof follows K. Takahashi's proof (see [**160**]) for the case of invertibility, but we present it here for the sake of completeness. First remark that if $\mathcal{N}(S|R)$ contains a noncompact operator, then $\dim(\mathcal{N}\left(\begin{bmatrix} S & R \end{bmatrix}\right)) = \infty$.

Let $\begin{bmatrix} G \\ H \end{bmatrix} : \mathcal{K} \to \mathcal{H} \oplus \mathcal{K}$ be left invertible such that

$$\mathcal{R}\left(\begin{bmatrix} G \\ H \end{bmatrix}\right) = \mathcal{N}\left(\begin{bmatrix} S & R \end{bmatrix}\right). \tag{2.5}$$

We consider 3 cases:

Case 1: R is left invertible: Evidently, since $R_l^{-1}SG + H = 0$, we get that G is left invertible. Now, there exists an operator $X \in \mathcal{B}(\mathcal{H}, \mathcal{K})$ such that $H - XG$ is invertible (for example let $X = G_l^{-1} - R_l^{-1}S$). Now, by

$$\begin{bmatrix} I & G \\ X & H \end{bmatrix} = \begin{bmatrix} I & 0 \\ X & I \end{bmatrix} \begin{bmatrix} I & G \\ 0 & H - XG \end{bmatrix},$$

we have that $\begin{bmatrix} I & G \\ X & H \end{bmatrix}$ is invertible. From

$$\begin{bmatrix} S & R \end{bmatrix} \begin{bmatrix} I & G \\ X & H \end{bmatrix} = \begin{bmatrix} S + RX & 0 \end{bmatrix},$$

we get that $S+RX$ is right semi-Fredholm. Hence the case when R is left invertible is solved.

Case 2: S is not left semi-Fredholm: Let E be the spectral measure of R^*R. For $x \in \mathcal{K}$, let E_x be the positive measure given by $E_x(S) = (E(S)x, x)$ for a Borel subset S of $\mathbb{C}$. First we prove that there exists $\epsilon > 0$ such that $\mathcal{K}_\epsilon = \mathcal{R}(E(\epsilon, \infty))$ is infinite dimensional. Let $R_\epsilon = RE(\epsilon, \infty)$. Then

$$\|(R - R_\epsilon)x\|^2 = \int \lambda \chi_{[0,\epsilon]} dE_{x_0}(\lambda) \leq \epsilon \|x\|^2,$$

where $x_0 = E[0, \epsilon]x$. If for each $\epsilon > 0$, $\mathcal{K}_\epsilon$ is finite dimensional, then we have that R is compact as a limit of finite-dimensional operators R_ϵ. Hence there exists $\epsilon_0 > 0$ such that $\mathcal{K}_{\epsilon_0} = \mathcal{R}(E(\epsilon_0, \infty))$ is infinite dimensional. Since the set of right semi-Fredholm operators is open in the set of bounded linear operators, we get that for small enough $\epsilon > 0$ such that $\epsilon < \epsilon_0$, the operator $\begin{bmatrix} S & R_\epsilon \end{bmatrix}$ is right semi-Fredholm; i.e., $\begin{bmatrix} S & R|_{\mathcal{K}_\epsilon} \end{bmatrix}$ is right semi-Fredholm, where $\mathcal{K}_\epsilon$ is an infinite-dimensional closed subspace since $\mathcal{K}_{\epsilon_0} \subseteq \mathcal{K}_\epsilon$. Also, we have that

$$\|R_\epsilon x\|^2 = (R^* R x_0, x_0) = \int \lambda \chi_{(\epsilon,\infty)} dE_{x_0}(\lambda) \geq \epsilon \|x_0\|^2,$$

where $x_0 = E(\epsilon, \infty)x$, implying that $R|_{\mathcal{K}_\epsilon}$ is left invertible. Furthermore, we have $\dim \mathcal{N}(\begin{bmatrix} S & R|_{\mathcal{K}_\epsilon} \end{bmatrix}) = \infty$. In fact, if $\dim(\mathcal{N}(\begin{bmatrix} S & R|_{\mathcal{K}_\epsilon} \end{bmatrix})) < \infty$, then $\begin{bmatrix} S & R|_{\mathcal{K}_\epsilon} \end{bmatrix}$ is left semi-Fredholm, which contradicts the assumption that S is not left semi-Fredholm. Indeed, from $\dim(\mathcal{N}(\begin{bmatrix} S & R|_{\mathcal{K}_\epsilon} \end{bmatrix})) < \infty$ we can conclude that $\dim(\mathcal{N}(S)) < \infty$ and that $\dim(\mathcal{R}(S) \cap \mathcal{R}(R|_{\mathcal{K}_\epsilon})) < \infty$ which, by the fact that $\begin{bmatrix} S & R|_{\mathcal{K}_\epsilon} \end{bmatrix}$ is right semi-Fredholm, implies that $\mathcal{R}(S)$ is closed. Hence S is left semi-Fredholm. Now, by Case 1 of this proof there exists $F_1 \in \mathcal{B}(\mathcal{L}, \mathcal{K}_\epsilon)$ such that $S + (R|_{\mathcal{K}_\epsilon})F_1$ is invertible. Define $F = JF_1$ where $J : \mathcal{K}_\epsilon \to \mathcal{K}$ is the inclusion mapping. Then $S + RF$ is right semi-Fredholm.

Case 3: S is left semi-Fredholm: It is sufficient to show that there exists a closed infinite-dimensional subspace $\mathcal{K}_1$ of $\mathcal{K}$ such that $R_1 = R|_{\mathcal{K}_1}$ is left invertible and $\begin{bmatrix} S & R_1 \end{bmatrix}$ is a right semi-Fredholm operator with $\dim(\mathcal{N}(\begin{bmatrix} S & R_1 \end{bmatrix})) = \infty$. By assumption there exists noncompact $D \in \mathcal{N}(S|R)$; i.e., $SD + RT = 0$, for some $T \in \mathcal{B}(\mathcal{K})$. Thus $\mathcal{R}\left(\begin{bmatrix} D \\ T \end{bmatrix}\right) \subseteq \mathcal{N}(\begin{bmatrix} S & R \end{bmatrix})$ which by (2.5) implies that $\mathcal{R}(D) \subseteq \mathcal{R}(G)$, so G is noncompact. Also, from (2.5) we have that $SG + RH = 0$ which, together with the fact that S is left semi-Fredholm, implies that RH is noncompact. Now similarly as in the previous case using the spectral measure of $(RH)^*(RH)$, there exists an infinite-dimensional subspace $\mathcal{K}_0$ of $\mathcal{K}$ such that $RH|_{\mathcal{K}_0}$ is left invertible. Let $\mathcal{K}_1 = H[\mathcal{K}_0] \oplus \mathcal{R}(H)^\perp$. To prove that $\mathcal{K}_1$ is closed it is sufficient to prove that $H[\mathcal{K}_0]$ is closed. Since $RH[K_0]$ is closed it follows that $R^{-1}[RH[\mathcal{K}_0]] = H[\mathcal{K}_0] \oplus \mathcal{N}(R)$ is closed as an inverse image of a closed subspace. Hence $H[\mathcal{K}_0]$ is a closed subspace. Now we have that $\mathcal{K}_1$ is closed. Let us show that it satisfies the required conditions. First, we prove that $P^{\mathtt{rst}}_{\mathcal{R}(S)^\perp} R|_{\mathcal{R}(H)^\perp} : \mathcal{R}(H)^\perp \to \mathcal{R}(S)^\perp$ is injective and right semi-Fredholm. Suppose that for $h \in \mathcal{R}(H)^\perp$, $P^{\mathtt{rst}}_{\mathcal{R}(S)^\perp} Rh = 0$. Then $Rh \in \mathcal{R}(S)$, so there exists $t \in \mathcal{K}$ such that $Rh = St$. Now $\begin{bmatrix} t \\ -h \end{bmatrix} \in \mathcal{N}(\begin{bmatrix} S & R \end{bmatrix})$ implying by (2.5) that $h \in \mathcal{R}(H)$. Hence, $h = 0$ and we have that $P^{\mathtt{rst}}_{\mathcal{R}(S)^\perp} R|_{\mathcal{R}(H)^\perp} : \mathcal{R}(H)^\perp \to \mathcal{R}(S)^\perp$ is injective. Right semi-Fredholmness of $P^{\mathtt{rst}}_{\mathcal{R}(S)^\perp} R|_{\mathcal{R}(H)^\perp} : \mathcal{R}(H)^\perp \to \mathcal{R}(S)^\perp$ is established as follows: Since $\begin{bmatrix} S & R \end{bmatrix}$ is right semi-Fredholm we have that $P^{\mathtt{rst}}_{\mathcal{R}(S)^\perp} R$ is of finite codimension in $\mathcal{R}(S)^\perp$. Since $\mathcal{R}(R|_{\mathcal{R}(H)}) \subseteq \mathcal{R}(S)$ (follows by (2.5)) we have $\mathcal{R}(P^{\mathtt{rst}}_{\mathcal{R}(S)^\perp} R|_{\mathcal{R}(H)^\perp}) = \mathcal{R}(P^{\mathtt{rst}}_{\mathcal{R}(S)^\perp} R)$ implying that $P^{\mathtt{rst}}_{\mathcal{R}(S)^\perp} R|_{\mathcal{R}(H)^\perp} : \mathcal{R}(H)^\perp \to \mathcal{R}(S)^\perp$ is right semi-Fredholm.

Now, by the right semi-Fredholmness of $P^{\mathtt{rst}}_{\mathcal{R}(S)^\perp} R|_{\mathcal{R}(H)^\perp} : \mathcal{R}(H)^\perp \to \mathcal{R}(S)^\perp$, having in mind that $\mathcal{R}(R|_{\mathcal{R}(H)^\perp}) \subseteq \mathcal{R}(R_1)$ and that $\begin{bmatrix} S & R \end{bmatrix}$ is right semi-Fredholm, we get that $\begin{bmatrix} S & R_1 \end{bmatrix}$ is right semi-Fredholm, where $R_1 = R|_{\mathcal{K}_1}$.

The left invertibility of $RH|_{\mathcal{K}_0}$ implies that of $R|_{H[\mathcal{K}_0]} : H[\mathcal{K}_0] \to \mathcal{R}(S)$, which together with the left invertibility of $P^{\mathtt{rst}}_{\mathcal{R}(S)^\perp} R|_{\mathcal{R}(H)^\perp}$ implies that R_1 is left invertible. Since

$$\mathcal{R}\left(\begin{bmatrix} G \\ H \end{bmatrix}|_{\mathcal{K}_0}\right) \subseteq \mathcal{N}(\begin{bmatrix} S & R_1 \end{bmatrix})$$

and since $\mathcal{K}_0$ is infinite dimensional and $\begin{bmatrix} G \\ H \end{bmatrix}$ is left invertible, it follows that $\dim(\mathcal{N}(\begin{bmatrix} S & R_1 \end{bmatrix})) = \infty$. □

In the next theorem, we present necessary and sufficient conditions for the existence of $X \in \mathcal{B}(\mathcal{H}, \mathcal{K})$ such that $S + RX$ is a Fredholm operator:

THEOREM 2.30. *Let $S \in \mathcal{B}(\mathcal{H})$ and $R \in \mathcal{B}(\mathcal{K}, \mathcal{H})$. The following hold:*

(1) *If R is compact, there exists $X \in \mathcal{B}(\mathcal{H}, \mathcal{K})$ such that $S + RX$ is Fredholm if and only if S is a Fredholm operator.*
(2) *If R is noncompact, there exists $X \in \mathcal{B}(\mathcal{H}, \mathcal{K})$ such that $S + RX$ is Fredholm if and only if $\begin{bmatrix} S & R \end{bmatrix} : \mathcal{H} \oplus \mathcal{K} \to \mathcal{H}$ is right semi-Fredholm and $\mathcal{N}(S|R)$ contains a noncompact operator.*

PROOF. (1) If R is compact and for some $X \in \mathcal{B}(\mathcal{H},\mathcal{K})$, $S+RX$ is Fredholm, then we have that S is a Fredholm operator. In the other direction take $X=0$.

(2) Suppose that R is noncompact, $\begin{bmatrix} S & R \end{bmatrix} : \mathcal{H}\oplus\mathcal{K} \to \mathcal{H}$ is right semi-Fredholm, and $\mathcal{N}(S|R)$ contains a noncompact operator. We consider 3 cases:

Case 1: $\mathcal{R}(R)$: Then $\mathcal{M} = S^{-1}[\mathcal{R}(R)]$ is closed and since $\mathcal{N}(S|R)$ contains a noncompact operator, we can conclude that $\mathcal{M}$ is infinite dimensional. In that case operators S and R have the following representations:

$$S = \begin{bmatrix} S_1 & S_2 \\ 0 & S_3 \end{bmatrix} : \begin{bmatrix} \mathcal{M} \\ \mathcal{M}^\perp \end{bmatrix} \to \begin{bmatrix} \mathcal{R}(R) \\ \mathcal{R}(R)^\perp \end{bmatrix},$$

$$R = \begin{bmatrix} R_1 & 0 \\ 0 & 0 \end{bmatrix} : \begin{bmatrix} \mathcal{N}(R)^\perp \\ \mathcal{N}(R) \end{bmatrix} \to \begin{bmatrix} \mathcal{R}(R) \\ \mathcal{R}(R)^\perp \end{bmatrix},$$

where S_3 is injective and R_1 is invertible. Also since $\begin{bmatrix} S & R \end{bmatrix}$ is right semi-Fredholm, it follows that S_3 is right semi-Fredholm. Hence S_3 is a Fredholm operator. Since $\mathcal{R}(R)$ is infinite dimensional, there exists an invertible $T: \mathcal{M} \to \mathcal{R}(R)$.

For

$$X = \begin{bmatrix} R_1^{-1}(T-S_1) & -R_1^{-1}S_2 \\ 0 & 0 \end{bmatrix} : \begin{bmatrix} \mathcal{M} \\ \mathcal{M}^\perp \end{bmatrix} \to \begin{bmatrix} \mathcal{N}(R)^\perp \\ \mathcal{N}(R) \end{bmatrix},$$

we have that

$$S+RX = \begin{bmatrix} T & 0 \\ 0 & S_3 \end{bmatrix} : \begin{bmatrix} \mathcal{M} \\ \mathcal{M}^\perp \end{bmatrix} \to \begin{bmatrix} \mathcal{R}(R) \\ \mathcal{R}(R)^\perp \end{bmatrix};$$

i.e., $S+RX$ is a Fredholm operator.

Case 2: S is not left semi-Fredholm: As in the proof of Theorem 2.29, let E be the spectral measure of R^*R. We know that there exists $\epsilon > 0$ such that $\mathcal{K}_\epsilon = \mathcal{R}(E(\epsilon,\infty))$ is closed and infinite dimensional, $R_\epsilon = RE(\epsilon,\infty)$ is left invertible, and $\begin{bmatrix} S & R_\epsilon \end{bmatrix}$ is right semi-Fredholm; i.e., $\begin{bmatrix} S & R|_{\mathcal{K}_\epsilon} \end{bmatrix}$ is right semi-Fredholm. Furthermore, we have $\dim \mathcal{N}(\begin{bmatrix} S & R|_{\mathcal{K}_\epsilon} \end{bmatrix}) = \infty$ which, since $R|_{\mathcal{K}_\epsilon}$ is left invertible, implies that $\mathcal{M} = S^{-1}[R|_{\mathcal{K}_\epsilon}]$ is closed and infinite dimensional. Now we consider Case 1 where instead of R we consider $R|_{\mathcal{K}_\epsilon}$.

Case 3: S is left semi-Fredholm: Using Case 1, it is sufficient to show that there exists an infinite-dimensional subspace $\mathcal{K}_0$ such that $R|_{\mathcal{K}_0}$ has closed infinite-dimensional range. Since there exists a noncompact $G \in \mathcal{N}(S|R)$, we have that $SG + RH = 0$ for some $H \in \mathcal{B}(\mathcal{K})$. Since S is left semi-Fredholm we have that SG is noncompact, so RH is noncompact. This means that there exists a closed infinite-dimensional subspace $\mathcal{T} \subseteq \mathcal{R}(RH)$. Let $\mathcal{K}_1 = R^{-1}[\mathcal{T}]$. Then clearly $\mathcal{K}_1$ is closed and infinite dimensional and $\mathcal{R}(R|_{\mathcal{K}_1}) = \mathcal{T}$ is closed.

The reverse implication follows directly from Theorem 2.29. □

From Theorems 2.25 and 2.28–2.30, we get the following corollaries:

COROLLARY 2.31. Let $S \in \mathcal{B}(\mathcal{H})$, and let $R \in \mathcal{B}(\mathcal{K},\mathcal{H})$ be a noncompact operator. Then the following are equivalent:

(1) There exists $X \in \mathcal{B}(\mathcal{H},\mathcal{K})$ such that $S+RX$ is right semi-Fredholm.
(2) There exists $X \in \mathcal{B}(\mathcal{H},\mathcal{K})$ such that $S+RX$ is Fredholm.

COROLLARY 2.32. Let $S \in \mathcal{B}(\mathcal{H})$, let $R \in \mathcal{B}(\mathcal{K},\mathcal{H})$ be a noncompact operator, and let $\begin{bmatrix} S & R \end{bmatrix} : \mathcal{H} \oplus \mathcal{K} \to \mathcal{H}$ be right invertible. Then the following are equivalent:

(1) There exists $X \in \mathcal{B}(\mathcal{H},\mathcal{K})$ such that $S+RX$ is invertible.
(2) There exists $X \in \mathcal{B}(\mathcal{H},\mathcal{K})$ such that $S+RX$ is right invertible.
(3) There exists $X \in \mathcal{B}(\mathcal{H},\mathcal{K})$ such that $S+RX$ is right semi-Fredholm.

2.7. Dense range of an operator $A+CX$

In [**55**] the problem of characterization of all the pairs of operators $A \in \mathcal{B}(\mathcal{H},\mathcal{K})$ and $C \in \mathcal{B}(\mathcal{L},\mathcal{K})$ for which the operator $A+CX$ is injective for some $X \in \mathcal{B}(\mathcal{H},\mathcal{L})$ was completely solved. In view of the main result of the previous section, to sort of complete this line of investigation it naturally remains to answer the question for which operators $A \in \mathcal{B}(\mathcal{H},\mathcal{K})$ and $C \in \mathcal{B}(\mathcal{L},\mathcal{K})$ there exists $X \in \mathcal{B}(\mathcal{H},\mathcal{L})$ such that the operator $A+CX$ is with dense range. Unfortunately, we have not been able to fully solve this problem, but we have rather given some partial answers to it. In particular, we show that if $A \notin \mathcal{F}_+(\mathcal{H},\mathcal{K})$, then there always exists such an operator X. Also, we give necessary and sufficient conditions for the existence of such an operator in the case when $A \in \mathcal{F}(\mathcal{H},\mathcal{K})$.

As before, the claim that there exists an operator $X \in \mathcal{B}(\mathcal{H},\mathcal{L})$ such that $A+CX$ is with dense range amounts to saying that $\mathcal{R}(A+T)$ is dense for some $T \in \mathcal{B}(\mathcal{H},\mathcal{K})$ with $\mathcal{R}(T) \subseteq \mathcal{R}(C)$. Clearly a necessary condition for this to hold is that the subspace $\mathcal{R}(A)+\mathcal{R}(C)$ is dense in $\mathcal{K}$ or, equivalently, that $\mathcal{N}(A^*)\cap\mathcal{N}(C^*) = \{0\}$.

A simple observation combined with Theorem 2.19 yields the following proposition.

PROPOSITION 2.33. Let $A \in \mathcal{B}(\mathcal{H},\mathcal{K})$ and $C \in \mathcal{B}(\mathcal{L},\mathcal{K})$ be such that $\mathcal{R}(A)$ is closed and $\mathcal{R}(C)$ is finite dimensional. The following are equivalent:

(1) There exists an operator $T \in \mathcal{B}(\mathcal{H},\mathcal{K})$ such that $\mathcal{R}(T) \subseteq \mathcal{R}(C)$ and $\mathcal{R}(A+T)$ is dense in $\mathcal{K}$.
(2) There exists an operator $T \in \mathcal{B}(\mathcal{H},\mathcal{K})$ such that $\mathcal{R}(T) \subseteq \mathcal{R}(C)$ and $\mathcal{R}(A+T)$ is surjective.
(3) $\mathcal{R}(A)+\mathcal{R}(C) = \mathcal{K}$ and $d(A) \leq \alpha(A)$.

PROOF. By Lemma 1.64 the subspace $\mathcal{R}(A+T)$ is closed for every $T \in \mathcal{B}(\mathcal{H},\mathcal{K})$ such that $\mathcal{R}(T) \subseteq \mathcal{R}(C)$. For this reason (1) and (2) say the same thing. To see that (3) is equivalent to these, in view of Theorem 2.19, it suffices to show that if there exists an operator $T \in \mathcal{B}(\mathcal{H},\mathcal{K})$ such that $\mathcal{R}(T) \subseteq \mathcal{R}(C)$ and $\mathcal{R}(A+T)$ is surjective, then it must be that $d(A) \leq \alpha(A)$. By Theorem 2.19 the inequality $\alpha(A) < d(A)$ would imply the existence of an infinite-dimensional subspace $M \subseteq \mathcal{H}$ such that $A[M] \subseteq \mathcal{R}(C)$. Given that $\alpha(A) < \infty$, we can infer from this that $M \cap \mathcal{N}(A)^\perp$ must also be of infinite dimension, which would mean that $A[M]$, and thus $\mathcal{R}(C)$ as well is infinite dimensional—a contradiction. □

As a second partial result we have the following conclusion which immediately follows from Theorem 2.20.

THEOREM 2.34. *Let $A \in \mathcal{B}(\mathcal{H}, \mathcal{K})$ and $C \in \mathcal{B}(\mathcal{L}, \mathcal{K})$ be such that $\mathcal{R}(A) + \mathcal{R}(C)$ is dense in $\mathcal{K}$. If $A^{-1}[\mathcal{R}(C)]$ contains a closed infinite-dimensional subspace, then there exists an operator $T \in \mathcal{B}(\mathcal{H}, \mathcal{K})$ such that $\mathcal{R}(T) \subseteq \mathcal{R}(C)$ and $\mathcal{R}(A+T)$ is dense in $\mathcal{K}$.*

Due to the next theorem, in our further considerations we need only consider the case when A is left semi-Fredholm.

THEOREM 2.35. *Let $A \in \mathcal{B}(\mathcal{H}, \mathcal{K})$ and $C \in \mathcal{B}(\mathcal{L}, \mathcal{K})$ be such that $\mathcal{R}(A) + \mathcal{R}(C)$ is dense in $\mathcal{K}$. If $\beta(A^*) = \infty$, then there exists an operator $T \in \mathcal{B}(\mathcal{H}, \mathcal{K})$ such that $\mathcal{R}(T) \subseteq \mathcal{R}(C)$ and $\mathcal{R}(A+T)$ is dense in $\mathcal{K}$.*

PROOF. From $\beta(A^*) = \infty$ it follows that there exists a closed infinite-dimensional subspace $F \subseteq \mathcal{H}$ such that $\mathcal{R}(A^*) \cap F = \{0\}$ (use Lemma 2.5 if $\mathcal{R}(A^*)$ is not closed). Thus we can fix an injective operator $Y \in \mathcal{B}(\mathcal{L}, \mathcal{H})$ such that $\mathcal{R}(Y) \subseteq F$. We claim that $A^* + YC^*$ is one-to-one. Indeed, suppose $x \in \mathcal{N}(A^*)^\perp$ and $y \in \mathcal{N}(A^*)$ are such that $(A^* + YC^*)(x+y) = 0$. $\mathcal{R}(A^*) \cap F = \{0\}$ then gives $A^*x = 0$ and $YC^*(x+y) = 0$. But $x \in \mathcal{N}(A^*)^\perp$ and Y is injective, so $x = 0$ and $C^*y = 0$. Given that $y \in \mathcal{N}(A^*)$, it now follows from $\mathcal{N}(A^*) \cap \mathcal{N}(C^*) = \{0\}$ that $y = 0$. □

The following auxiliary lemma is needed in the proof of Theorem 2.37.

LEMMA 2.36. *Let $A \in \mathcal{F}_+(\mathcal{H}, \mathcal{K})$ be such that* $\operatorname{ind}(A) < 0$. *If $C \in \mathcal{B}(\mathcal{L}, \mathcal{K})$ is compact, then there is no $X \in \mathcal{B}(\mathcal{H}, \mathcal{L})$ such that $\mathcal{R}(A + CX)$ is dense in $\mathcal{K}$.*

PROOF. Suppose towards a contradiction that $\mathcal{R}(A + CX)$ is dense in $\mathcal{K}$ for some $X \in \mathcal{B}(\mathcal{H}, \mathcal{L})$. If $\mathcal{H}$ is finite dimensional, then $\mathcal{R}(A+CX)$ is finite dimensional and $\mathcal{R}(A + CX) = \mathcal{K}$, so $\dim \mathcal{K} < \infty$. But by Proposition 2.33 we must then have $\operatorname{ind}(A) \geq 0$, contrary to our assumption. Thus $\mathcal{H}$ must be infinite dimensional. Since $\alpha(A) < \infty$, this also implies that $\dim \mathcal{K} = \infty$. C is compact, so the operator $A + CX$ is semi-Fredholm and $\operatorname{ind}(A + CX) < 0$, by Theorem 2.3. But this means that $\mathcal{R}(A + CX)$ is closed and that also $\mathcal{R}(A + CX) \neq \mathcal{K}$, which contradicts our assumption again. □

In view of Theorem 2.19, the following result completely answers our question in the case when $A \in \mathcal{F}(\mathcal{H}, \mathcal{K})$.

THEOREM 2.37. *Suppose $A \in \mathcal{F}(\mathcal{H}, \mathcal{K})$ and $C \in \mathcal{B}(\mathcal{L}, \mathcal{K})$. Then the following are equivalent:*

(1) *There exists an operator $X \in \mathcal{B}(\mathcal{H}, \mathcal{L})$ such that $\mathcal{R}(A + CX)$ is dense in $\mathcal{K}$.*
(2) *There exists an operator $X \in \mathcal{B}(\mathcal{H}, \mathcal{L})$ such that $A + CX$ is surjective.*
(3) *$\mathcal{R}(A) + \mathcal{R}(C) = \mathcal{K}$ and either* $\operatorname{ind}(A) \geq 0$ *or $\mathcal{R}(A) \cap \mathcal{R}(C)$ contains a closed infinite-dimensional subspace.*

PROOF. (1) ⇒ (2): $\mathcal{R}(A) + \mathcal{R}(C)$ is dense in $\mathcal{K}$, but it is also closed, being of finite codimension, so in fact $\mathcal{R}(A) + \mathcal{R}(C) = \mathcal{K}$. If $\operatorname{ind}(A) \geq 0$, then (2) follows directly from Theorem 2.24. Thus assume that $\operatorname{ind}(A) < 0$.

By Lemma 2.36 we have that C is not compact, so there exists a closed infinite-dimensional subspace $F \subseteq \mathcal{R}(C)$. From $d(A) < \infty$, it follows that $F_1 = F \cap \mathcal{R}(A)$ is closed and infinite dimensional as well. Thus the subspace $F_2 = A^{-1}[F_1]$ is closed and infinite dimensional; clearly $A[F_2] \subseteq \mathcal{R}(C)$. Now (2) follows from Theorem 2.21.

Since the implication (2) $\Rightarrow$ (1) is trivial, it remains to see that (2) $\Leftrightarrow$ (3). But this is easily seen to be a direct consequence of Theorem 2.19, in view of Theorem 2.4 and the fact that $\alpha(A) < \infty$. □

We end this discussion with a sufficient condition for our question to have an affirmative answer.

THEOREM 2.38. *Let $A \in \mathcal{B}(\mathcal{H},\mathcal{K})$ and $C \in \mathcal{B}(\mathcal{L},\mathcal{K})$ be such that $\mathcal{R}(A)+\mathcal{R}(C)$ is dense in $\mathcal{K}$. If $C \in \mathcal{F}_-(\mathcal{L},\mathcal{K})$ and $\dim(\mathcal{R}(A)) = \infty$, then there is some $X \in \mathcal{B}(\mathcal{H},\mathcal{L})$ such that $A + CX$ is with dense range.*

PROOF. There exists $X \in \mathcal{B}(\mathcal{H},\mathcal{L})$ such that $A + CX$ is with dense range if and only if there exists $Y \in \mathcal{B}(\mathcal{L},\mathcal{H})$ such that $A^* + YC^*$ is injective. We establish the existence of such an operator Y.

Since $\alpha(C^*) < \infty$, the subspace $\mathcal{N}(A^*) \oplus \mathcal{N}(C^*)$ is closed and consequently we have $\mathcal{K} = \mathcal{N}(A^*) \oplus \mathcal{N}(C^*) \oplus U$, where $U = (\mathcal{N}(A^*) \oplus \mathcal{N}(C^*))^\perp$. Since $\mathcal{R}(C^*)$ is closed, it now follows that the subspaces $V_1 = C^*[\mathcal{N}(A^*)]$ and $V_2 = C^*[U]$ are both closed and we have $\mathcal{R}(C^*) = V_1 \oplus V_2$. The closed subspace U is infinite dimensional, since $\alpha(C^*) < \infty$ and $\dim(\mathcal{R}(A^*)) = \infty$. Thus there are closed subspaces $S_1, S_2 \subseteq U$ such that $\dim S_1 = \dim V_1$, $\dim S_2 = \infty$, and $U = S_1 \oplus S_2$. Let $P \in \mathcal{B}(U, S_2)$ be the projector onto S_2 parallel to S_1. The subspace S_2 being of infinite dimension, we can fix an injective operator $B_0 \in \mathcal{B}(U, S_2)$ and set $B = P + B_0$. The restriction $C_0 \in \mathcal{B}(U, V_2)$ of the operator C^* is invertible and for $D_2 = BC_0^{-1} \in \mathcal{B}(V_2, S_2)$ we now clearly have $D_2C^*x \neq Px$ for every nonzero $x \in U$. Since $\dim S_1 = \dim V_1$, there is an injective operator $D_1 \in \mathcal{B}(V_1, S_1)$. Define $D \in \mathcal{B}(\mathcal{L},\mathcal{K})$ by

$$D = \begin{bmatrix} D_1 & 0 & 0 \\ 0 & D_2 & 0 \\ 0 & 0 & 0 \end{bmatrix} : \begin{bmatrix} V_1 \\ V_2 \\ \mathcal{R}(C^*)^\perp \end{bmatrix} \to \begin{bmatrix} S_1 \\ S_2 \\ \mathcal{N}(A^*) \oplus \mathcal{N}(C^*) \end{bmatrix}.$$

For $Y = -A^*D$ we claim that the operator $A^* + YC^*$ is one-to-one. To see this let $z \in \mathcal{N}(A^*)$, $y \in \mathcal{N}(C^*)$, and $x \in U$ be such that $(A^* + YC^*)(z + y + x) = 0$, i.e., such that $w = (1 - DC^*)(z + y + x) \in \mathcal{N}(A^*)$. From $z + y + x = DC^*z + DC^*x + w$ and $DC^*z = D_1C^*z \in S_1$, $DC^*x = D_2C^*x \in S_2$, it now follows that $y = 0$ and $x = D_2C^*x + D_1C^*z$ and consequently $D_2C^*x = Px$. As observed before this implies that $x = 0$. But then we have $D_1C^*z = 0$ whence, by injectivity of D_1, $C^*z = 0$. Finally, $\mathcal{N}(A^*) \cap \mathcal{N}(C^*) = \{0\}$ gives $z = 0$. □

2.8. Closed range of an operator $A + CX$

In this section we describe the set of all pairs of operators $(A, C) \in \mathcal{B}(\mathcal{H}) \times \mathcal{B}(\mathcal{H})$ such that there exists an operator $X \in \mathcal{B}(\mathcal{H})$ for which $A + CX$ is a closed range operator. As far as we know there are even no partial results on this subject when the property of closedness of range is considered. As an application we get some results related to the completion of the operator matrix of the form $\begin{bmatrix} A & C \\ ? & B \end{bmatrix} \in \mathcal{B}(\mathcal{H} \oplus \mathcal{H})$ to a closed range operator.

Now we list two auxiliary results that will be used throughout the section.

THEOREM 2.39. *Let $S, R \in \mathcal{B}(\mathcal{H})$ be such that $\mathcal{R}(S) + \mathcal{R}(R)$ and $\mathcal{R}(S) \cap \mathcal{R}(R)$ are closed; then $\mathcal{R}(S)$ and $\mathcal{R}(R)$ are closed.*

PROOF. Let P be the orthogonal projector onto $\mathcal{R}(S)\cap\mathcal{R}(R)$. The proof follows from the fact that $\mathcal{R}(S)+\mathcal{R}(R)=\mathcal{R}((I-P)R)\oplus\mathcal{R}(S)$ and Theorem 2.3 from [**77**]. □

Notice that the next result is a generalization of Lemma 2.5.

THEOREM 2.40 ([**111**]). *Let $A\in\mathcal{B}(\mathcal{H})$ be such that $\mathcal{R}(A)$ is not closed. Then there exists a closed subspace $\mathcal{M}$ of $\mathcal{H}$ such that $\mathcal{M}\subseteq\overline{\mathcal{R}(A)}$ and $\mathcal{R}(A)\cap\mathcal{M}=\{0\}$.*

The following theorem presents the main result of this section.

THEOREM 2.41. *Let $A,C\in\mathcal{B}(\mathcal{H})$. Then there exists an operator $X\in\mathcal{B}(\mathcal{H})$ such that $A+CX$ is a closed range operator if and only if one of the following conditions holds:*

(1) *$\mathcal{R}(A)$ is closed.*
(2) *C is a noncompact operator.*
(3) *C is a compact operator and there exists a closed subspace F of $\mathcal{R}(A)$ such that $\mathcal{R}(A)=F+\mathcal{R}(A)\cap\mathcal{R}(C)$.*

The proof of Theorem 2.41 will be given in a form of a series of results to be presented below.

The case when $\mathcal{R}(A)$ is closed is trivial since then we can take $X=0$.

In the following theorem we consider the case when $C\in\mathcal{B}(\mathcal{H})$ is a noncompact operator.

THEOREM 2.42. *Let $A\in\mathcal{B}(\mathcal{H})$, and let $C\in\mathcal{B}(\mathcal{H})$ be a noncompact operator. Then there exists an operator $X\in\mathcal{B}(\mathcal{H})$ such that $A+CX$ is a closed range operator.*

PROOF. Let $P\in\mathcal{B}(\mathcal{H})$ be an orthogonal projector whose infinite-dimensional range is contained in $\mathcal{R}(C)$. By R. G. Douglas's result (see [**68**]), we have $P=CY$, for some $Y\in\mathcal{B}(\mathcal{H})$. We next construct an operator $Z\in\mathcal{B}(\mathcal{H})$ such that $A+PZ$ has closed range. If the range of P has finite codimension, we may take $Z=-A$ to conclude that $A+PZ$ has finite rank and thus has closed range. Otherwise, decompose $\mathcal{H}=\mathcal{R}(P)\oplus\mathcal{N}(P)$. Then

$$P=\begin{bmatrix} I & 0\\ 0 & 0\end{bmatrix}:\begin{bmatrix}\mathcal{R}(P)\\ \mathcal{N}(P)\end{bmatrix}\rightarrow\begin{bmatrix}\mathcal{R}(P)\\ \mathcal{N}(P)\end{bmatrix},$$

$$A=\begin{bmatrix} A_1 & A_2\\ A_3 & A_4\end{bmatrix}:\begin{bmatrix}\mathcal{R}(P)\\ \mathcal{N}(P)\end{bmatrix}\rightarrow\begin{bmatrix}\mathcal{R}(P)\\ \mathcal{N}(P)\end{bmatrix}.$$

Choose $Z\in\mathcal{B}(\mathcal{H})$ given by

$$Z=\begin{bmatrix} Z_1 & Z_2\\ Z_3 & Z_4\end{bmatrix}:\begin{bmatrix}\mathcal{R}(P)\\ \mathcal{N}(P)\end{bmatrix}\rightarrow\begin{bmatrix}\mathcal{R}(P)\\ \mathcal{N}(P)\end{bmatrix}$$

such that the first row of

$$A+PZ=\begin{bmatrix} A_1+Z_1 & A_2+Z_2\\ A_3 & A_4\end{bmatrix}:\begin{bmatrix}\mathcal{R}(P)\\ \mathcal{N}(P)\end{bmatrix}\rightarrow\begin{bmatrix}\mathcal{R}(P)\\ \mathcal{N}(P)\end{bmatrix}$$

is an invertible operator from $\mathcal{R}(P)\oplus\mathcal{N}(P)$ to $\mathcal{R}(P)$. Then $A+PZ$ has closed range. Taking $X=YZ$ we complete the proof. □

Now, let us consider the last case: when $C \in \mathcal{B}(\mathcal{H})$ is a compact operator.

THEOREM 2.43. *Let $A \in \mathcal{B}(\mathcal{H})$, and let $C \in \mathcal{B}(\mathcal{H})$ be a compact operator. Then the following are equivalent:*

(i) *There exists $X \in \mathcal{B}(\mathcal{H})$ such that $A + CX$ has closed range,*
(ii) *There exists a closed subspace F of $\mathcal{R}(A)$ such that $\mathcal{R}(A) = F + \mathcal{R}(A) \cap \mathcal{R}(C)$.*

PROOF. (i)⇒(ii): Suppose that there exists $X \in \mathcal{B}(\mathcal{H})$ such that $A + CX$ has closed range and let $\mathcal{N} = \mathcal{N}(A + CX)^{\perp}$. Then $(A + CX)|_{\mathcal{N}}$ is left invertible and since $CX|_{\mathcal{N}}$ is a compact operator, it follows that $A|_{\mathcal{N}}$ is left semi-Fredholm. Let $F = A(\mathcal{N})$. Then F is closed and we can check that $\mathcal{R}(A) = F + \mathcal{R}(A) \cap \mathcal{R}(C)$.

(ii)⇒(i): Let F be a closed subspace of $\mathcal{R}(A)$ such that $\mathcal{R}(A) = F + \mathcal{R}(A) \cap \mathcal{R}(C)$. Then $\mathcal{N}(A)^{\perp} = A^{-1}[F] \cap \mathcal{N}(A)^{\perp} + A^{-1}[\mathcal{R}(A) \cap \mathcal{R}(C)] \cap \mathcal{N}(A)^{\perp}$. Now since $A^{-1}[F] \cap \mathcal{N}(A)^{\perp}, A^{-1}[\mathcal{R}(A) \cap \mathcal{R}(C)] \cap \mathcal{N}(A)^{\perp}$ are operator ranges, by Theorem 2.7 there exist closed subspaces $M_1 \subseteq A^{-1}[F] \cap \mathcal{N}(A)^{\perp}$ and $M_2 \subseteq A^{-1}[\mathcal{R}(A) \cap \mathcal{R}(C)] \cap \mathcal{N}(A)^{\perp}$ such that $\mathcal{N}(A)^{\perp} = M_1 \oplus M_2$. Now, $\mathcal{R}(A) = A(M_1) \oplus A(M_2)$ and we can check that $A(M_1)$ is a closed subspace since $A|_{A^{-1}[F] \cap \mathcal{N}(A)^{\perp}}$ is left invertible. Now, define the operator $K \in \mathcal{B}(\mathcal{H})$ by

$$K = \begin{bmatrix} 0 & 0 & -A_2 \end{bmatrix} : \begin{bmatrix} \mathcal{N}(A) \\ M_1 \\ M_2 \end{bmatrix} \to \mathcal{H},$$

where

$$A = \begin{bmatrix} 0 & A_1 & A_2 \end{bmatrix} : \begin{bmatrix} \mathcal{N}(A) \\ M_1 \\ M_2 \end{bmatrix} \to \mathcal{H}.$$

Evidently, $K \in \mathcal{B}(\mathcal{H})$, $\mathcal{R}(K) \subseteq \mathcal{R}(C)$, and $\mathcal{R}(A + K) = A(M_1)$ is closed. Now for some $X \in \mathcal{B}(\mathcal{H})$, $K = CX$. □

Notice that the compactness is not needed in the implication (ii)⇒(i) of Theorem 2.43. Now, we are ready to prove Theorem 2.41:

Proof of Theorem 2.41: The proof in the case when $C \in \mathcal{B}(\mathcal{H})$ is noncompact follows directly from Theorem 2.42. Assume $C \in \mathcal{B}(\mathcal{H})$ is compact. If $\operatorname{codim}_{\mathcal{R}(A)}(\mathcal{R}(A) \cap \mathcal{R}(C)) < \infty$, then clearly $\mathcal{R}(A) = F + \mathcal{R}(A) \cap \mathcal{R}(C)$, for a finite-dimensional subspace F, so by Theorem 2.43, $\mathcal{R}(A + CX)$ is closed for a suitable X. If on the other hand $\operatorname{codim}_{\mathcal{R}(A)}(\mathcal{R}(A) \cap \mathcal{R}(C)) = \infty$, then such an operator X exists if and only if there exists a closed, necessarily infinite-dimensional, subspace F of $\mathcal{R}(A)$ such that $\mathcal{R}(A) = F + \mathcal{R}(A) \cap \mathcal{R}(C)$; notice that in this case A must be a noncompact operator.

Now, we present an alternative characterization in the case when $C \in \mathcal{B}(\mathcal{H})$ is a compact operator.

THEOREM 2.44. *Let $A \in \mathcal{B}(\mathcal{H})$ be a nonclosed range operator, and let $C \in \mathcal{B}(\mathcal{H})$ be a compact operator. Then the following are equivalent:*

(i) *There exists $X \in \mathcal{B}(\mathcal{H})$ such that $A + CX$ has closed range.*
(ii) *There exists a closed infinite-dimensional subspace M of $A^{-1}[\mathcal{R}(C)] \cap \mathcal{N}(A)^{\perp}$ such that $\mathcal{R}(A) = F \oplus A(M)$, for some closed subspace F.*

PROOF. (ii)⇒(i): This is evident by Theorem 2.43.

(i)⇒(ii): By the proof of the part (ii)⇒(i) from Theorem 2.43, we know there are closed subspaces $M_1 \subseteq A^{-1}[F] \cap \mathcal{N}(A)^{\perp}$ and $M_2 \subseteq A^{-1}[\mathcal{R}(A) \cap \mathcal{R}(C)] \cap \mathcal{N}(A)^{\perp}$ such that $\mathcal{R}(A) = A(M_1) \oplus A(M_2)$, where $A(M_1)$ is closed. Obviously M_2 must be infinite dimensional because if M_2 is finite dimensional, then $A(M_2)$ is finite dimensional, which together with $\mathcal{R}(A) = A(M_1) \oplus A(M_2)$ implies that $\mathcal{R}(A)$ is closed, which is not true. □

Given $A \in \mathcal{B}(\mathcal{H})$ and $C \in \mathcal{B}(\mathcal{H})$, it can easily be seen that the set $\mathcal{N}(A|C)$ contains a noncompact operator if and only if there is a closed infinite-dimensional subspace $F \subseteq \mathcal{H}$ such that $A[F] \subseteq \mathcal{R}(C)$. Indeed, if $G \in \mathcal{N}(A|C)$ is noncompact, then there is a closed infinite-dimensional subspace $F \subseteq \mathcal{R}(G) \subseteq \mathcal{H}$; $A[F] \subseteq \mathcal{R}(C)$ then follows from $\mathcal{R}(AG) \subseteq \mathcal{R}(C)$. If, on the other hand, $F \subseteq \mathcal{H}$ is a closed infinite-dimensional subspace such that $A[F] \subseteq \mathcal{R}(C)$, then we can take some $A \in \mathcal{B}(\mathcal{H})$ with $\mathcal{R}(A) = F$ (both $\mathcal{H}$ and F being separable infinite-dimensional Hilbert spaces); clearly, $\mathcal{R}(AG) \subseteq \mathcal{R}(C)$ is satisfied.

In view of the preceding observation, the reader is invited to notice that the condition that $\mathcal{N}(A)^{\perp} \cap A^{-1}[\mathcal{R}(C)]$ contains a closed infinite-dimensional subspace might as well have been replaced by the equivalent condition that $\mathcal{N}(A_0 \mid C)$ contains a noncompact operator, where $A_0 = A\,|_{\mathcal{N}(A)^{\perp}}$.

COROLLARY 2.45. Let $A \in \mathcal{B}(\mathcal{H})$ be a nonclosed range operator and $C \in \mathcal{B}(\mathcal{H})$ a compact operator. If there exists $X \in \mathcal{B}(\mathcal{H})$ such that $A+CX$ has closed range, then there exists $X \in \mathcal{B}(\mathcal{H})$ such that $A+CX$ is injective.

PROOF. If there exists $X \in \mathcal{B}(\mathcal{H})$ such that $A+CX$ has closed range, then by Theorem 2.44 , there exists a closed infinite-dimensional subspace M of $A^{-1}[\mathcal{R}(C)] \cap \mathcal{N}(A)^{\perp}$, which is by Theorem 2.18 a sufficient condition for the existence of $X \in \mathcal{B}(\mathcal{H})$ such that $A+CX$ is injective. □

COROLLARY 2.46. Let $A \in \mathcal{B}(\mathcal{H})$ be a nonclosed range operator and $C \in \mathcal{B}(\mathcal{H})$ a compact operator. If $\dim(\mathcal{R}(A) \cap \mathcal{R}(C)) < \infty$, then there is no $X \in \mathcal{B}(\mathcal{H})$ such that $A+CX$ has closed range.

Now let us discuss this problem in terms of operator ranges: We know that operator ranges are of four types: closed subspaces, ranges of compact operators, class 2(a), and class 2(b). Class 2(a) is the class of all operator ranges which contain a closed infinite-dimensional subspace $\mathcal{N}$ and that any such $\mathcal{N}$ is contained in another such space $\mathcal{N}'$ with $\dim(\mathcal{N}' \ominus \mathcal{N}) = \infty$. An operator range $\mathcal{R}$ is of class 2(b) if it is not closed and there exists a closed infinite-dimensional subspace $\mathcal{N} \subset \mathcal{R}$ that is not itself of infinite deficiency in another closed subspace $\mathcal{N}' \subset \mathcal{R}$. In Theorem 2.8 of [**77**] we can find the following characterization of operator ranges of class 2(b):

THEOREM 2.47. *An operator range $\mathcal{R}$ is of class* 2(b) *if and only if there exists a closed infinite-dimensional subspace $\mathcal{M}$ and a nonclosed compact operator range $\mathcal{R}_0 \subseteq \mathcal{M}^{\perp}$ such that $\mathcal{R} = \mathcal{M} + \mathcal{R}_0$.*

In the special case of positive operators we have the following characterization:

THEOREM 2.48. *Let $A \in \mathcal{B}(\mathcal{H})$ be a positive operator with nonclosed range. Then $\mathcal{R}(A)$ is of class* 2(b) *if and only if* $A = \begin{bmatrix} K & 0 \\ 0 & B \end{bmatrix} : \begin{bmatrix} \mathcal{M} \\ \mathcal{M}^\perp \end{bmatrix} \to \begin{bmatrix} \mathcal{M} \\ \mathcal{M}^\perp \end{bmatrix}$, *where $K \in \mathcal{B}(\mathcal{M})$ is compact, $B \in \mathcal{B}(\mathcal{M}^\perp)$ is invertible with infinite rank, and $\mathcal{M}$ is a closed subspace of $\mathcal{H}$.*

Now using Theorem 2.41 we get the next result:

THEOREM 2.49. *Let $A \in \mathcal{B}(\mathcal{H})$ be a nonclosed range operator and $C \in \mathcal{B}(\mathcal{H})$ a compact operator such that* $\dim_{\mathcal{R}(A)}(\mathcal{R}(A) \cap \mathcal{R}(C)) = \infty$. *If there exists an operator $X \in \mathcal{B}(\mathcal{H})$ such that $A + CX$ is a closed range operator, then $\mathcal{R}(A)$ is of class* 2(b).

PROOF. By Theorem 2.43 there exists a closed subspace F such that $\mathcal{R}(A) = F + \mathcal{R}(A) \cap \mathcal{R}(C)$, implying $F_0 = P_{F^\perp}(\mathcal{R}(A) \cap \mathcal{R}(C)) = F^\perp \cap \mathcal{R}(A)$. As an intersection of two operator ranges, $\mathcal{R}(A) \cap \mathcal{R}(C)$ is an operator range; i.e., $\mathcal{R}(A) \cap \mathcal{R}(C) = \mathcal{R}(S)$ for a suitable operator S. From $\mathcal{R}(S) \subseteq \mathcal{R}(C)$ it follows that S is compact; hence so is $P_{F^\perp}S$. Clearly $\mathcal{R}(A) = F + F_0$, $F_0 = \mathcal{R}(P_{F^\perp}S) \subseteq F^\perp$, and F_0 is nonclosed since otherwise $\mathcal{R}(A)$ would be closed. It remains to apply Theorem 2.47. □

From Theorem 2.41, it is clear that in the case when $\mathcal{R}(A)$ is not closed and C is compact, the existence of X such that $A + CX$ is a closed-range operator depends only on the ranges of A and C, so without loss of generality we can suppose that A and C are positive operators. In this case if $\dim_{\mathcal{R}(A)}(\mathcal{R}(A) \cap \mathcal{R}(C)) = \infty$, by Theorem 2.48 we can suppose that, in the case when such an operator X exists, we have that $A = B \oplus K$, where K is compact and B is invertible with infinite rank.

CHAPTER 3

Completions of operator matrices M_X

In this chapter we consider a particular case of completions of matrices given by

$$M_X = \begin{bmatrix} A & C \\ X & B \end{bmatrix} : \begin{bmatrix} \mathcal{H} \\ \mathcal{K} \end{bmatrix} \to \begin{bmatrix} \mathcal{H} \\ \mathcal{K} \end{bmatrix}.$$

More precisely, given operators $A \in \mathcal{B}(\mathcal{H})$, $B \in \mathcal{B}(\mathcal{K})$, and $C \in \mathcal{B}(\mathcal{K}, \mathcal{H})$, the question arises under what circumstances there exists an operator $X \in \mathcal{B}(\mathcal{H}, \mathcal{K})$ such that M_X belongs to a certain prescribed class of operators. Often completions to some sort of invertibility were considered and in these cases various types of spectra of M_X have been analyzed.

In Section 2.2 we indicated that the research over completions of M_X was first initiated by K. Takahashi in his seminal work [**160**]. The considerations of the present chapter are also rooted in [**160**], as the problem considered therein also covers the situations in which M_X is invertible for some $X \in \mathcal{B}(\mathcal{H}, \mathcal{K})$. The same problem, but using a different approach, was solved in [**170**]. Afterwards, the following questions were considered:

- When is M_X Fredholm for all $X \in \mathcal{B}(\mathcal{H}, \mathcal{K})$? (See [**35**].)
- When does there exist $X \in \mathcal{B}(\mathcal{H}, \mathcal{K})$ such that M_X is left (right) semi-Fredholm? (See [**92**].)
- When does there exist $X \in \mathcal{B}(\mathcal{H}, \mathcal{K})$ such that M_X is injective under the additional assumption that $\mathcal{R}(C)$ is closed? (See [**133**].)

In the case when $\mathcal{R}(C)$ is closed, we will describe the following sets:

$$\bigcap_{X \in \mathcal{B}(\mathcal{H},\mathcal{K})} \sigma_p(M_X), \qquad \bigcap_{X \in \mathcal{B}(\mathcal{H},\mathcal{K})} \sigma_r(M_X), \qquad \bigcap_{X \in \mathcal{B}(\mathcal{H},\mathcal{K})} \sigma_c(M_X).$$

In this chapter, we present two completely different approaches to the problem of completions of the operator matrix M_X to invertibility. We give some particular solutions of the problem of completion to injectivity, that is still an open one, to which partial solutions in the special cases when $\mathcal{R}(C)$ is closed were given in [**35**]. We also describe all of the operators $X \in \mathcal{B}(\mathcal{H}, \mathcal{K})$ for which M_X is injective in the particular cases considered. Next we will address the problem of the completion of the operator matrix M_X to an operator whose range is not dense that was considered in [**59**]. We will give a complete answer to it and then use the result to describe the intersection $\bigcap_{X \in \mathcal{B}(\mathcal{H},\mathcal{K})} \rho_p(M_X)$. Some results on the residual spectrum of the operator matrix M_X will be given as well. Additionally, we will pay attention to the completion problem to right invertibility, which was considered in [**90**]. Here we will present a much shorter proof than in the original source, with the intention of pointing out the connection between the completion problem for this type of operator matrices and the existence of an operator X for which $A + CX$ is of an appropriate type. In the last two sections we consider the completion of the operator matrix M_X to the right and left semi-Fredholm and Fredholm operators.

Given $A \in \mathcal{B}(\mathcal{H})$, $B \in \mathcal{B}(\mathcal{K})$, and $C \in \mathcal{B}(\mathcal{K},\mathcal{H})$, by $S_I(A,B,C)$ we denote the set of all $X \in \mathcal{B}(\mathcal{H},\mathcal{K})$ such that M_X is injective. Throughout this chapter all Hilbert spaces under consideration are assumed to be infinite dimensional and separable.

3.1. Completions of operator matrices M_X to invertible operator

In this section we present two completely different approaches to the problem of completions of the operator matrix

$$M_X = \begin{bmatrix} A & C \\ X & B \end{bmatrix} : \begin{bmatrix} \mathcal{H} \\ \mathcal{K} \end{bmatrix} \to \begin{bmatrix} \mathcal{H} \\ \mathcal{K} \end{bmatrix},$$

to the invertibility. One is given by K. Takahashi who in his paper [**160**] completely answered this question using the answer to the question of the existence of $X \in \mathcal{B}(\mathcal{H},\mathcal{K})$ such that $S + RX$ is invertible, where $S \in \mathcal{B}(\mathcal{H})$ and $R \in \mathcal{B}(\mathcal{K},\mathcal{H})$ are given operators. The other one is given by X. Yao [**170**] who used the spectral measure properties to give a necessary and sufficient conditions for the existence of $X \in \mathcal{B}(\mathcal{H},\mathcal{K})$ such that M_X is invertible.

First, we give K. Takahashi's characterization:

THEOREM 3.1. *Let $A \in \mathcal{B}(\mathcal{H})$, $B \in \mathcal{B}(\mathcal{K})$, and $C \in \mathcal{B}(\mathcal{K},\mathcal{H})$. Assume that the operators $\begin{bmatrix} A & C \end{bmatrix} : \mathcal{H} \oplus \mathcal{K} \to \mathcal{H}$ and $\begin{bmatrix} B^* & C^* \end{bmatrix} : \mathcal{K} \oplus \mathcal{H} \to \mathcal{K}$ are right invertible. The following are equivalent:*

(i) *There exists $X \in \mathcal{B}(\mathcal{H},\mathcal{K})$ such that M_X is invertible.*
(ii) *There exists $X \in \mathcal{B}(\mathcal{H},\mathcal{K})$ such that M_X is Weyl.*
(iii) *The operator M_0 $(= M(A,B,C;0))$ is Weyl or both $\mathcal{N}(A|C)$ and $\mathcal{N}(B^*|C^*)$ contain noncompact operators.*

PROOF. The implication (i)⇒(ii) is trivial.

(ii)⇒(iii): Let X be such that M_X is a Weyl operator. It must be $\mathcal{N}\left(\begin{bmatrix} A & C \end{bmatrix}\right) \neq \{0\}$. Indeed, if $\mathcal{N}\left(\begin{bmatrix} A & C \end{bmatrix}\right) = \{0\}$, then $\mathcal{N}(M_X) = \{0\}$ and therefore M_X is invertible because $\operatorname{ind}(M_X) = 0$. On the other hand, the condition $\mathcal{N}\left(\begin{bmatrix} A & C \end{bmatrix}\right) = \{0\}$ also implies $\mathcal{R}(M_X) \cap (\{0\} \oplus \mathcal{K}) = \{0\}$, which is a contradiction. Let $\mathcal{K}_1$ be a Hilbert space with $\dim \mathcal{K}_1 = \dim(\mathcal{N}\left(\begin{bmatrix} A & C \end{bmatrix}\right))$, and let $\begin{bmatrix} G \\ H \end{bmatrix} : \mathcal{K}_1 \to \mathcal{H} \oplus \mathcal{K}$ be a left invertible operator such that

$$\mathcal{N}\left(\begin{bmatrix} A & C \end{bmatrix}\right) = \mathcal{R}\left(\begin{bmatrix} G \\ H \end{bmatrix}\right). \tag{3.1}$$

Since $\begin{bmatrix} A & C \end{bmatrix}$ is right invertible, there exists a (left invertible) operator

$$\begin{bmatrix} Y \\ Z \end{bmatrix} : \mathcal{H} \to \mathcal{H} \oplus \mathcal{K}$$

such that $AY + CZ = I$. Let

$$W = \begin{bmatrix} Y & G \\ Z & H \end{bmatrix} : \mathcal{H} \oplus \mathcal{K}_1 \to \mathcal{H} \oplus \mathcal{K}. \tag{3.2}$$

Clearly W is invertible and satisfies

$$M_X W = \begin{bmatrix} I & 0 \\ XY + BZ & XG + BH \end{bmatrix}. \tag{3.3}$$

Hence $XG + BH$ is Weyl and $\dim \mathcal{K}_1 = \dim \mathcal{K}$.

Now suppose that $\mathcal{N}(A|C)$ consists of compact operators only. Then, since (3.1) implies $GV \in \mathcal{N}(A|C)$ where $V : \mathcal{K} \to \mathcal{K}_1$ is unitary, the operator G is compact and therefore it follows that BH is Weyl. This, together with the identity (3.3) with $X = 0$, shows that M_0 is Fredholm with $\operatorname{ind}(M_0) = 0$. In the case when $\mathcal{N}(B^*|C^*)$ consists of compact operators only, we can also show by applying the above argument to the adjoint M_X^* that M_0 is Fredholm with $\operatorname{ind}(M_0) = 0$.

(iii)⇒(i): First, suppose that M_0 is Weyl. Let $\begin{bmatrix} G \\ H \end{bmatrix}$ and W be as in the proof of the implication (ii)⇒(iii). Then the identity (3.3) with $X = 0$ shows that BH is Weyl and $\dim \mathcal{K}_1 = \dim \mathcal{K}$. Thus we may suppose $\mathcal{K}_1 = \mathcal{K}$. We also see that $\begin{bmatrix} G \\ BH \end{bmatrix}$ is left invertible, so that $((BH)^*, G^*)$ is right invertible. Indeed, since $AG = -CH$ by (3.1), we have

$$\begin{aligned} G^*G + (BH)^*(BH) &\geq \delta((AG)^*(AG) + (BH)^*(BH)) \\ &= \delta H^*(C^*C + B^*B)H, \end{aligned}$$

for some $\delta > 0$. Hence it follows from the left invertibility of the operators $\begin{bmatrix} C \\ B \end{bmatrix}$ and $\begin{bmatrix} G \\ H \end{bmatrix}$ that $\begin{bmatrix} G \\ BH \end{bmatrix}$ is left invertible. Now, applying Lemma 2.11 to the right invertible pair $\begin{bmatrix} (BH)^* & G^* \end{bmatrix}$, we obtain an operator $X \in \mathcal{B}(\mathcal{H}, \mathcal{K})$ such that $BH + XG$ is invertible. Then it follows from (3.3) that M_X is invertible.

Next suppose that $\mathcal{N}(A|C)$ and $\mathcal{N}(B^*|C^*)$ contain noncompact operators. Then $\dim(\mathcal{N}([\begin{matrix} A & C \end{matrix}])) = \infty = \dim \mathcal{K}$ and we can take a left invertible operator $\begin{bmatrix} G \\ H \end{bmatrix} : \mathcal{K} \to \mathcal{H} \oplus \mathcal{K}$ whose range is equal to $\mathcal{N}([\begin{matrix} A & C \end{matrix}])$. Note that G is not compact because $\mathcal{N}(A|C)$ contains a noncompact operator. As shown above, $\begin{bmatrix} (BH)^* & G^* \end{bmatrix}$ is right invertible. For a noncompact operator $D \in \mathcal{N}(B^*|C^*)$, we have $B^*D = C^*E$ for some $E \in \mathcal{B}(\mathcal{H})$ and so

$$(BH)^*D = H^*C^*E = -G^*A^*E,$$

which shows that $\mathcal{N}((BH)^*|G^*)$ contains the noncompact operator D. Therefore, by Theorem 2.10(2) there exists an operator X such that $BH + XG$ is invertible and, as in the first part, we can conclude that M_X is invertible. □

Now we present a different characterization of the existence of $X \in \mathcal{B}(\mathcal{H}, \mathcal{K})$ to be M_X invertible for a given $A \in \mathcal{B}(\mathcal{H})$, $B \in \mathcal{B}(\mathcal{K})$, and $C \in \mathcal{B}(\mathcal{K}, \mathcal{H})$ which is given by X. Yao [**170**]. The approach taken in [**170**] is completely different based on the spectral measure techniques and geometrical structure of operator.

THEOREM 3.2. *Let (A, B, C) be a given triple of operators, where $A \in \mathcal{B}(\mathcal{H})$, $B \in \mathcal{B}(\mathcal{K}, \mathcal{H})$, and $C \in \mathcal{B}(\mathcal{K}, \mathcal{H})$. Suppose that $\begin{bmatrix} A & C \end{bmatrix}$ and $\begin{bmatrix} B^* & C^* \end{bmatrix}$ are right invertible and $\mathcal{R}(C)$ is closed. Then there exists $X \in \mathcal{B}(\mathcal{H}, \mathcal{K})$ such that M_X is invertible if and only if* $\dim(\mathcal{K} \ominus B(\mathcal{N}(C))) = \dim(\mathcal{H} \ominus A^*(\mathcal{N}(C^*)))$.

PROOF. If $\mathcal{R}(C)$ is closed, then M_X has the operator matrix

$$M_X = \begin{bmatrix} A_1 & C_0 & 0 \\ A_2 & 0 & 0 \\ X & B_1 & B_2 \end{bmatrix} : \begin{bmatrix} \mathcal{H} \\ \mathcal{R}(C^*) \\ \mathcal{N}(C) \end{bmatrix} \to \begin{bmatrix} \mathcal{R}(C) \\ \mathcal{N}(C^*) \\ \mathcal{K} \end{bmatrix}.$$

It is clear that $C_0 \in \mathcal{B}(\mathcal{R}(C^*), \mathcal{R}(C))$ is an invertible operator. Observing that two operators

$$T_1 = \begin{bmatrix} I & 0 & 0 \\ 0 & I & 0 \\ -B_1 C_0^{-1} & 0 & I \end{bmatrix} \quad \text{and} \quad T_2 = \begin{bmatrix} I & 0 & 0 \\ -C_0^{-1} A_1 & I & 0 \\ 0 & 0 & I \end{bmatrix}$$

are invertible and

$$\begin{aligned} T_1 M_X T_2 &= \begin{bmatrix} I & 0 & 0 \\ 0 & I & 0 \\ -B_1 C_0^{-1} & 0 & I \end{bmatrix} \begin{bmatrix} A_1 & C_0 & 0 \\ A_2 & 0 & 0 \\ X & B_1 & B_2 \end{bmatrix} \begin{bmatrix} I & 0 & 0 \\ -C_0^{-1} A_1 & I & 0 \\ 0 & 0 & I \end{bmatrix} \\ &= \begin{bmatrix} 0 & C_0 & 0 \\ A_2 & 0 & 0 \\ -B_1 C_0^{-1} A_1 + X & 0 & B_2 \end{bmatrix}, \end{aligned}$$

there exists an operator $X \in \mathcal{B}(\mathcal{H}, \mathcal{K})$ such that M_X is invertible if and only if there exists an operator $Y \in \mathcal{B}(\mathcal{H}, \mathcal{K})$ such that the operator

$$\tilde{M}_Y = \begin{bmatrix} A_2 & 0 \\ Y & B_2 \end{bmatrix} : \begin{bmatrix} \mathcal{H} \\ \mathcal{N}(C) \end{bmatrix} \to \begin{bmatrix} \mathcal{N}(C^*) \\ \mathcal{K} \end{bmatrix}$$

is invertible. By the hypotheses that $\begin{bmatrix} A & C \end{bmatrix}$ and $\begin{bmatrix} B^* & C^* \end{bmatrix}$ are right invertible, we have that $\mathcal{R}(A_2) = \mathcal{N}(C^*)$ and $\mathcal{R}(B_2^*) = \mathcal{N}(C)$, where $B_2 \in \mathcal{B}(\mathcal{N}(C), \mathcal{K})$ is left invertible. By Theorem 1.8 there exists an operator $Y \in \mathcal{B}(\mathcal{H}, \mathcal{K})$ such that $\tilde{M}_Y$ is invertible if and only if $\dim(\mathcal{N}(A_2)) = \dim(\mathcal{N}(B_2^*))$. Moreover, observe that $\mathcal{N}(A_2) = \mathcal{H} \ominus A^*(\mathcal{N}(C^*))$, since $\mathcal{R}(A_2) = \mathcal{N}(C^*)$, and $\mathcal{N}(B_2^*) = \mathcal{K} \ominus B(\mathcal{N}(C))$; that is, there exists an operator $X \in \mathcal{B}(\mathcal{H}, \mathcal{K})$ such that M_X is invertible if and only if $\dim(\mathcal{K} \ominus B(\mathcal{N}(C))) = \dim(\mathcal{H} \ominus A^*(\mathcal{N}(C^*)))$. □

To consider the case that $\mathcal{R}(C)$ is not closed, we begin with some well-known facts: Namely, if $\mathcal{R}(C)$ is not closed, let $C = QU$ be the polar decomposition of C, where $Q = (CC^*)^{\frac{1}{2}}$. Let E_λ be a spectral measure of Q. For $\varepsilon > 0$, define Q_ϵ and C_ϵ by

$$Q_\epsilon = \int_\epsilon^{\|C\|} \lambda dE_\lambda \quad \text{and} \quad C_\epsilon = Q_\epsilon U. \tag{3.4}$$

Since Q_ϵ is left invertible, it is clear that $\mathcal{R}(C_\epsilon)$ is closed.

DEFINITION 3.3. Let (A, B, C) be a given triple of operators, where $A \in \mathcal{B}(\mathcal{H})$, $B \in \mathcal{B}(\mathcal{K}, \mathcal{H})$, and $C \in \mathcal{B}(\mathcal{K}, \mathcal{H})$. Suppose that $\begin{bmatrix} A & C \end{bmatrix}$ and $\begin{bmatrix} B^* & C^* \end{bmatrix}$ are right invertible. The right invertible index $d(A, B, C)$ of the triple (A, B, C) is defined by

$$d(A, B, C) = \max\{\gamma : AA^* + CC^* \geq \gamma I_{\mathcal{H}},\ B^*B + C^*C \geq \gamma I_{\mathcal{K}}\}.$$

Evidently, $\begin{bmatrix} A & C \end{bmatrix}$ and $\begin{bmatrix} B^* & C^* \end{bmatrix}$ are right invertible if and only if $d(A,B,C) > 0$. Also, $AA^* + CC^* \geq d(A,B,C)I_{\mathcal{H}}$ and $B^*B + C^*C \geq d(A,B,C)I_{\mathcal{K}}$.

Now, we consider our problem in the case when $\mathcal{R}(C)$ is not closed:

THEOREM 3.4. *Let (A,B,C) be a given triple of operators, where $A \in \mathcal{B}(\mathcal{H})$, $B \in \mathcal{B}(\mathcal{K},\mathcal{H})$, and $C \in \mathcal{B}(\mathcal{K},\mathcal{H})$. Suppose that $\begin{bmatrix} A & C \end{bmatrix}$ and $\begin{bmatrix} B^* & C^* \end{bmatrix}$ are right invertible and $\mathcal{R}(C)$ is not closed. Then there exists $X \in \mathcal{B}(\mathcal{H},\mathcal{K})$ such that M_X is invertible if and only if there exists $\epsilon > 0$ with $\frac{1}{4}d(A,B,C) > \epsilon^2$ such that $\dim(\mathcal{K} \ominus B(\mathcal{N}(C_\epsilon))) = \dim(\mathcal{H} \ominus A^*(\mathcal{N}(C_\epsilon^*)))$, where C_ϵ is defined by (3.4).*

PROOF. Suppose that there exists $X \in \mathcal{B}(\mathcal{H},\mathcal{K})$ such that M_X is invertible. Since the set of invertible operators is open there exists ϵ_0 such that for $\epsilon \in (0,\epsilon_0)$, operator $M_X(\epsilon) = \begin{bmatrix} A & C_\epsilon \\ X & B \end{bmatrix}$ is invertible since $\|M_X - M_X(\epsilon)\| \leq \epsilon$. Since $\mathcal{R}(C_\epsilon)$ is closed by Theorem 3.2 we have that $M_X(\epsilon)$ is invertible if and only if $\dim(\mathcal{K} \ominus B(\mathcal{N}(C_\epsilon))) = \dim(\mathcal{H} \ominus A^*(\mathcal{N}(C_\epsilon^*)))$. Evidently, there exists such $\epsilon > 0$ with $\frac{1}{4}d(A,B,C) > \epsilon^2$.

Conversely, suppose that there exists $\epsilon > 0$ with $\frac{1}{4}d(A,B,C) > \epsilon^2$ such that $\dim(\mathcal{K} \ominus B(\mathcal{N}(C_\epsilon))) = \dim(\mathcal{H} \ominus A^*(\mathcal{N}(C_\epsilon^*)))$ and let $C_\epsilon^0 = C - C_\epsilon = \int_0^{\epsilon-0} \lambda dE_\lambda U$. Define $M_Y(\epsilon)$ by

$$M_Y(\epsilon) = \begin{bmatrix} A_1(\epsilon) & C_\epsilon & 0 \\ A_2(\epsilon) & 0 & C_\epsilon^0 \\ Y & B_1(\epsilon) & B_1(\epsilon) \end{bmatrix} : \begin{bmatrix} \mathcal{H} \\ \mathcal{R}(C_\epsilon^*) \\ \mathcal{N}(C_\epsilon) \end{bmatrix} \to \begin{bmatrix} \mathcal{R}(C_\epsilon) \\ \mathcal{N}(C_\epsilon^*) \\ \mathcal{K} \end{bmatrix}.$$

Since $\begin{bmatrix} A & C \end{bmatrix}$ and $\begin{bmatrix} B^* & C^* \end{bmatrix}$ are right invertible, we have that

$$AA^* + CC^* = \begin{bmatrix} A_1(\epsilon)A_1^*(\epsilon) + C_\epsilon C_\epsilon^* & A_1(\epsilon)A_2^*(\epsilon) \\ A_2(\epsilon)A_1(\epsilon)^* & A_2(\epsilon)A_2^*(\epsilon) + C_\epsilon^{0*}C_\epsilon^0 \end{bmatrix} \geq d(A,B,C)I_{\mathcal{H}}$$

and

$$B^*B + C^*C = \begin{bmatrix} B_1^*(\epsilon)B_1(\epsilon) + C_\epsilon^* C_\epsilon & B_1^*(\epsilon)B_2(\epsilon) \\ B_2^*(\epsilon)B_1(\epsilon) & B_2^*(\epsilon)B_2(\epsilon) + C_\epsilon^0 C_\epsilon^{0*} \end{bmatrix} \geq d(A,B,C)I_{\mathcal{K}}.$$

Hence,

$$A_2(\epsilon)A_2^*(\epsilon) + C_\epsilon^{0*}C_\epsilon^0 \geq d(A,B,C)I_{\mathcal{N}(C_\epsilon^*)}$$

and

$$B_2^*(\epsilon)B_2(\epsilon) + C_\epsilon^0 C_\epsilon^{0*} \geq d(A,B,C)I_{\mathcal{N}(C_\epsilon)}.$$

Since $\|C_\epsilon^0\| \leq \epsilon$ and $\frac{1}{4}d(A,B,C) > \epsilon^2$, we have that

$$A_2(\epsilon)A_2^*(\epsilon) > \frac{3}{4}d(A,B,C)I_{\mathcal{N}(C_\epsilon^*)}$$

and

$$B_2^*(\epsilon)B_2(\epsilon) > \frac{3}{4}d(A,B,C)I_{\mathcal{N}(C_\epsilon)}$$

which implies that $A_2(\epsilon)$ and $B_2^*(\epsilon)$ are right invertible. Now, there exists $Y(\epsilon) \in \mathcal{B}(\mathcal{H} \ominus A^*(\mathcal{N}(C_\epsilon^*)), \mathcal{K} \ominus B(\mathcal{N}(C_\epsilon)))$ such that

$$\tilde{M}_Y(\epsilon) = \begin{bmatrix} A_2(\epsilon) & 0 \\ \tilde{Y}(\epsilon) & B_2(\epsilon) \end{bmatrix} : \begin{bmatrix} \mathcal{H} \\ \mathcal{N}(C_\epsilon) \end{bmatrix} \to \begin{bmatrix} \mathcal{N}(C_\epsilon^*) \\ \mathcal{K} \end{bmatrix}$$

is invertible and $\tilde{M}_Y(\epsilon)\tilde{M}_Y^*(\epsilon) > \frac{3}{4}d(A,B,C)I_{\mathcal{N}(C_\epsilon^*)\oplus\mathcal{K}}$, where $\tilde{Y}(\epsilon)$ is defined by

$$\tilde{Y}(\epsilon) = \begin{bmatrix} 0 & 0 \\ 0 & Y(\epsilon) \end{bmatrix} : \begin{bmatrix} A^*(\mathcal{N}(C_\epsilon^*)) \\ \mathcal{H} \ominus A^*(\mathcal{N}(C_\epsilon^*)) \end{bmatrix} \to \begin{bmatrix} B(\mathcal{N}(C_\epsilon)) \\ \mathcal{K} \ominus B(\mathcal{N}(C_\epsilon)) \end{bmatrix}.$$

Evidently, $\tilde{M}_Y(\epsilon)$ is invertible, so $M_Y(\epsilon)$ is invertible too. □

It is clear that Theorem 3.4 can be reformulated in the following way:

THEOREM 3.5. *Let (A, B, C) be a given triple of operators, where $A \in \mathcal{B}(\mathcal{H})$, $B \in \mathcal{B}(\mathcal{K},\mathcal{H})$, and $C \in \mathcal{B}(\mathcal{K},\mathcal{H})$. Suppose that $\begin{bmatrix} A & C \end{bmatrix}$ and $\begin{bmatrix} B^* & C^* \end{bmatrix}$ are right invertible and $\mathcal{R}(C)$ is not closed. Then there exists $X \in \mathcal{B}(\mathcal{H},\mathcal{K})$ such that M_X is invertible if and only if there exists $\epsilon > 0$ small enough such that $\dim(\mathcal{K} \ominus B(\mathcal{N}(C_\epsilon))) = \dim(\mathcal{H} \ominus A^*(\mathcal{N}(C_\epsilon^*)))$ for all $\epsilon \in (0, \epsilon_0)$.*

Now, we consider a special case:

THEOREM 3.6. *Let (A, B, C) be a given triple of operators, where $A \in \mathcal{B}(\mathcal{H})$, $B \in \mathcal{B}(\mathcal{K},\mathcal{H})$, and $C \in \mathcal{B}(\mathcal{K},\mathcal{H})$. Suppose that $\begin{bmatrix} A & C \end{bmatrix}$ and $\begin{bmatrix} B^* & C^* \end{bmatrix}$ are right invertible and 0 is an accumulation point of $\sigma_e(CC^*)$. Then there exists $X \in \mathcal{B}(\mathcal{H},\mathcal{K})$ such that M_X is invertible.*

PROOF. That 0 is an accumulation point of $\sigma_e(CC^*)$ implies that $\mathcal{R}(C)$ is not closed and that for small enough $\epsilon > 0$,

$$\dim(\mathcal{K} \ominus B(\mathcal{N}(C_\epsilon))) = \dim(\mathcal{H} \ominus A^*(\mathcal{N}(C_\epsilon^*))).$$

In fact $\begin{bmatrix} A & C \end{bmatrix}$ and $\begin{bmatrix} B^* & C^* \end{bmatrix}$ are right invertible implies that there exists $\epsilon_0 > 0$ such that for each $\epsilon \in (0, \epsilon_0)$ operators A^* and B are injective on $\mathcal{N}(C_\epsilon^*)$ and $\mathcal{N}(C_\epsilon)$, respectively. That 0 is an accumulation point of $\sigma_e(CC^*)$ implies that for each $\epsilon \in (0, \epsilon_0)$ there exists $\delta \in (0, \epsilon)$ such that

$$\begin{aligned} \dim(\mathcal{N}(C_\delta)) &= \dim(\mathcal{N}(C_\epsilon) \ominus \mathcal{N}(C_\delta)) = \dim(\mathcal{N}(C_\delta^*)) \\ &= \dim(\mathcal{N}(C_\epsilon^*) \ominus \mathcal{N}(C_\delta^*)) = \infty. \end{aligned}$$ □

THEOREM 3.7. *Let (A, B, C) be a given triple of operators, where $A \in \mathcal{B}(\mathcal{H})$, $B \in \mathcal{B}(\mathcal{K},\mathcal{H})$, and $C \in \mathcal{B}(\mathcal{K},\mathcal{H})$. Then:*

(1) *If $\mathcal{R}(C)$ is closed, then*

$$\bigcap_{X \in \mathcal{B}(\mathcal{H},\mathcal{K})} \sigma(M_X) = \Psi(A,B,C) \cup \Phi(A,B,C).$$

(2) *If $\mathcal{R}(C)$ is not closed, then $\bigcap_{X\in\mathcal{B}(\mathcal{H},\mathcal{K})} \sigma(M_X) = \Psi(A,B,C) \bigcup \chi(A,B,C)$, where*

$$\Psi(A,B,C) = \Big\{\lambda\colon \mathcal{R}(A-\lambda) + \mathcal{R}(C) \neq \mathcal{H} \text{ or } \mathcal{R}(B^* - \overline{\lambda}) + \mathcal{R}(C^*) \neq \mathcal{K}\Big\},$$

$$\begin{aligned} \Phi(A,B,C) = \Big\{\lambda\colon &\mathcal{R}(A-\lambda) + \mathcal{R}(C) = \mathcal{H},\ \mathcal{R}(B^* - \overline{\lambda}) + \mathcal{R}(C^*) = \mathcal{K}, \\ &\dim(\mathcal{K} \ominus (B-\lambda)(\mathcal{N}(C))) \neq \dim(\mathcal{H} \ominus (A-\lambda)^*(\mathcal{N}(C^*)))\Big\}, \end{aligned}$$

$$\chi(A,B,C) = \Big\{\lambda\colon \text{there exist infinitely many } \epsilon_n > 0 \text{ such that } \epsilon_n \to 0 \text{ and } \lambda \in \Phi(A,B,C_{\epsilon_n})\Big\}.$$

3.2. Completions of M_X to injective and nondense range operator

After K. Takahashi's result in [**160**], there have been a number of follow-ups to his work, treating completions problems for various forms of operator matrices (see [**13, 30, 35, 67, 90, 92, 94, 95, 111, 133, 155, 160, 162, 163, 170**]). In those few papers [**35, 72, 90, 92, 133**] to appear after K. Takahashi's result in [**160**] that treat the problem of completing M_X to a given type of operators, the focus of investigation was, among other things, as follows:

- the question of when M_X is Fredholm for all $X \in \mathcal{B}(\mathcal{H}, \mathcal{K})$, i.e., a description of $\bigcup_{X\in\mathcal{B}(\mathcal{H},\mathcal{K})} \sigma_e(M_X)$,
- the question of existence of an operator $X \in \mathcal{B}(\mathcal{H}, \mathcal{K})$ such that M_X is left (right) semi-Fredholm, i.e., a description of $\bigcap_{X\in\mathcal{B}(\mathcal{H},\mathcal{K})} \sigma_{le}(M_X)$ and $\bigcap_{X\in\mathcal{B}(\mathcal{H},\mathcal{K})} \sigma_{re}(M_X)$,
- the question of existence of an operator $X \in \mathcal{B}(\mathcal{H}, \mathcal{K})$ such that M_X is injective under the additional assumption that $\mathcal{R}(C)$ is closed, i.e., a description of $\bigcap_{X\in\mathcal{B}(\mathcal{H},\mathcal{K})} \sigma_p(M_X)$; also, in the case when $\mathcal{R}(C)$ is closed the following sets are described: $\bigcap_{X\in\mathcal{B}(\mathcal{H},\mathcal{K})} \sigma_{rs}(M_X)$, $\bigcap_{X\in\mathcal{B}(\mathcal{H},\mathcal{K})} \sigma_c(M_X)$.

Thus, whether M_X can be completed to injectivity is still an open problem. In the case when $\mathcal{R}(C)$ is closed the problem has been completely solved by A. Chen et al. in [**35**]. More precisely, from Theorem 4 of [**35**], which is one of the main results of that paper, we have the following important corollary.

THEOREM 3.8. *Suppose that $A \in \mathcal{B}(\mathcal{H})$, $B \in \mathcal{B}(\mathcal{K})$ and assume $C \in \mathcal{B}(\mathcal{K}, \mathcal{H})$ is such that $\mathcal{R}(C)$ is closed. Then M_X is injective for some $X \in \mathcal{B}(\mathcal{H}, \mathcal{K})$ if and only if B_1 is injective and either $\mathcal{R}(B_1)$ is closed or $\alpha(A_1) \le d(B_1)$, where $A_1 = P^{\mathtt{rst}}_{\mathcal{R}(C)^\perp} A \in \mathcal{B}(\mathcal{H}, \mathcal{R}(C)^\perp)$ and $B_1 = B\,|_{\mathcal{N}(C)} \in \mathcal{B}(\mathcal{N}(C), \mathcal{K})$.*

However, in the case when $\mathcal{R}(C)$ is not closed, classical algebraic techniques alone do not seem to be sufficient to solve the problem, which is probably one of the reasons why it is still open. In this section we first use standard algebraic methods to reformulate the problem, under the assumption that $\mathcal{R}(B)$ is closed (the case when $\mathcal{R}(B)$ is not closed is dealt with in a rather straightforward manner as we will show below), so as to give it a form to which the results of the previous section can be applied. This leads us to necessary conditions for it to have a positive answer. Then, using a different approach, we give two conditions that guarantee the existence of some X for which M_X is injective. Hopefully the progress in solving this problem that we will thus have made will give new insight into it and be of further help to finally arrive at its complete solution.

As already noted, in the case when $\mathcal{R}(B)$ is not closed the problem is easily solved by the following lemma.

LEMMA 3.9. *Let $A \in \mathcal{B}(\mathcal{H})$, $B \in \mathcal{B}(\mathcal{K})$, and $C \in \mathcal{B}(\mathcal{K}, \mathcal{H})$ be given operators, such that $\mathcal{R}(B)$ is not closed. There exists $X \in \mathcal{B}(\mathcal{H}, \mathcal{K})$ such that M_X is injective if and only if $\mathcal{N}(C) \cap \mathcal{N}(B) = \{0\}$.*

PROOF. It is immediate that for the operator M_X to be injective for some $X \in \mathcal{B}(\mathcal{H}, \mathcal{K})$ it is necessary that $\mathcal{N}(C) \cap \mathcal{N}(B) = \{0\}$ holds. If $\mathcal{R}(B)$ is not closed, then the assumption $\mathcal{N}(C) \cap \mathcal{N}(B) = \{0\}$ alone is sufficient for M_X to be injective for some X. Indeed, in this case there is a closed subspace $F \subseteq \mathcal{K}$ of infinite dimension such that $F \cap \mathcal{R}(B) = \{0\}$, so we can fix an injective operator $X \in \mathcal{B}(\mathcal{H}, \mathcal{K})$ such that $\mathcal{R}(X) \subseteq F$. To see that M_X is injective, suppose $Ax + Cy = 0$, $Xx + By = 0$. From $\mathcal{R}(X) \cap \mathcal{R}(B) = \{0\}$ it follows that $Xx = 0$, $By = 0$. Injectivity of X now yields $x = 0$ and thus $y \in \mathcal{N}(B) \cap \mathcal{N}(C)$; i.e., $y = 0$. $\square$

Thus assume $\mathcal{R}(B)$ is closed. To state the problem in a form suitable for application of the results of the previous section, we begin by representing M_X as follows:

$$M_X = \begin{bmatrix} C_2 & C_1 & A \\ 0 & B_1 & X_1 \\ 0 & 0 & X_2 \end{bmatrix} : \begin{bmatrix} \mathcal{N}(B) \\ \mathcal{N}(B)^\perp \\ \mathcal{H} \end{bmatrix} \to \begin{bmatrix} \mathcal{H} \\ \mathcal{R}(B) \\ \mathcal{R}(B)^\perp \end{bmatrix}. \tag{3.5}$$

We can easily check that the problem of completing the operator matrix M_X to injectivity, under the assumption $\mathcal{N}(C) \cap \mathcal{N}(B) = \{0\}$, can be reduced to the equivalent problem of the existence of $X_1 \in \mathcal{B}(\mathcal{H}, \mathcal{R}(B))$ and $X_2 \in \mathcal{B}(\mathcal{H}, \mathcal{R}(B)^\perp)$ such that

$$M_{X_1,X_2} = \begin{bmatrix} C_2 & A - C_1 B_1^{-1} X_1 \\ 0 & X_2 \end{bmatrix} : \begin{bmatrix} \mathcal{N}(B) \\ \mathcal{H} \end{bmatrix} \to \begin{bmatrix} \mathcal{H} \\ \mathcal{R}(B)^\perp \end{bmatrix} \tag{3.6}$$

is injective. M_{X_1,X_2} is injective if and only if

$$\begin{bmatrix} C_2 & (A - C_1 B_1^{-1} X_1)|_{\mathcal{N}(X_2)} \end{bmatrix} : \begin{bmatrix} \mathcal{N}(B) \\ \mathcal{N}(X_2) \end{bmatrix} \to \mathcal{H}$$

is injective. The above discussion thus shows that M_X is injective for some $X \in \mathcal{B}(\mathcal{H}, \mathcal{K})$ if and only if $\mathcal{N}(C) \cap \mathcal{N}(B) = \{0\}$ and there exists a subspace $\mathcal{H}_0$ of $\mathcal{H}$ such that $\operatorname{codim}(\mathcal{H}_0) \le \dim(\mathcal{R}(B)^\perp)$ and an operator $Y \in \mathcal{B}(\mathcal{N}(B)^\perp, \mathcal{H}_0)$ such that

$$M_{Y,\mathcal{H}_0} = \begin{bmatrix} C_2 & A_0 - C_1 Y \end{bmatrix} : \begin{bmatrix} \mathcal{N}(B) \\ \mathcal{H}_0 \end{bmatrix} \to \mathcal{H} \tag{3.7}$$

is injective, where $A_0 = A|_{\mathcal{H}_0}$. Since

$$\begin{bmatrix} C_2 & A_0 - C_1 Y \end{bmatrix} = \begin{bmatrix} C_2 & A_0 \end{bmatrix} + \begin{bmatrix} 0 & C_1 \end{bmatrix} \begin{bmatrix} 0 & 0 \\ 0 & -Y \end{bmatrix},$$

we can now directly apply Theorem 2.18 to deduce the following necessary condition for the considered completion problem to have a positive answer:

THEOREM 3.10. *Let $A \in \mathcal{B}(\mathcal{H})$, $B \in \mathcal{B}(\mathcal{K})$, and $C \in \mathcal{B}(\mathcal{K}, \mathcal{H})$ be given operators, such that $\mathcal{R}(B)$ is closed. There exists $X \in \mathcal{B}(\mathcal{H}, \mathcal{K})$ such that M_X is injective only if $\mathcal{N}(C) \cap \mathcal{N}(B) = \{0\}$ and for some subspace $\mathcal{H}_0 \subseteq \mathcal{H}$ such that* $\operatorname{codim}(\mathcal{H}_0) \le \dim(\mathcal{R}(B)^\perp)$ *one of the following conditions is satisfied:*

(1*) $\operatorname{codim}_{\mathcal{R}(C_1)} [(\mathcal{R}(A_0) + \mathcal{R}(C_2)) \cap \mathcal{R}(C_1)] \ge \dim(\mathcal{N}([\, C_2 \;\; A_0 \,]))$.

(2*) $\mathcal{N}([\, C_2 \;\; A_0 \,])^\perp \cap (C_2\, A_0)^{-1} [(\mathcal{R}(A_0) + \mathcal{R}(C_2)) \cap \mathcal{R}(C_1)]$ *contains a closed infinite-dimensional subspace,*

where $C_1 = C\,|_{\mathcal{N}(B)^\perp}$, $C_2 = C\,|_{\mathcal{N}(B)}$, $A_0 = A|_{\mathcal{H}_0}$.

The next theorem shows in particular that if B is right invertible, the condition (1$*$) is also sufficient, and if B is not right invertible, it gives a special condition under which the existence of $X \in \mathcal{B}(\mathcal{H}, \mathcal{K})$ such that M_X is injective is guaranteed:

THEOREM 3.11. *Let $A \in \mathcal{B}(\mathcal{H})$, $B \in \mathcal{B}(\mathcal{K})$, and $C \in \mathcal{B}(\mathcal{K}, \mathcal{H})$ be given operators and let $C_1 = C\,|_{\mathcal{N}(B)^\perp}$, $C_2 = C\,|_{\mathcal{N}(B)}$. If $\mathcal{N}(B) \cap \mathcal{N}(C) = \{0\}$ and*

$$\operatorname{codim}_{\mathcal{R}(C_1)} (\mathcal{R}(C_1) \cap (\mathcal{R}(A) + \mathcal{R}(C_2))) \ge \dim(\mathcal{N}([\, C_2 \;\; A_0 \,])),$$

then there exists $X \in \mathcal{B}(\mathcal{H}, \mathcal{K})$ such that M_X is injective.

PROOF. We assume that $\mathcal{R}(B)$ is closed, for otherwise M_X can always be completed to injectivity, as already noted. In this case M_X can be represented by (3.5). In order to prove the existence of $X \in \mathcal{B}(\mathcal{H}, \mathcal{K})$ such that M_X is injective, we first recall the discussion preceding Theorem 3.10 and observe that the operator C_2 must be injective. We thus need to prove that there exists a subspace $\mathcal{H}_0$ of $\mathcal{H}$ such that $\operatorname{codim}(\mathcal{H}_0) \leq \dim(\mathcal{R}(B)^\perp)$ and an operator $T \in \mathcal{B}(\mathcal{H}_0, \mathcal{H})$ such that $\mathcal{R}(T) \subseteq \mathcal{R}(C_1)$, A_0+T is injective, and $\mathcal{R}(A_0+T) \cap \mathcal{R}(C_2) = \{0\}$, where $A_0 = A|_{\mathcal{H}_0}$. Actually, we will show that one can take $\mathcal{H}_0 = \mathcal{H}$.

Put $M = \mathcal{R}(C_1) \cap (\mathcal{R}(A) + \mathcal{R}(C_2))$. We distinguish two cases.

Case 1: $\operatorname{codim}_{\mathcal{R}(C_1)} M = \infty$. By (i) of Lemma 2.1 there is an infinite-dimensional operator range $L \subseteq \mathcal{R}(C_1)$ such that $M \cap L = \{0\}$. By (ii) of the same lemma we can fix an injective $T \in \mathcal{B}(\mathcal{H}_1, \mathcal{H}_2)$ such that $\mathcal{R}(T) \subseteq L$. As $\mathcal{R}(T) \subseteq \mathcal{R}(C_1)$ is satisfied, we only need to check that $(A+T)x \in \mathcal{R}(C_2)$ implies $x = 0$, for every $x \in \mathcal{H}$. But if $Ax + Tx \in \mathcal{R}(C_2)$, then $Tx \in (\mathcal{R}(A) + \mathcal{R}(C_2)) \cap L = M \cap L = \{0\}$, so $Tx = 0$ and $x = 0$, since T is one-to-one.

Case 2: $\infty > \operatorname{codim}_{\mathcal{R}(C_1)} M \geq \dim(\mathcal{N}([\ C_2 \quad A_0\ ]))$. Let $F \subseteq \mathcal{R}(C_1)$ be any subspace such that $M \oplus F = \mathcal{R}(C_1)$. If $\pi \in \mathcal{B}(\mathcal{K} \oplus \mathcal{H}, \mathcal{H})$ is the orthogonal projector onto $\mathcal{H}$, then, letting $P = \pi\left[\mathcal{N}([\ C_2 \quad A_0\ ])\right]$, we have

$$\dim P \leq \dim(\mathcal{N}([\ C_2 \quad A_0\ ])) \leq \dim F < \infty \tag{3.8}$$

by our assumption. In particular, P is a closed subspace so $\mathcal{H} = P \oplus P^\perp$. Using (3.8) we can pick an injective $T_0 \in \mathcal{B}(P, F)$ and then consider the operator $T \in \mathcal{B}(\mathcal{H})$ defined by

$$T = \begin{bmatrix} 0 & T_0 \\ 0 & 0 \end{bmatrix} : \begin{bmatrix} P^\perp \\ P \end{bmatrix} \to \begin{bmatrix} F \\ F^\perp \end{bmatrix}. \tag{3.9}$$

Clearly $\mathcal{R}(T) = \mathcal{R}(T_0) \subseteq F \subseteq \mathcal{R}(C)$. Let us show that $(A+T)x \in \mathcal{R}(C_2)$ implies $x = 0$, for every $x \in \mathcal{H}$. Suppose $a \in P^\perp, b \in P$ are such that $(A+T)(a+b) \in \mathcal{R}(C_2)$. Since $T(a+b) = T_0 b$ this means that $T_0 b \in M \cap F = \{0\}$, so $T_0 b = 0$ and consequently $b = 0$, by injectivity of T_0. It now follows that $Aa \in \mathcal{R}(C_2)$ so, noticing that $P = A^{-1}[\mathcal{R}(C_2)]$, we conclude that $a \in P \cap P^\perp$; i.e., $a = 0$. □

Yet another sufficient condition is provided by the following theorem.

THEOREM 3.12. *Let $A \in \mathcal{B}(\mathcal{H})$, $B \in \mathcal{B}(\mathcal{K})$, and $C \in \mathcal{B}(\mathcal{K}, \mathcal{H})$ be given operators and let $C_1 = C|_{\mathcal{N}(B)^\perp}$, $C_2 = C|_{\mathcal{N}(B)}$. If $\mathcal{N}(B) \cap \mathcal{N}(C) = \{0\}$, $\dim(\mathcal{R}(A) \cap \mathcal{R}(C_2)) < \infty$, and there exists a closed infinite-dimensional subspace $F \subseteq \mathcal{N}(A)^\perp \cap A^{-1}[\mathcal{R}(C_1)]$, then there exists $X \in \mathcal{B}(\mathcal{H}, \mathcal{K})$ such that M_X is injective.*

PROOF. We assume that $\mathcal{R}(B)$ is closed and that M_X is consequently given by (3.5). As in the proof of Theorem 3.11, it suffices to establish the existence of an operator $T \in \mathcal{B}(\mathcal{H})$ such that $\mathcal{R}(T) \subseteq \mathcal{R}(C_1)$, $A+T$ is injective, and $\mathcal{R}(A+T) \cap \mathcal{R}(C_2) = \{0\}$.

Set $M = \mathcal{R}(C_1) \cap (\mathcal{R}(A) + \mathcal{R}(C_2))$. Let $Q \subseteq F$ be a closed infinite-dimensional subspace such that $\operatorname{codim}_Q F = \infty$, and by $Q_0 = Q^\perp \cap F$ denote the orthogonal complement of Q in F. Put $F_0 = F^\perp \cap \mathcal{N}(A)^\perp$ and $P = \mathcal{N}(A)^\perp \cap A^{-1}[\mathcal{R}(C_2)]$; clearly $\dim P < \infty$. Write $I_{\mathcal{N}(A)^\perp} = \pi_1 + \pi_2 + \pi_3$, where $\pi_1, \pi_2, \pi_3 \in \mathcal{B}\left(\mathcal{N}(A)^\perp\right)$, and $\mathcal{R}(\pi_1) \subseteq F_0$, $\mathcal{R}(\pi_2) \subseteq Q_0$, and $\mathcal{R}(\pi_3) \subseteq Q$. The subspaces $P_1 = \pi_2[P]$ and $P_2 = \pi_3[P]$ being finite dimensional, we can find closed infinite-dimensional subspaces $K \subseteq Q_0$ and $L \subseteq Q$ such that $K \cap P_1 = \{0\}$ and $L \cap P_2 = \{0\}$.

Let K_1 and K_2 be any two closed infinite-dimensional subspaces of K such that $K_1 \cap K_2 = \{0\}$. Similarly, let L_1 and L_2 be any two closed infinite-dimensional subspaces of L such that $L_1 \cap L_2 = \{0\}$.

Let $W_1 \in \mathcal{B}(\mathcal{N}(A), Q_0)$ and $W_2 \in \mathcal{B}(F_0, Q_0)$ be any two injective operators such that $\mathcal{R}(W_1) \subseteq K_1$ and $\mathcal{R}(W_2) \subseteq K_2$. Let $W_3 \in \mathcal{B}(Q_0, Q)$ be any injective operator with $\mathcal{R}(W_3) \subseteq L_1$. We can find $W_4 \in \mathcal{B}(Q)$ such that $\mathcal{R}(I_Q + W_4) \subseteq L_2$ and $I_Q + W_4$ is one-to-one. Define $W \in \mathcal{B}(\mathcal{H})$ by

$$W = \begin{bmatrix} 0 & 0 & 0 & 0 \\ 0 & 0 & 0 & 0 \\ W_1 & W_2 & 0 & 0 \\ 0 & 0 & W_3 & W_4 \end{bmatrix} : \begin{bmatrix} \mathcal{N}(A) \\ F_0 \\ Q_0 \\ Q \end{bmatrix} \to \begin{bmatrix} \mathcal{N}(A) \\ F_0 \\ Q_0 \\ Q \end{bmatrix}.$$

We claim that for every $x \in \mathcal{H}$, if $x \neq 0$, then $(I+W)x \notin A^{-1}[\mathcal{R}(C_2)]$. Let $x = a+b+c+d$, where $a \in \mathcal{N}(A)$, $b \in F_0$, $c \in Q_0$, $d \in Q$ and suppose that $x + Wx \in A^{-1}[\mathcal{R}(C_2)]$. Since $A^{-1}[\mathcal{R}(C_2)] = \mathcal{N}(A) \oplus P$ and $P \subseteq \mathcal{N}(A)^{\perp}$, from

$$x + Wx = a+b+c+d+Wa+Wb+Wc+Wd$$

and

$$b+c+d+Wa+Wb+Wc+Wd \in \mathcal{N}(A)^{\perp},$$

we conclude that actually $p = b+c+d+Wa+Wb+Wc+Wd \in P$. From the definition of the operator W we see that $c+Wa+Wb \in Q_0$ and $Wc+d+Wd \in Q$ and, since $b \in F_0$, that consequently $c+Wa+Wb = \pi_2(p) \in P_2$ and $Wc+d+Wd = \pi_3(p) \in P_3$. On the other hand, again from the definition of W, we have that $Wc+d+Wd \in L$ from which it follows that $Wc+d+Wd = 0$. Now $Wc = W_3c \in L_1$ and $(I+W)d = (I_Q+W_4)\,d \in L_2$, so $W_3c = 0$ and $(I_Q+W_4)\,d = 0$. But both W_3 and $(I_Q+W_4)\,d$ are injective so $c = d = 0$. We thus conclude that $Wa+Wb \in P_2$, but by construction of W we have $Wa+Wb \in K$, so $Wa+Wb = 0$. From $Wa = W_1a \in K_1$ and $Wb = W_2b \in K_2$ we further deduce that $W_1a = 0$ and $W_2b = 0$, whence finally $a = b = 0$, since W_1 and W_2 are injective.

We have thus shown that if $x \in \mathcal{H}$ is a nonzero vector, then $(A+AW)x \notin \mathcal{R}(C_2)$. Letting $T = AW$, this means that $\mathcal{R}(A+T) \cap \mathcal{R}(C_2) = \{0\}$ and that $A+T$ is one-to-one. Clearly $\mathcal{R}(W) \subseteq F \subseteq A^{-1}[\mathcal{R}(C)]$ so $\mathcal{R}(T) \subseteq \mathcal{R}(C_1)$. □

In order to make a comparison between our results and the result of A. Chen et al. (Theorem 3.8 in this book or see [**35**]), as well as to illustrate the interplay between the various conditions appearing therein, we give the following example of a particular set of circumstances in which from one of the conditions in Theorem 3.12 we can deduce the conditions in Theorem 3.8.

EXAMPLE 3.13. Suppose that $A \in \mathcal{B}(\mathcal{H})$, $B \in \mathcal{B}(\mathcal{K})$, and $C \in \mathcal{B}(\mathcal{K}, \mathcal{H})$ are such that $\mathcal{N}(B) \subseteq \mathcal{N}(C)^{\perp}$. Assume that the subspace $S = \mathcal{N}(A)^{\perp} \cap A^{-1}[\mathcal{R}(C_1)]$ is infinite dimensional (which is an even weaker condition than the corresponding one appearing in Theorem 3.12). We show that if $\mathcal{R}(B_1)$ is closed, then $d(B_1) = \infty$, ensuring that $\alpha(A_1) \leq d(B_1)$, where A_1 and B_1 are as in Theorem 3.8.

Let $V \subseteq \mathcal{N}(C)^{\perp}$ be the orthogonal complement of $\mathcal{N}(B)$ in $\mathcal{N}(C)^{\perp}$ and let C_1 be as in Theorem 3.12. Under our assumptions we have $\mathcal{N}(B)^{\perp} = \mathcal{N}(C) \oplus V$. Here $\dim(S) = \infty$ and $S \subseteq \mathcal{N}(A)^{\perp}$ imply $\dim(A[S]) = \infty$ and consequently

$\dim(\mathcal{R}(C_1)) = \infty$. Since $C[V] = \mathcal{R}(C_1)$, it follows that $\dim(V) = \infty$. But then it must also be that $\dim(B[V]) = \infty$. From $B[V] \cap \mathcal{R}(B_1) = \{0\}$ it now follows that $\operatorname{codim}(\mathcal{R}(B_1)) = \infty$. Thus either $\mathcal{R}(B_1)$ is not closed or $d(B_1) = \infty$. Thus all the conditions from Theorem 3.8 are met.

Closely related to the completion problem considered above is the one of completing $\begin{bmatrix} A & C \\ ? & B \end{bmatrix}$ to a left invertible operator. This problem is completely solved by G. Hai et al. in Theorem 2.7 of [**90**]. To state the result, for $A \in \mathcal{B}(\mathcal{H})$ and $C \in \mathcal{B}(\mathcal{K}, \mathcal{H})$ let us recall that

$$\mathcal{N}(A|C) = \{G \in \mathcal{B}(\mathcal{K}, \mathcal{H}) \,|\, \mathcal{R}(AG) \subseteq \mathcal{R}(C)\}.$$

One then has the following:

THEOREM 3.14 ([**90**]). *Let $A \in \mathcal{B}(\mathcal{H})$, $B \in \mathcal{B}(\mathcal{K})$, and $C \in \mathcal{B}(\mathcal{K}, \mathcal{H})$ be given operators. Then M_X if a left invertible operator for some $X \in \mathcal{B}(\mathcal{H}, \mathcal{K})$ if and only if $\mathcal{R}(B^*) + \mathcal{R}(C^*) = \mathcal{K}$ and one of the following conditions holds:*

(i) *$\mathcal{N}(B^*|C^*)$ contains a noncompact operator.*
(ii) *$M_0 = M(A, B, C; 0)$ is a left semi-Fredholm operator and*

$$\alpha(M_0) \leq d(B) + \dim\left(\mathcal{R}(B^*) \cap \mathcal{R}(C^*|_{\mathcal{N}(A^*)})\right).$$

Given $A \in \mathcal{B}(\mathcal{H})$ and $C \in \mathcal{B}(\mathcal{K}, \mathcal{H})$, it can easily be seen that the set $\mathcal{N}(A|C)$ contains a noncompact operator if and only if there is a closed infinite-dimensional subspace $F \subseteq \mathcal{H}$ such that $A[F] \subseteq \mathcal{R}(C)$. Indeed, if $G \in \mathcal{N}(A|C)$ is noncompact, then there is a closed infinite-dimensional subspace $F \subseteq \mathcal{R}(G) \subseteq \mathcal{H}$; $A[F] \subseteq \mathcal{R}(C)$ then follows from $\mathcal{R}(AG) \subseteq \mathcal{R}(C)$. If on the other hand, $F \subseteq \mathcal{H}$ is a closed infinite-dimensional subspace such that $A[F] \subseteq \mathcal{R}(C)$, then there we can take some $A \in \mathcal{B}(\mathcal{H})$ with $\mathcal{R}(A) = F$ (both $\mathcal{H}$ and F being separable infinite-dimensional Hilbert spaces); clearly, $\mathcal{R}(AG) \subseteq \mathcal{R}(C)$ is satisfied.

In view of the preceding observation, the reader is invited to notice an interesting common feature appearing in both Theorem 3.12 and Theorem 3.14: the condition that $\mathcal{N}(A)^{\perp} \cap A^{-1}[\mathcal{R}(C_1)]$ contains a closed infinite-dimensional subspace might as well have been replaced by the equivalent condition that $\mathcal{N}(A_0 \mid C_1)$ contains a noncompact operator, where $A_0 = A\,|_{\mathcal{N}(A)^{\perp}}$.

Below we present two examples of nonempty classes of triples (A, B, C) of operators for which there are injective completions of $\begin{bmatrix} A & C \\ ? & B \end{bmatrix}$, but there are no injective completions having closed range (left invertible completion). We will apply both of the results we have obtained, as well as those of G. Hai et al. (Theorem 2.7 in [**90**]). In the first example, triples (A, B, C) are exhibited such that existence of injective completions M_X follows from Theorem 3.12, but for which no left invertible completions are possible, as will be seen by an application of Theorem 3.14.

EXAMPLE 3.15. Write $\mathcal{H} = M \oplus M^{\perp}$, where M is a closed subspace of infinite dimension and $M \neq \mathcal{H}$. Let $C \in \mathcal{B}(\mathcal{K}, \mathcal{H})$ be a compact operator with infinite-dimensional range (it is well known that such operators exist; see, e.g., [**5**]). Note that $\mathcal{R}(C)$ is not closed. Since $\mathcal{N}(C)^{\perp}$ is closed and infinite dimensional, we can fix an injective $T \in \mathcal{B}(M, \mathcal{K})$ such that $\mathcal{R}(T) = \mathcal{N}(C)^{\perp}$. If $P_M \in \mathcal{B}(\mathcal{H}, M)$ is the orthogonal projector onto the subspace M, set $A = CTP_M \in \mathcal{B}(\mathcal{H})$. It is obvious that then we must have $\mathcal{N}(A) = M^\perp \neq \{0\}$, $\mathcal{N}(A)^\perp = M$, and $\mathcal{R}(A) = \mathcal{R}(C)$. Let $B \in \mathcal{B}(\mathcal{K})$ be any invertible operator. Set $C_1 = C\,|_{\mathcal{N}(B)^\perp}$, $C_2 = C\,|_{\mathcal{N}(B)}$. We have $\mathcal{N}(B) = \{0\}$, so $C_2 = 0$ and $C_1 = C$. Taking $F = \mathcal{N}(A)^\perp$ in Theorem 3.12, we see that for the triple (A, B, C) all the conditions of the theorem are met. Thus there are injective completions of $\begin{bmatrix} A & C \\ ? & B \end{bmatrix}$. However, none of those can have closed range, which can be seen from Theorem 3.14, as follows:

(i) of Theorem 3.14 cannot hold. For, suppose $\mathcal{N}(B^*|C^*)$ contains a non-compact operator. Then, by the remark above, there is a closed infinite-dimensional subspace $F \subseteq \mathcal{K}$ such that $B^*[F] \subseteq \mathcal{R}(C^*)$. But B^* is invertible, so $B^*[F]$ is a closed infinite-dimensional subspace of $\mathcal{R}(C^*)$, implying that C^*, and hence C as well, is not compact, contradicting our choice of the operator C.

(ii) of Theorem 3.14 cannot hold either. This is because the range of the operator $\begin{bmatrix} A & C \\ 0 & B \end{bmatrix}$ is not closed: indeed, since $\mathcal{R}(A)$ is not closed, $\mathcal{R}(B) = \mathcal{K}$, and $\mathcal{N}(B) = \{0\}$, this follows directly from Theorem 2.5 of [**67**].

Note that, in the previous example, Theorem 3.11 could not have been used to ensure existence of injective completions whereas

$$\begin{aligned}\operatorname{codim}_{\mathcal{R}(C_1)}(\mathcal{R}(C_1) \cap (\mathcal{R}(A) + \mathcal{R}(C_2))) &= \operatorname{codim}_{\mathcal{R}(C)}(\mathcal{R}(C)) = 0 \\ &< \alpha(A) = \dim(\mathcal{N}(\begin{bmatrix} C_2 & A \end{bmatrix})).\end{aligned}$$

In the second example, triples (A, B, C) are exhibited such that existence of injective completions M_X follows from Theorem 3.11, but for which no left invertible completions are possible, as will be seen by an application of Theorem 3.14.

EXAMPLE 3.16. Write $\mathcal{H} = L_1 \oplus L_2$, where L_1 and L_2 are closed subspaces of infinite dimension. Let $C \in \mathcal{B}(\mathcal{K}, \mathcal{H})$ be any compact operator with infinite-dimesional range such that $\mathcal{R}(C) \subseteq L_2$. Again, $\mathcal{R}(C)$ is not closed. Let $A \in \mathcal{B}(\mathcal{H})$ be any operator with nonclosed range such that $\mathcal{R}(A) \subseteq L_1$. Finally, let $B \in \mathcal{B}(\mathcal{K})$ be any invertible operator. Set $C_1 = C\,|_{\mathcal{N}(B)^\perp}$, $C_2 = C\,|_{\mathcal{N}(B)}$.

As before $\mathcal{N}(B) = \{0\}$, so $C_2 = 0$ and $C_1 = C$. Here also we have $\mathcal{R}(A) \cap \mathcal{R}(C) = \{0\}$ and thus

$$\operatorname{codim}_{\mathcal{R}(C_1)}(\mathcal{R}(C_1) \cap (\mathcal{R}(A) + \mathcal{R}(C_2))) = \operatorname{codim}_{\mathcal{R}(C)}(\{0\}) = \infty,$$

so that for the triple (A, B, C) all the conditions of Theorem 3.11 are met. Hence there are injective completions of $\begin{bmatrix} A & C \\ ? & B \end{bmatrix}$. Using Theorem 3.14 as in the previous example, we can see that none of those have closed range.

Note that, in the previous example, to conclude that injective completions exist it would not have been possible to use Theorem 3.12. Indeed, if $F \subseteq \mathcal{N}(A)^\perp = \mathcal{H}$ is a subspace of infinite dimension, then $A[F]$ is infinite dimensional as well, by injectivity of A. But then from $\mathcal{R}(A) \cap \mathcal{R}(C) = \{0\}$, it follows that $A[F]$ cannot be contained in $\mathcal{R}(C)$.

3.3. Completions of operator matrices M_X to right and left invertible operator

This section is concerned with the completion problem of operator matrices M_X to a right invertibile operator. This problem was already considered in [**90**]. Here we present a much shorter proof with the intention of pointing out the connection between the completion problem for this type of operator matrices and the existence of an operator X for which $A + CX$ is an operator of an appropriate type.

First we state an auxiliary result:

THEOREM 3.17. *Let* $A \in \mathcal{B}(\mathcal{H})$, $B \in \mathcal{B}(\mathcal{K})$, *and* $C \in \mathcal{B}(\mathcal{K}, \mathcal{H})$. *Here* M_X *is right invertible for some* $X \in \mathcal{B}(\mathcal{H}, \mathcal{K})$ *if and only if* $\begin{bmatrix} A & C \end{bmatrix} : \mathcal{H} \oplus \mathcal{K} \to \mathcal{H}$ *is right invertible,* $\dim(\mathcal{N}(\begin{bmatrix} A & C \end{bmatrix})) = \infty$, *and* $XG + BH$ *is a right invertible operator, where* $\begin{bmatrix} G \\ H \end{bmatrix} : \mathcal{K} \to \mathcal{H} \oplus \mathcal{K}$ *is a left invertible operator such that*

$$\mathcal{R}\left(\begin{bmatrix} G \\ H \end{bmatrix}\right) = \mathcal{N}\left(\begin{bmatrix} A & C \end{bmatrix}\right). \tag{3.10}$$

PROOF. If M_X is right invertible for some $X \in \mathcal{B}(\mathcal{H}, \mathcal{K})$, then there exists $W = \begin{bmatrix} Y & G \\ Z & H \end{bmatrix} : \mathcal{H} \oplus \mathcal{K} \to \mathcal{H} \oplus \mathcal{K}$ such that

$$M_X W = \begin{bmatrix} A & C \\ X & B \end{bmatrix} \begin{bmatrix} Y & G \\ Z & H \end{bmatrix} = \begin{bmatrix} I & 0 \\ 0 & I \end{bmatrix}. \tag{3.11}$$

Evidently, we have that $\begin{bmatrix} A & C \end{bmatrix}$ is right invertible, $\begin{bmatrix} G \\ H \end{bmatrix}$ is left invertible, and $\dim(\mathcal{N}(\begin{bmatrix} A & C \end{bmatrix})) = \infty$, since $\mathcal{R}\left(\begin{bmatrix} G \\ H \end{bmatrix}\right) \subseteq \mathcal{N}(\begin{bmatrix} A & C \end{bmatrix})$.

Now, let $\begin{bmatrix} G \\ H \end{bmatrix} : \mathcal{K} \to \mathcal{H} \oplus \mathcal{K}$ be a left invertible operator such that

$$\mathcal{R}\left(\begin{bmatrix} G \\ H \end{bmatrix}\right) = \mathcal{N}\left(\begin{bmatrix} A & C \end{bmatrix}\right). \tag{3.12}$$

Since $\begin{bmatrix} A & C \end{bmatrix}$ is right invertible, there exists an (left invertible) operator

$$\begin{bmatrix} E \\ F \end{bmatrix} : \mathcal{H} \to \mathcal{H} \oplus \mathcal{K}$$

such that $AE + CF = I$. Let

$$W = \begin{bmatrix} E & G \\ F & H \end{bmatrix} : \mathcal{H} \oplus \mathcal{K} \to \mathcal{H} \oplus \mathcal{K}. \tag{3.13}$$

Clearly W is invertible and satisfies

$$M_X W = \begin{bmatrix} I & 0 \\ XE + BF & XG + BH \end{bmatrix}. \tag{3.14}$$

Hence $XG + BH$ is right invertible. The reverse implication follows similarly. □

The following theorem gives certain necessary and sufficient conditions for the completion of M_X to a right invertible operator.

THEOREM 3.18. *Let $A \in \mathcal{B}(\mathcal{H})$, $B \in \mathcal{B}(\mathcal{K})$, and $C \in \mathcal{B}(\mathcal{K}, \mathcal{H})$. Then there exists $X \in \mathcal{B}(\mathcal{H}, \mathcal{K})$ such that M_X is right invertible if and only if $\begin{bmatrix} A & C \end{bmatrix} : \mathcal{H} \oplus \mathcal{K} \to \mathcal{H}$ is right invertible, $\dim(\mathcal{N}(\begin{bmatrix} A & C \end{bmatrix})) = \infty$, and one of the following conditions is satisfied:*

(i) *$\mathcal{N}(A|C)$ contains a noncompact operator.*
(ii) *The operator M_0 $(= M(A, B, C; 0))$ is right semi-Fredholm and*

$$d(M_0) \leq \alpha(A) + \dim\left(\mathcal{R}(A) \cap \mathcal{R}(C|_{\mathcal{N}(B)})\right).$$

PROOF. If M_X is a right invertible operator for some $X \in \mathcal{B}(\mathcal{H}, \mathcal{K})$, then evidently $\begin{bmatrix} A & C \end{bmatrix}$ is right invertible and $\dim(\mathcal{N}(\begin{bmatrix} A & C \end{bmatrix})) = \infty$. Let W be defined by (3.13). If G is noncompact, then evidently (i) holds. So, let us suppose that G is a compact operator. Since $XG + BH$ is right invertible we have that BH is right semi-Fredholm and that $\mathrm{ind}(BH) \geq 0$. Since

$$M_0 W = \begin{bmatrix} I & 0 \\ BF & BH \end{bmatrix} \tag{3.15}$$

it follows that M_0 is right semi-Fredholm and $\mathrm{ind}(M_0) = \mathrm{ind}(BH) \geq 0$. Hence, $d(M_0) \leq \alpha(M_0)$. Also, we can suppose that $\mathcal{N}(C) \cap \mathcal{N}(B) = \{0\}$ since if this is not the case we will consider

$$M'_X = \begin{bmatrix} A & C' \\ X & B' \end{bmatrix} : \begin{bmatrix} \mathcal{H} \\ \mathcal{K}' \end{bmatrix} \to \begin{bmatrix} \mathcal{H} \\ \mathcal{K} \end{bmatrix},$$

where $\mathcal{K}' = (\mathcal{N}(B) \cap \mathcal{N}(C))^{\perp}$, $B' = B|_{\mathcal{K}'}$, and $C' = C|_{\mathcal{K}'}$ with the remark that M_X is right invertible if and only if M'_X is right invertible. Hence,

$$d(M_0) \leq \alpha(M_0) \leq \alpha(A) + \dim\left(\mathcal{R}(A) \cap \mathcal{R}(C|_{\mathcal{N}(B)})\right).$$

For the opposite direction if (i) holds, then there exists $\begin{bmatrix} G \\ H \end{bmatrix} : \mathcal{K} \to \mathcal{H} \oplus \mathcal{K}$ such that $AG + CH = 0$ and G is noncompact. Since G^* is noncompact, then by Theorem 2.27 there exists $X_0 \in \mathcal{B}(\mathcal{H}, \mathcal{K})$ such that $(BH)^* + G^* X_0^*$ is left invertible; i.e., $X_0 G + BH$ is right invertible. Define

$$W = \begin{bmatrix} E & G \\ F & H \end{bmatrix} : \mathcal{H} \oplus \mathcal{K} \to \mathcal{H} \oplus \mathcal{K},$$

where $\begin{bmatrix} E \\ F \end{bmatrix} : \mathcal{H} \to \mathcal{H} \oplus \mathcal{K}$ is a right inverse of $\begin{bmatrix} A & C \end{bmatrix}$. Then

$$M_{X_0} W = \begin{bmatrix} I & 0 \\ X_0 E + BF & X_0 G + BH \end{bmatrix}$$

is a right invertible operator, so M_{X_0} is right invertible.

If (ii) holds, then let W be defined by (3.13). By (3.15) we have that BH is right semi-Fredholm and $d(M_0) = d(BH)$, $\alpha(M_0) = \alpha(BH)$. □

3.4. Completions of operator matrices M_X to semi-Fredholm operator

In this section we present results concerning completions of operator matrices M_X to right and left semi-Fredholm operators. In the literature we can already find some particular results as the one proved by G. Hai et al. [**92**] that concerns the completion of operator matrices M_X to right semi-Fredholm operator. For the proof of this result we need the following lemma whose proof can be found in [**69**]:

LEMMA 3.19 ([**69**]). *If $\mathcal{H}$ is an infinite-dimensional Hilbert space and T is an operator on $\mathcal{H}$, then T is compact if and only if the range of T contains no closed infinite-dimensional subspaces.*

THEOREM 3.20 ([**92**]). *Let $A \in \mathcal{B}(\mathcal{H}_1)$, $B \in \mathcal{B}(\mathcal{H}_2)$, and $C \in \mathcal{B}(\mathcal{H}_2, \mathcal{H}_1)$ be given operators. Then M_X is a right semi-Fredholm operator for some $X \in \mathcal{B}(\mathcal{H}_1, \mathcal{H}_2)$ if and only if $\begin{bmatrix} A & C \end{bmatrix} : \mathcal{H}_1 \oplus \mathcal{H}_2 \to \mathcal{H}_1$ is a right semi-Fredholm operator and one of the following conditions holds:*

(i) *$\mathcal{N}(A|C)$ contains a noncompact operator.*

(ii) *$M_0 = M(A, B, C; 0)$ is a right semi-Fredholm operator.*

PROOF. If the condition (ii) holds, the sufficiency is clear. Now we suppose that $\mathcal{N}(A|C)$ contains a noncompact operator. From Lemma 3.19, we have that there exists a closed infinite-dimensional subspace $\mathcal{M}$ such that $\mathcal{R}(A|_{\mathcal{M}}) \subseteq \mathcal{R}(C)$, and therefore $C^+AP_{\mathcal{M}} : \mathcal{H}_1 \to \mathcal{H}_2$ is a bounded linear operator. Since $\mathcal{H}_1$ and $\mathcal{H}_2$ are separable, it follows that there exists a right invertible operator $T \in \mathcal{B}(\mathcal{H}_1, \mathcal{H}_2)$ such that $\mathcal{N}(T)^\perp = \mathcal{M}$. Define an operator $X \in \mathcal{B}(\mathcal{H}_1, \mathcal{H}_2)$ by

$$X = T + BC^+AP_{\mathcal{M}}.$$

Then M_X is a right semi-Fredholm operator. Indeed, let $\mathcal{H}' = \mathcal{R}(A) + \mathcal{R}(C)$. It is obvious that $\mathcal{R}(M_X) \subseteq \mathcal{H}' \oplus \mathcal{H}_2$. On the other hand, for any $u \in \mathcal{H}'$ and $v \in \mathcal{H}_2$, since $\mathcal{R}(A)+\mathcal{R}(C) = \mathcal{H}'$ and $\mathcal{R}(A|_{\mathcal{M}}) \subseteq \mathcal{R}(C)$, it follows that there exist $x_1 \in \mathcal{M}^\perp$ and $y_1 \in \mathcal{H}_2$ such that $Ax_1 + Cy_1 = u$. Also, by the right invertibility of T, there exists $x_2 \in \mathcal{M}$ such that $Tx_2 = v - By_1$. Let $x_0 = x_1 + x_2$ and $y_0 = y_1 - C^+Ax_2$. From the definition of X we infer that

$$\begin{bmatrix} A & C \\ X & B \end{bmatrix} \begin{bmatrix} x_0 \\ y_0 \end{bmatrix} = \begin{bmatrix} u \\ v \end{bmatrix},$$

which means that $\mathcal{R}(M_X) = \mathcal{H}' \oplus \mathcal{H}_2$. This, together with the right semi-Fredholmness of $\begin{bmatrix} A & C \end{bmatrix} : \mathcal{H}_1 \oplus \mathcal{H}_2 \to \mathcal{H}_1$, shows that M_X is a right semi-Fredholm operator.

On the other side suppose that M_X is a right semi-Fredholm operator for some $X \in \mathcal{B}(\mathcal{H}_1, \mathcal{H}_2)$. It follows that there exist

$$\begin{bmatrix} Y & F \\ T & E \end{bmatrix} \in \mathcal{B}(\mathcal{H}_1, \mathcal{H}_2)$$

and finite rank operators $K_{ij} : H_j \to \mathcal{H}_i$, $i, j = 1, 2$, such that

$$\begin{bmatrix} A & C \\ X & B \end{bmatrix} \begin{bmatrix} Y & F \\ T & E \end{bmatrix} = \begin{bmatrix} I_{\mathcal{H}_1} & 0 \\ 0 & I_{\mathcal{H}_2} \end{bmatrix} + \begin{bmatrix} K_{11} & K_{12} \\ K_{21} & K_{22} \end{bmatrix}. \tag{3.16}$$

Thus, $AY + CT = I_{\mathcal{H}_1} + K_{11}$ and $XF + BE = I_{\mathcal{H}_2} + K_{22}$. So from the perturbation theory of the Fredholm operator we have that both $AY + CT$ and $XF + BE$ are Fredholm operators. Therefore $\begin{bmatrix} A & C \end{bmatrix} : \mathcal{H}_1 \oplus \mathcal{H}_2 \to \mathcal{H}_1$ is a right semi-Fredholm operator and $\begin{bmatrix} F \\ E \end{bmatrix} : \mathcal{H}_2 \to \mathcal{H}_1 \oplus \mathcal{H}_2$ is a left semi-Fredholm operator. On the other hand, from (3.16) we get that $AF + CE = K_{12}$ is a finite rank operator. So $\mathcal{N}\left(\begin{bmatrix} A & C \end{bmatrix}\right)$ is infinite dimensional. Hence there exists a left invertible operator $\begin{bmatrix} G \\ H \end{bmatrix} : \mathcal{H}_2 \to \mathcal{H}_1 \oplus \mathcal{H}_2$ whose range is equal to $\mathcal{N}\left(\begin{bmatrix} A & C \end{bmatrix}\right)$. It is clear that $AG + CH = 0$.

If G is a noncompact operator, it follows from $AG + CH = 0$ that $\mathcal{R}(AG) \subseteq \mathcal{R}(C)$. Therefore A and C satisfy (i).

If G is a compact operator, let $\begin{bmatrix} W \\ Z \end{bmatrix} : \mathcal{H}_1 \to \mathcal{H}_1 \oplus \mathcal{H}_2$ be a generalized inverse of $\begin{bmatrix} A & C \end{bmatrix} : \mathcal{H}_1 \oplus \mathcal{H}_2 \to \mathcal{H}_1$. Then

$$\mathcal{R}\left(\begin{bmatrix} W \\ Z \end{bmatrix}\right) = \left(\mathcal{N}\left(\begin{bmatrix} A & C \end{bmatrix}\right)\right)^{\perp} \quad \text{and} \quad AW + CZ = I_{\mathcal{H}_1} - P_{\mathcal{H}'^{\perp}},$$

where $\mathcal{H}' = \mathcal{R}(A) + \mathcal{R}(C)$. Put

$$L = \begin{bmatrix} W & G \\ Z & H \end{bmatrix} : \mathcal{H}_1 \oplus \mathcal{H}_2 \to \mathcal{H}_1 \oplus \mathcal{H}_2.$$

Clearly, L is a Fredholm operator with $d(L) = 0$. Thus

$$M_X L = \begin{bmatrix} A & C \\ X & B \end{bmatrix} \begin{bmatrix} W & G \\ Z & H \end{bmatrix} = \begin{bmatrix} I_{\mathcal{H}_1} - P_{\mathcal{H}'^{\perp}} & 0 \\ XW + BZ & XG + BH \end{bmatrix} \tag{3.17}$$

is a right semi-Fredholm operator. Since $P_{\mathcal{H}'^{\perp}}$ is a finite rank operator (by the right semi-Fredholmness of $\begin{bmatrix} A & C \end{bmatrix}$) and G is compact, it follows that

$$\begin{bmatrix} I_{\mathcal{H}_1} & 0 \\ XW + BZ & BH \end{bmatrix}$$

is a right semi-Fredholm operator, and therefore BH is a right semi-Fredholm operator. Take $X = 0$ in (3.17). Then it follows from (3.17) and the right semi-Fredholmness of BH that $M_0 L$ is a right semi-Fredholm operator. Thus

$$M_0 = \begin{bmatrix} A & C \\ 0 & B \end{bmatrix} : \mathcal{H}_1 \oplus \mathcal{H}_2 \to \mathcal{H}_1 \oplus \mathcal{H}_2$$

is a right semi-Fredholm operator. □

The following is the dual statement of Theorem 3.20 in the case of left semi-Fredholmness:

THEOREM 3.21 ([**92**]). *Let $A \in \mathcal{B}(\mathcal{H}_1)$, $B \in \mathcal{B}(\mathcal{H}_2)$, and $C \in \mathcal{B}(\mathcal{H}_2, \mathcal{H}_1)$ be given operators. Then M_X is a left semi-Fredholm operator for some $X \in \mathcal{B}(\mathcal{H}_1, \mathcal{H}_2)$ if and only if $\begin{bmatrix} B^* & C^* \end{bmatrix} : \mathcal{H}_2 \oplus \mathcal{H}_1 \to \mathcal{H}_2$ is a right semi-Fredholm operator and one of the following conditions holds:*

(i) *$\mathcal{N}(B^*|C^*)$ contains a noncompact operator.*
(ii) *$M_0 = M(A, B, C; 0)$ is a left semi-Fredholm operator.*

As a corollary of Theorem 3.20, we have

COROLLARY 3.22 ([**92**]). Let $A \in \mathcal{B}(\mathcal{H}_1)$, $B \in \mathcal{B}(\mathcal{H}_2)$, and $C \in \mathcal{B}(\mathcal{H}_2, \mathcal{H}_1)$ be given operators. If C is compact, then M_X is a right semi-Fredholm operator for some $X \in \mathcal{B}(\mathcal{H}_2, \mathcal{H}_1)$ if and only if A is a right semi-Fredholm operator and one of the following conditions holds:

(i) B is a right semi-Fredholm operator.
(ii) $\alpha(A) = \infty$.

PROOF. If A is right semi-Fredholm and C is a compact operator, it follows that $\begin{bmatrix} A & C \end{bmatrix} : \mathcal{H}_1 \oplus \mathcal{H}_2 \to \mathcal{H}_1$ is a right semi-Fredholm operator. If B is a right semi-Fredholm operator, from the right semi-Fredholmness of A we have that $M_0 = M(A, B, C; 0)$ is a right semi-Fredholm operator. If $\alpha(A) = \infty$, then $\mathcal{N}(A|C)$ contains a noncompact operator. In fact, since $\alpha(A) = \infty$, there exists a left invertible operator $G \in \mathcal{B}(\mathcal{H}_2, \mathcal{H}_1)$ such that $\mathcal{R}(G) = \mathcal{N}(A)$. Clearly, G is a noncompact operator and $G \in \mathcal{N}(A|C)$. Using Theorem 3.20, we conclude that $M_X = M(A, B, C; X)$ is a right semi-Fredholm operator for some $X \in \mathcal{B}(\mathcal{H}_1, \mathcal{H}_2)$.

Suppose that M_X is a right semi-Fredholm operator for some $X \in \mathcal{B}(\mathcal{H}_1, \mathcal{H}_2)$. By the compactness of C,

$$M(A, B, 0; X) = \begin{bmatrix} A & 0 \\ X & B \end{bmatrix}$$

is a right semi-Fredholm operator. Therefore A is a right semi-Fredholm operator. On the other hand, from Theorem 3.20 we obtain that M_0 is a right semi-Fredholm operator or $\mathcal{N}(A|C)$ contains a noncompact operator.

If M_0 is a right semi-Fredholm operator, then clearly B is a right semi-Fredholm operator.

If $\mathcal{N}(A|C)$ contains a noncompact operator, then $\alpha(A) = \infty$. Assume to the contrary that $\alpha(A) < \infty$. Thus A is a Fredholm operator (because A is a right semi-Fredholm operator). If follows that there exist $S \in \mathcal{B}(\mathcal{H}_1)$ and a finite rank operator $F \in \mathcal{B}(\mathcal{H}_1)$ such that

$$SA = I_{\mathcal{H}_1} + F.$$

For any $G \in \mathcal{N}(A|C)$, there exists $H \in \mathcal{B}(\mathcal{H}_2)$ such that $AG = CH$. This, together with $SA = I_{\mathcal{H}_1} + F$, shows that

$$G = SCH - FG.$$

Since C and F are compact operators, if follows that G is compact. It is in contradiction to the fact that $\mathcal{N}(A|C)$ contains a noncompact operator. □

A similar idea can be used to complete the proof of the following corollary.

COROLLARY 3.23 ([**92**]). Let $A \in \mathcal{B}(\mathcal{H}_1)$, $B \in \mathcal{B}(\mathcal{H}_2)$, and $C \in \mathcal{B}(\mathcal{H}_2, \mathcal{H}_1)$ be given operators. If C is compact, then M_X is a left semi-Fredholm operator for some $X \in \mathcal{B}(\mathcal{H}_1, \mathcal{H}_2)$ if and only if B is a left semi-Fredholm operator and one of the following conditions holds:

(i) A is a left semi-Fredholm operator.
(ii) $d(B) = \infty$.

In particular, we have

COROLLARY 3.24 ([**30**]). Let $A \in \mathcal{B}(\mathcal{H}_1)$ and $B \in \mathcal{B}(\mathcal{H}_2)$ be given operators. An operator matrix $L_X = \begin{bmatrix} A & X \\ 0 & B \end{bmatrix}$ is a left semi-Fredholm operator for some $X \in \mathcal{B}(\mathcal{H}_2, \mathcal{H}_1)$ if and only if A is a left semi-Fredholm operator and

$$\begin{cases} \alpha(B) < \infty \text{ or } \alpha(B) = d(A) = \infty, & \text{if } \mathcal{R}(B) \text{ is closed,} \\ d(A) = \infty, & \text{if } \mathcal{R}(B) \text{ is not closed.} \end{cases}$$

The following results are immediate from the argument above.

COROLLARY 3.25 ([**92**]). For a given $A \in \mathcal{B}(\mathcal{H}_1)$, $B \in \mathcal{B}(\mathcal{H}_2)$, and $C \in \mathcal{B}(\mathcal{H}_2, \mathcal{H}_1)$,

$$\bigcap_{X \in \mathcal{B}(\mathcal{H}_1,\mathcal{H}_2)} \sigma_{re}(M_X) = \{\lambda \in \mathbb{C}\colon (A - \lambda I, C) \text{ is not right semi-Fredholm}\}$$
$$\cup \{\lambda \in \mathbb{C}\colon \lambda \in \sigma_{re}(M_0),\, \mathcal{N}(A - \lambda I/C) \subseteq \mathcal{K}(\mathcal{H}_2, \mathcal{H}_1)\},$$

$$\bigcap_{X \in \mathcal{B}(\mathcal{H}_1,\mathcal{H}_2)} \sigma_{le}(M_X) = \{\lambda \in \mathbb{C}\colon (B^* - \overline{\lambda} I, C^*) \text{ is not right semi-Fredholm}\}$$
$$\cup \{\lambda \in \mathbb{C}\colon \lambda \in \sigma_{le}(M_0),\, \mathcal{N}(B^* - \overline{\lambda} I/C^*) \subseteq \mathcal{K}(\mathcal{H}_1, \mathcal{H}_2)\}.$$

COROLLARY 3.26 ([**92**]). Let $A \in \mathcal{B}(\mathcal{H}_1)$, $B \in \mathcal{B}(\mathcal{H}_2)$, and $C \in \mathcal{B}(\mathcal{H}_2, \mathcal{H}_1)$ be given operators. If C is a compact operator, then

$$\bigcap_{X \in \mathcal{B}(\mathcal{H}_1,\mathcal{H}_2)} \sigma_{re}(M_X) = \sigma_{re}(A) \cup \{\lambda \in \mathbb{C}\colon \lambda \in \sigma_{re}(B),\, \alpha(A - \lambda I) < \infty\},$$

$$\bigcap_{X \in \mathcal{B}(\mathcal{H}_1,\mathcal{H}_2)} \sigma_{le}(M_X) = \sigma_{le}(B) \cup \{\lambda \in \mathbb{C}\colon \lambda \in \sigma_{le}(A),\, d(B - \lambda I) < \infty\}.$$

From Theorem 3.20 and Theorem 3.21, we get

COROLLARY 3.27 ([**92**]). Let $A \in \mathcal{B}(\mathcal{H}_1)$ and $C \in \mathcal{B}(\mathcal{H}_2, \mathcal{H}_1)$ be given operators. Assume that $\begin{bmatrix} A & C \end{bmatrix} : \mathcal{H}_2 \to \mathcal{H}_1$ is a right semi-Fredholm operator.

(i) If C is compact, then there exists $X \in \mathcal{B}(\mathcal{H}_1, \mathcal{H}_2)$ such that $A + CX$ is a right semi-Fredholm operator if and only if A is a right semi-Fredholm operator.

(ii) If C is not compact, then there exists $X \in \mathcal{B}(\mathcal{H}_1, \mathcal{H}_2)$ such that $A + CX$ is a right semi-Fredholm operator if and only if $\mathcal{N}(A|C)$ contains a noncompact operator.

PROOF. (i) It is clear.

(ii) Assume that $\mathcal{N}(A|C)$ contains a noncompact operator. Since

$$\begin{bmatrix} A + CX & 0 \\ 0 & I_{\mathcal{H}_2} \end{bmatrix} = \begin{bmatrix} I_{\mathcal{H}_1} & -C \\ 0 & I_{\mathcal{H}_2} \end{bmatrix} \begin{bmatrix} A & C \\ -X & I_{\mathcal{H}_2} \end{bmatrix} \begin{bmatrix} I_{\mathcal{H}_1} & 0 \\ X & I_{\mathcal{H}_2} \end{bmatrix}, \tag{3.18}$$

it follows that there exists $X \in \mathcal{B}(\mathcal{H}_1, \mathcal{H}_2)$ such that $A + CX$ is a right semi-Fredholm operator if and only if $M(A, I_{\mathcal{H}_2}, C; X)$ is a right semi-Fredholm operator for some $X \in \mathcal{B}(\mathcal{H}_1, \mathcal{H}_2)$. So we only need to prove that $M(A, I_{\mathcal{H}_2}, C; X)$ is a right semi-Fredholm operator for some $X \in \mathcal{B}(\mathcal{H}_1, \mathcal{H}_2)$. We infer that $M_X = M(A, I_{\mathcal{H}_2}, C; X)$ is a right semi-Fredholm operator for some $X \in \mathcal{B}(\mathcal{H}_1, \mathcal{H}_2)$ which follows by Theorem 3.20.

Conversely, assume that there exists $X \in \mathcal{B}(\mathcal{H}_1, \mathcal{H}_2)$ such that $A + CX$ is a right semi-Fredholm operator. By (3.18), $M_X = M(A, I_{\mathcal{H}_2}, C; -X)$ is a right semi-Fredholm operator. From Theorem 3.20 we obtain that $\mathcal{N}(A|C)$ contains a noncompact operator or $M_0 = M(A, I_{\mathcal{H}_2}, C; 0)$ is a right semi-Fredholm operator. In the case when $M(A, I_{\mathcal{H}_2}, C; 0)$ is a right semi-Fredholm operator, it is easy to show that A is a right semi-Fredholm operator. Thus $C_1 = CP_{\mathcal{N}(P_{\mathcal{R}(A)^\perp} C)}$

is a noncompact operator (because $\dim(\mathcal{N}(P_{\mathcal{R}(A)^\perp}C)^\perp) \leq d(A) < \infty$ and $C_1 + CP_{\mathcal{N}(P_{\mathcal{R}(A)^\perp}C)^\perp} = C$ is not compact). It is obvious that $P_{\mathcal{R}(A)^\perp}C_1 = 0$. This, together with the closedness of $\mathcal{R}(A)$, implies $\mathcal{R}(C_1) \subseteq \mathcal{R}(A)$. Therefore there exists $G \in \mathcal{B}(\mathcal{H}_2, \mathcal{H}_1)$ such that $C_1 = AG$ (Theorem 1 from [**68**]). Clearly, G is a noncompact operator and $\mathcal{R}(AG) \subseteq \mathcal{R}(C)$. This proves that $\mathcal{N}(A|C)$ contains a noncompact operator. □

COROLLARY 3.28 ([**92**]). Let $A \in \mathcal{B}(\mathcal{H}_1)$ and $C \in \mathcal{B}(\mathcal{H}_2, \mathcal{H}_1)$ be given operators. Assume that $\begin{bmatrix} A & C \end{bmatrix} : \mathcal{H}_1 \oplus \mathcal{H}_2 \to \mathcal{H}_1$ is a right semi-Fredholm operator.

(i) If C is compact, then there exists $X \in \mathcal{B}(\mathcal{H}_1, \mathcal{H}_2)$ such that $A + CX$ is a left semi-Fredholm operator if and only if A is a left semi-Fredholm operator.

(ii) If C is not compact, then there exists $X \in \mathcal{B}(\mathcal{H}_1, \mathcal{H}_2)$ such that $A + CX$ is a left semi-Fredholm operator.

PROOF. The proof follows from (3.18), Theorem 3.21, and the perturbation theory of the Fredholm operator. □

The next two examples illustrate the main results of this section:

EXAMPLE 3.29. Let $\mathcal{H}_1 = \mathcal{H}_2 = l_2$. Define $A \in \mathcal{B}(l_2)$ and $C \in \mathcal{B}(l_2)$ by

$$\begin{aligned} Ax &= (0, x_1, x_2, \dots), \\ Cx &= \left(x_1, \frac{1}{2}x_2, \frac{1}{3}x_3, \dots\right), \end{aligned}$$

where $x = (x_1, x_2, \dots) \in l_2$. Clearly, A is a right semi-Fredholm operator. It is easy to show that $I_{l_2} - A$ is not right semi-Fredholm. Since C is compact and $\alpha(A) = 0$, thus by Corollary 3.22, we have that

$$M_X = \begin{bmatrix} A & C \\ X & I_{l_2} - A \end{bmatrix}$$

is not a right semi-Fredholm operator for every $X \in \mathcal{B}(l_2)$.

EXAMPLE 3.30. Let $C \in \mathcal{B}(\mathcal{H}_2, \mathcal{H}_1)$ be an invertible operator. For any $A \in \mathcal{B}(\mathcal{H}_1)$ and $B \in \mathcal{B}(\mathcal{H}_2)$, it is easy to show that $\begin{bmatrix} A & C \end{bmatrix} : \mathcal{H}_1 \oplus \mathcal{H}_2 \to \mathcal{H}_1$ is a right semi-Fredholm operator and $\mathcal{N}(A|C)$ contains a noncompact operator. By Theorem 3.20, there exists $X \in \mathcal{B}(\mathcal{H}_1, \mathcal{H}_2)$ such that M_X is a right semi-Fredholm operator. In fact, define an operator $X \in \mathcal{B}(\mathcal{H}_1, \mathcal{H}_2)$ by

$$X = T + BC^{-1}A,$$

where $T \in \mathcal{B}(\mathcal{H}_1, \mathcal{H}_2)$ is a right semi-Fredholm operator. From the right semi-Fredholmness of T we get that there exists $S \in \mathcal{B}(\mathcal{H}_2, \mathcal{H}_1)$ such that $TS \in \mathcal{B}(\mathcal{H}_2)$ is a Fredholm operator. Therefore

$$\begin{bmatrix} A & C \\ X & B \end{bmatrix} \begin{bmatrix} 0 & S \\ C^{-1} & -C^{-1}AS \end{bmatrix} = \begin{bmatrix} I_{\mathcal{H}_1} & 0 \\ BC^{-1} & TS \end{bmatrix}$$

is a Fredholm operator. It follows that M_X is a right semi-Fredholm operator.

3.5. Completions of operator matrices M_X to Fredholm operator

In this section, we present an application of the previously presented results in the Chapter 2 to the problem of completions of the operator matrix

$$\begin{bmatrix} A & C \\ ? & B \end{bmatrix} : \begin{bmatrix} \mathcal{H} \\ \mathcal{K} \end{bmatrix} \to \begin{bmatrix} \mathcal{H} \\ \mathcal{K} \end{bmatrix}$$

to a Fredholm operator. Note that the problem of completions of this type of operator matrix to left and right semi-Fredholmness was considered in our Section 3.4 (originally published in [**92**]), while the completions to a Fredholm operator with index 0 (i.e., to a Weyl operator) was considered in [**160**]. Here we completely solve the completion problem to a Fredholm operator with arbitrary index using Theorem 2.30.

First we state one auxiliary result whose proof will be omitted since it is elementary:

LEMMA 3.31. *Let* $\begin{bmatrix} A & C \end{bmatrix} : \mathcal{H} \oplus \mathcal{K} \to \mathcal{H}$ *be a right semi-Fredholm operator such that* $\dim(\mathcal{N}(\begin{bmatrix} A & C \end{bmatrix})) = \infty$. *Let* $\begin{bmatrix} G \\ H \end{bmatrix} : \mathcal{K} \to \mathcal{H} \oplus \mathcal{K}$ *be a left invertible operator such that*

$$\mathcal{R}\left(\begin{bmatrix} G \\ H \end{bmatrix}\right) = \mathcal{N}(\begin{bmatrix} A & C \end{bmatrix}) \tag{3.19}$$

and let $\begin{bmatrix} E \\ F \end{bmatrix} : \mathcal{H} \to \mathcal{H} \oplus \mathcal{K}$ *be such that*

$$\begin{bmatrix} A & C \end{bmatrix} \begin{bmatrix} E \\ F \end{bmatrix} = I + K,$$

for some compact operator $K \in \mathcal{B}(\mathcal{H})$. *Then*

$$W = \begin{bmatrix} E & G \\ F & H \end{bmatrix}$$

is a Fredholm operator.

The main result of the section is the following:

THEOREM 3.32. *Let* $A \in \mathcal{B}(\mathcal{H})$, $B \in \mathcal{B}(\mathcal{K})$, *and* $C \in \mathcal{B}(\mathcal{K}, \mathcal{H})$. *Then* M_X *is Fredholm for some* $X \in \mathcal{B}(\mathcal{H}, \mathcal{K})$ *if and only if* $\begin{bmatrix} A & C \end{bmatrix} : \mathcal{H} \oplus \mathcal{K} \to \mathcal{H}$ *is right semi-Fredholm,* $\begin{bmatrix} B \\ C \end{bmatrix} : \mathcal{K} \to \mathcal{H} \oplus \mathcal{K}$ *is left semi-Fredholm, and either one of conditions* (a)–(b) *and either one of conditions* (c)–(d) *is satisfied:*

(a) $\mathcal{N}(A|C)$ *contains a noncompact operator.*
(b) M_0 *is a right semi-Fredholm operator.*
(c) $\mathcal{N}(B^*|C^*)$ *contains a noncompact operator.*
(d) M_0 *is a left semi-Fredholm operator.*

PROOF. If there exists $X \in \mathcal{B}(\mathcal{H}, \mathcal{K})$ such that M_X is Fredholm by Theorems 3.20 and 3.21, we have that $\begin{bmatrix} A & C \end{bmatrix} : \mathcal{H} \oplus \mathcal{K} \to \mathcal{H}$ is right semi-Fredholm, $\begin{bmatrix} B \\ C \end{bmatrix} : \mathcal{K} \to \mathcal{H} \oplus \mathcal{K}$ is left semi-Fredholm, and one of conditions (a)–(b) as well as one of conditions (c)–(d) is satisfied. On the other hand suppose that

$\begin{bmatrix} A & C \end{bmatrix} : \mathcal{H} \oplus \mathcal{K} \to \mathcal{H}$ is right semi-Fredholm, $\begin{bmatrix} B \\ C \end{bmatrix} : \mathcal{K} \to \mathcal{H} \oplus \mathcal{K}$ is left semi-Fredholm, and either one of conditions (a)–(b) and either one of conditions (c)–(d) is satisfied. We consider 4 possible cases:

Case 1: Suppose that (a) and (c) are satisfied and let $D \in \mathcal{N}(B^*|C^*)$ be a noncompact operator. Then $\dim(\mathcal{N}(\begin{bmatrix} A & C \end{bmatrix})) = \infty$, so take W as defined in Lemma 3.31. Then for any $X \in \mathcal{B}(\mathcal{H}, \mathcal{K})$, we have

$$M_X W = \begin{bmatrix} I + K & 0 \\ XE + BF & XG + BH \end{bmatrix}. \tag{3.20}$$

Since W is a Fredholm operator, we have that M_X is Fredholm if and only if $XG + BH$ is Fredholm. Since $B^*D = C^*E$, for some $E \in \mathcal{B}(\mathcal{K})$ and $AG + CH = 0$, we have that

$$H^*B^*D = H^*C^*E = -G^*A^*E;$$

i.e., $D \in \mathcal{N}((BH)^*|G^*)$. Also, we have that G must be noncompact. Now we have only to prove that $\begin{bmatrix} G \\ BH \end{bmatrix}$ is left semi-Fredholm: Evidently, $\mathcal{N}(H) \cap \mathcal{N}\left(\begin{bmatrix} G \\ BH \end{bmatrix}\right) = \{0\}$ since $\begin{bmatrix} G \\ H \end{bmatrix}$ is left invertible. Now, take $x \in \mathcal{N}(H)^\perp$. If $x \in \mathcal{N}\left(\begin{bmatrix} G \\ BH \end{bmatrix}\right)$, then $CHx = 0$ which implies that $Hx \in \mathcal{N}\left(\begin{bmatrix} C \\ B \end{bmatrix}\right)$. Since $\begin{bmatrix} B \\ C \end{bmatrix}$ is left semi-Fredholm, it follows that $\begin{bmatrix} G \\ BH \end{bmatrix}$ is left semi-Fredholm. Now, Theorem 2.30 implies the existence of X such that $XG + BH$ is Fredholm. By (3.20) it follows that M_X is Fredholm.

Case 2: Suppose that (a) and (d) are satisfied. Then $\dim(\mathcal{N}(\begin{bmatrix} A & C \end{bmatrix})) = \infty$, so take W as defined in Lemma 3.31. Then for any $X \in \mathcal{B}(\mathcal{H}, \mathcal{K})$, we have that (3.20) holds. By left semi-Fredholmness of M_0, Lemma 3.31, and the equality

$$M_0 W = \begin{bmatrix} I + K & 0 \\ BF & BH \end{bmatrix},$$

we get that BH is a left semi-Fredholm operator. So for any compact $X \in \mathcal{B}(\mathcal{H}, \mathcal{K})$ we have that $XG + BH$ is left semi-Fredholm. Since G is noncompact, by Theorem 2.30 there exists X such that $(BH)^* + G^*X^*$ is Fredholm; i.e., by (3.20) it follows that M_X is Fredholm.

Case 3: Suppose that (b) and (c) are satisfied. Here the proof is analogous to that in Case 2.

Case 4: Suppose that (b) and (d) are satisfied. If $\dim(\mathcal{N}(\begin{bmatrix} A & C \end{bmatrix})) < \infty$, then for any $X \in \mathcal{B}(\mathcal{H}, \mathcal{K})$ it follows that M_X is left semi-Fredholm. The existence of $X \in \mathcal{B}(\mathcal{H}, \mathcal{K})$ such that M_X is right semi-Fredholm follows from Theorem 3.20. So, for such X, we have that M_X is Fredholm. If $\dim(\mathcal{N}(\begin{bmatrix} A & C \end{bmatrix})) = \infty$, let W be defined as in Lemma 3.31. By Fredholmness of M_0, Lemma 3.31, and the equality

$$M_0 W = \begin{bmatrix} I + K & 0 \\ BF & BH \end{bmatrix},$$

we get that BH is a Fredholm operator. Now, for any compact operator $X \in \mathcal{B}(\mathcal{H}, \mathcal{K})$, by (3.20) we have that M_X is Fredholm. $\square$

3.6. Completions of operator matrices M_X to closed range operator

Call a triple (A, C, B) of operators closed range completable if there exists $X \in \mathcal{B}(\mathcal{H})$ such that $M_X = \begin{bmatrix} A & C \\ X & B \end{bmatrix}$ has closed range. The closed range completion problem seeks a characterization of those triples (A, C, B) which are completable. Now we consider how to solve the closed range completion problem in various special cases.

(A) As an application of Theorem 2.41, we characterize completable triples of the form (A, C, I):

THEOREM 3.33. *Let $A, C \in \mathcal{B}(\mathcal{H})$. There exists an operator $X \in \mathcal{B}(\mathcal{H})$ such that $\begin{bmatrix} A & C \\ X & I \end{bmatrix}$ is a closed range operator if and only if one of the following conditions holds:*

(1) *$\mathcal{R}(A)$ is closed.*
(2) *C is a noncompact operator.*
(3) *C is a compact operator and there exists a closed subspace F of $\mathcal{R}(A)$ such that $\mathcal{R}(A) = F + \mathcal{R}(A) \cap \mathcal{R}(C)$.*

Note that the same result holds for the operator matrix $\begin{bmatrix} A & C \\ X & B \end{bmatrix}$ in the case when $B \in \mathcal{B}(\mathcal{H})$ is an invertible operator since

$$\begin{bmatrix} I & 0 \\ 0 & B^{-1} \end{bmatrix} \begin{bmatrix} A & C \\ X & B \end{bmatrix} = \begin{bmatrix} A & C \\ B^{-1}X & I \end{bmatrix}.$$

(B) The next theorem characterizes completable triples of the form $(A, C, 0)$.

THEOREM 3.34. *Let $A, C \in \mathcal{B}(\mathcal{H})$. There exists an operator $X \in \mathcal{B}(\mathcal{H})$ such that $\begin{bmatrix} A & C \\ X & 0 \end{bmatrix}$ is a closed range operator if and only if $\overline{\mathcal{R}(C)} \subseteq \mathcal{R}(A) + \mathcal{R}(C)$.*

PROOF. Suppose first that $\overline{\mathcal{R}(C)} \subseteq \mathcal{R}(A) + \mathcal{R}(C)$ which is equivalent with $\overline{\mathcal{R}(C)} = \mathcal{R}(A) \cap \overline{\mathcal{R}(C)} + \mathcal{R}(C)$. If $\mathcal{R}(A) + \mathcal{R}(C)$ is closed, then clearly for $X = 0$ we have that M_0 has a closed range. If $\mathcal{R}(C)$ is closed, then for any $X \in \mathcal{B}(\mathcal{H})$,

$$M_X = \begin{bmatrix} A_1 & C_1 & 0 \\ A_2 & 0 & 0 \\ X & 0 & 0 \end{bmatrix} : \begin{bmatrix} \mathcal{H} \\ \mathcal{N}(C)^\perp \\ \mathcal{N}(C) \end{bmatrix} \to \begin{bmatrix} \mathcal{R}(C) \\ \mathcal{R}(C)^\perp \\ \mathcal{H} \end{bmatrix},$$

where C_1 is invertible. Evidently for invertible $X : \mathcal{H} \to \mathcal{H}$ we get that $\mathcal{R}(M_X)$ has closed range.

Now suppose that neither one of $\mathcal{R}(A) + \mathcal{R}(C)$ and $\mathcal{R}(C)$ is closed. Let $\mathcal{M} = \{x \colon Ax \in \overline{\mathcal{R}(C)}\}$. Evidently $\mathcal{M}$ is a closed subspace. Since $\mathcal{R}(C)$ is not closed, we conclude that $\dim \mathcal{M} = \infty$. Also, notice that $\operatorname{codim} \mathcal{M} = \infty$. If this is not the case, then $\mathcal{R}(A) = \mathcal{R}(A) \cap \overline{\mathcal{R}(C)} + \mathcal{H}_1$, where $\mathcal{H}_1 = \{Ax \colon x \in \mathcal{M}^\perp\}$ is finite dimensional. Since $\overline{\mathcal{R}(C)} = \mathcal{R}(A) \cap \overline{\mathcal{R}(C)} + \mathcal{R}(C)$ we have $\overline{\mathcal{R}(C)} + \mathcal{H}_1 = \mathcal{R}(A) + \mathcal{R}(C)$ and we get that $\mathcal{R}(A) + \mathcal{R}(C)$ is a closed subspace. This is in contradiction to the fact that $\mathcal{R}(A) + \mathcal{R}(C)$ is not closed.

Let us define $X \in \mathcal{B}(\mathcal{H})$ as

$$Xx = \begin{cases} 0, & x \in \mathcal{M}, \\ X_1 x, & x \in \mathcal{M}^\perp, \end{cases}$$

where $X_1 : \mathcal{M}^\perp \to \mathcal{H}$ is invertible. We show that for such X, $\mathcal{R}(M_X)$ is closed.

Now M_X can be represented by

$$M_X = \begin{bmatrix} A_1 & A_2 & C_1 & 0 \\ A_3 & 0 & 0 & 0 \\ X_1 & 0 & 0 & 0 \end{bmatrix} : \begin{bmatrix} \mathcal{M}^\perp \\ \mathcal{M} \\ \mathcal{N}(C)^\perp \\ \mathcal{N}(C) \end{bmatrix} \to \begin{bmatrix} \overline{\mathcal{R}(C)} \\ \mathcal{R}(C)^\perp \\ \mathcal{H} \end{bmatrix},$$

where X_1 is invertible and A_3 is injective. Evidently since X_1 is invertible it follows that $\mathcal{R}(M_X)$ is closed if and only if $\mathcal{R}(A_2)+\mathcal{R}(C_1)$ is closed. Since $\mathcal{R}(A)\cap\overline{\mathcal{R}}(C) = \mathcal{R}(A_2)$, it follows that $\mathcal{R}(A_2) + \mathcal{R}(C_1) = \overline{\mathcal{R}(C)}$.

For the converse, let us suppose that there exists an operator $X \in \mathcal{B}(\mathcal{H})$ such that M_X has closed range and let $y \in \overline{\mathcal{R}(C)}$. Then $\begin{bmatrix} y \\ 0 \end{bmatrix} \in \overline{\mathcal{R}(M_X)}$ which implies that $y \in \mathcal{R}(A) + \mathcal{R}(C)$. □

(C) Using the fact that an operator has closed range if and only if the same is true for its adjoint, we get that $(0, C, B)$ is closed range completable if and only if $\overline{\mathcal{R}(C^*)} \subseteq \mathcal{R}(B^*) + \mathcal{R}(C^*)$.

(D) Similarly we characterize completable triples of the form (I, C, B).

(E) (A, I, B) is always closed range completable since

$$\begin{bmatrix} I & 0 \\ -B & I \end{bmatrix}\begin{bmatrix} A & I \\ X & B \end{bmatrix}\begin{bmatrix} I & 0 \\ -A & I \end{bmatrix} = \begin{bmatrix} 0 & I \\ X - BA & 0 \end{bmatrix}. \tag{3.21}$$

(F) Theorems 2.5 and 2.6 from [**67**] address closed range completability of $(A, 0, B)$.

(G) Since multiplication of a row or column of M_X by an invertible operator does not affect whether its range is closed, parts (**B**), (**C**), and (**E**) above can be extended to cover all triples (A, C, B) where at least one of the given components A, B, C is invertible.

(H) Suppose C is a closed range operator. Then we can represent M_X as

$$M_X = \begin{bmatrix} A_1 & A_2 & C_1 & 0 \\ A_3 & A_4 & 0 & 0 \\ X_1 & X_2 & B_1 & B_2 \\ X_3 & X_4 & B_3 & B_4 \end{bmatrix},$$

where C_1 is invertible. Subtraction of multiples of the first row from the third and fourth rows and of the third column from the first and second columns reduces the question of range closure to a three-by-three matrix whose first entry is 0, which by blocking is reduced to range closure of a matrix covered by part (**C**).

CHAPTER 4

Completions of operator matrices $M_{(X,Y)}$

In this chapter several results are presented concerning the problem of when operator matrices of the form

$$\begin{bmatrix} A & C \\ ? & ? \end{bmatrix}$$

can be completed to obtain an operator that belongs to a particular class of operators, in the case when $A \in \mathcal{B}(\mathcal{X})$ and $C \in \mathcal{B}(\mathcal{Y}, \mathcal{X})$ are given. We denote by $M_{(X,Y)}$ the operator on $\mathcal{Z} = \mathcal{X} \oplus \mathcal{Y}$ of the form

$$M_{(X,Y)} = \begin{bmatrix} A & C \\ X & Y \end{bmatrix} : \begin{bmatrix} \mathcal{X} \\ \mathcal{Y} \end{bmatrix} \to \begin{bmatrix} \mathcal{X} \\ \mathcal{Y} \end{bmatrix},$$

for $X \in \mathcal{B}(\mathcal{X}, \mathcal{Y})$ and $Y \in \mathcal{B}(\mathcal{Y})$.

4.1. Completions of operator matrices $M_{(X,Y)}$ to invertibility, right and left invertibility

Let $\mathcal{X}$ and $\mathcal{Y}$ be Banach spaces, and let $A \in \mathcal{B}(\mathcal{X})$ and $C \in \mathcal{B}(\mathcal{Y}, \mathcal{X})$ be given operators. In this section we consider a necessary and sufficient condition for $\begin{bmatrix} A & C \\ X & Y \end{bmatrix}$ to be invertible (respectively, right, left invertible), for some $X \in \mathcal{B}(\mathcal{X}, \mathcal{Y})$ and $Y \in \mathcal{B}(\mathcal{Y})$. Throughout this section, let $\mathcal{Z}$ be a Banach space, and let $\mathcal{Z} = \mathcal{X} \oplus \mathcal{Y}$, for some closed and complementary subspaces $\mathcal{X}$ and $\mathcal{Y}$. It is obvious that $\mathcal{X}$ and $\mathcal{Y}$ are Banach spaces. If $\mathcal{Z}$ is a Hilbert space, then we always assume that $\mathcal{X}$ and $\mathcal{Y}$ are closed and mutually orthogonal subspaces of $\mathcal{Z}$, so in this case $\mathcal{Z} = \mathcal{X} \oplus \mathcal{Y}$ denotes the orthogonal sum.

For given operators $A \in \mathcal{B}(\mathcal{X})$ and $C \in \mathcal{B}(\mathcal{Y}, \mathcal{X})$, in this section we are interested to find $X \in \mathcal{B}(\mathcal{X}, \mathcal{Y})$ and $Y \in \mathcal{B}(\mathcal{Y})$ such that $M_{(X,Y)}$ is invertible. When $\mathcal{X}$ and $\mathcal{Y}$ are separable Hilbert spaces, the invertibility, the right invertibility, and the left invertibility of $M_{(X,Y)}$ were considered in [**34, 93**]. In a Banach space a closed subspace need not have a complement and the fact that a closed subspace is the null space or the range of a bounded operator does not imply that it is complemented. Therefore more assumptions are required to ensure the invertibility and the left invertibility of $M_{(X,Y)}$ when $\mathcal{X}$ and $\mathcal{Y}$ are Banach spaces than when they are Hilbert spaces. Recently, [**34**, Theorem 1.1] concerning the right invertibility of $M_{(X,Y)}$ was generalized to the case where $\mathcal{X}$ and $\mathcal{Y}$ are Banach spaces. In this section, we present the results from [**95**] that concern the invertibility and the left invertibility of $M_{(X,Y)}$ and extend [**34**, Theorem 2.1] and [**93**, Theorem 1] to a Banach space setting.

THEOREM 4.1 ([**95**]). *Let $\mathcal{X}$ and $\mathcal{Y}$ be Banach spaces, and let $A \in \mathcal{B}(\mathcal{X})$ and $C \in \mathcal{B}(\mathcal{Y}, \mathcal{X})$ be given operators. The following are equivalent:*

(i) *$M_{(X,Y)}$ is an invertible operator for some $X \in \mathcal{B}(\mathcal{X}, \mathcal{Y})$ and $Y \in \mathcal{B}(\mathcal{Y})$.*
(ii) *$\begin{bmatrix} A & C \end{bmatrix} : \mathcal{X} \oplus \mathcal{Y} \to \mathcal{X}$ is a right invertible operator and $\mathcal{N}\left(\begin{bmatrix} A & C \end{bmatrix}\right) \cong \mathcal{Y}$.*
(iii) *$\mathcal{R}(A) + \mathcal{R}(C) = \mathcal{X}$, $\mathcal{N}\left(\begin{bmatrix} A & C \end{bmatrix}\right)$ is a complemented subspace in $\mathcal{X} \oplus \mathcal{Y}$, and $\mathcal{N}(\begin{bmatrix} A & C \end{bmatrix}) \cong \mathcal{Y}$.*

In the case of separable Hilbert spaces, the following result from [**93**] is a direct consequence of Theorem 4.1.

COROLLARY 4.2 ([**93**]). Let $\mathcal{H}$ and $\mathcal{K}$ be separable Hilbert spaces and let $A \in \mathcal{B}(\mathcal{H})$ and $C \in \mathcal{B}(\mathcal{K}, \mathcal{H})$. The following are equivalent:

(i) $M_{(X,Y)}$ is an invertible operator for some $X \in \mathcal{B}(\mathcal{H}, \mathcal{K})$ and $Y \in \mathcal{B}(\mathcal{K})$.
(ii) $\begin{bmatrix} A & C \end{bmatrix} : \mathcal{H} \oplus \mathcal{K} \to \mathcal{H}$ is right invertible and $\dim(\mathcal{N}\left(\begin{bmatrix} A & C \end{bmatrix}\right)) = \dim \mathcal{K}$.
(iii) $\mathcal{R}(A) + \mathcal{R}(C) = \mathcal{H}$ and $\dim(\mathcal{N}\left(\begin{bmatrix} A & C \end{bmatrix}\right)) = \dim \mathcal{K}$.

THEOREM 4.3 ([**95**]). *Let $\mathcal{X}$ and $\mathcal{Y}$ be Banach spaces, and let $A \in \mathcal{B}(\mathcal{X})$ and $C \in \mathcal{B}(\mathcal{Y}, \mathcal{X})$ be given operators.*

(i) *If $\mathcal{N}(\begin{bmatrix} A & C \end{bmatrix})$ is complemented in $\mathcal{X} \oplus \mathcal{Y}$ and $M_{(X,Y)}$ is left invertible for some $X \in \mathcal{B}(\mathcal{X}, \mathcal{Y})$ and $Y \in \mathcal{B}(\mathcal{Y})$, then $\mathcal{N}(\begin{bmatrix} A & C \end{bmatrix}) \preceq \mathcal{Y}$.*
(ii) *If $\begin{bmatrix} A & C \end{bmatrix}$ is generalized invertible and $\mathcal{N}(\begin{bmatrix} A & C \end{bmatrix}) \preceq \mathcal{Y}$, then $M_{(X,Y)}$ is left invertible for some $X \in \mathcal{B}(\mathcal{X}, \mathcal{Y})$ and $\mathcal{Y} \in \mathcal{B}(\mathcal{Y})$.*
(iii) *If $\mathcal{X} \oplus \mathcal{Y} \preceq \mathcal{Y}$, then $M_{(X,Y)}$ is left invertible for some $X \in \mathcal{B}(\mathcal{X}, \mathcal{Y})$ and $Y \in \mathcal{B}(\mathcal{Y})$.*

In the case of separable Hilbert spaces, we have the following corollary:

COROLLARY 4.4 ([**34**, Theorem 2.1]). Let $\mathcal{H}$ and $\mathcal{K}$ be separable Hilbert spaces. Let $A \in \mathcal{B}(\mathcal{H})$ and $C \in \mathcal{B}(\mathcal{K}, \mathcal{H})$.

(i) If $\mathcal{K}$ is infinite dimensional, then $M_{(X,Y)}$ is left invertible for some $X \in \mathcal{B}(\mathcal{H}, \mathcal{K})$ and $Y \in \mathcal{B}(\mathcal{K})$.
(ii) If $\mathcal{K}$ is finite dimensional, then $M_{(X,Y)}$ is left invertible for some $X \in \mathcal{B}(\mathcal{H}, \mathcal{K})$ and $Y \in \mathcal{B}(\mathcal{K})$ if and only if $\mathcal{R}(A)$ is a closed subspace and $\dim(\mathcal{N}(\begin{bmatrix} A & C \end{bmatrix})) \leq \dim \mathcal{K}$.

The next result gives a description for the set $\bigcap_{X \in \mathcal{B}(\mathcal{X},\mathcal{Y}),\, Y \in \mathcal{B}(\mathcal{Y})} \sigma(M_{(X,Y)})$:

COROLLARY 4.5 ([**95**]). Let $\mathcal{X}$ and $\mathcal{Y}$ be Banach spaces and let $A \in \mathcal{B}(\mathcal{X})$ and $C \in \mathcal{B}(\mathcal{Y}, \mathcal{X})$. Then

$$\begin{aligned}\bigcap_{X \in \mathcal{B}(\mathcal{X},\mathcal{Y}),\, Y \in \mathcal{B}(\mathcal{Y})} \sigma(M_{(X,Y)}) =& \{\lambda \in \mathbb{C} \colon [A - \lambda I_{\mathcal{X}} \quad C] \text{ is not right invertible}\} \\ & \cup \{\lambda \in \mathbb{C} \colon \mathcal{N}([A - \lambda I_{\mathcal{X}} \quad C]) \cong \mathcal{Y} \text{ does not hold}\}.\end{aligned}$$

COROLLARY 4.6 ([**95**]). Let $\mathcal{X}$ and $\mathcal{Y}$ be Banach spaces and let $A \in \mathcal{B}(\mathcal{X})$ and $C \in \mathcal{B}(\mathcal{Y}, \mathcal{X})$.

(i) If $\mathcal{X} \oplus \mathcal{Y} \preceq \mathcal{Y}$, then

$$\bigcap_{X \in \mathcal{B}(\mathcal{X},\mathcal{Y}),\, Y \in \mathcal{B}(\mathcal{Y})} \sigma_l(M_{(X,Y)}) = \emptyset.$$

(ii) If $\mathcal{X} \oplus \mathcal{Y} \preceq \mathcal{Y}$ does not hold, then

$$\begin{aligned}
&\{\lambda \in \mathbb{C} \colon \mathcal{N}([A - \lambda I_{\mathcal{X}} \;\; C]) \text{ is complemented in } \mathcal{X} \oplus \mathcal{Y}\} \\
&\quad \cap \{\lambda \in \mathbb{C} \colon \mathcal{N}([A - \lambda I_{\mathcal{X}} \;\; C]) \preceq \mathcal{Y} \text{ does not hold}\} \\
&\quad \subseteq \bigcap_{X \in \mathcal{B}(\mathcal{X},\mathcal{Y}),\, Y \in \mathcal{B}(\mathcal{Y})} \sigma_l(M_{(X,Y)}) \\
&\quad \subseteq \{\lambda \in \mathbb{C} \colon \mathcal{N}([A - \lambda I_{\mathcal{X}} \;\; C]) \preceq \mathcal{Y} \text{ does not hold}\} \\
&\qquad \cup \{\lambda \in \mathbb{C} \colon \mathcal{N}([A - \lambda I_{\mathcal{X}} \;\; C]) \text{ is not generalized invertible}\}.
\end{aligned}$$

Now, we assume that $\mathcal{H}_1$ and $\mathcal{H}_2$ are separable Hilbert spaces and $A \in \mathcal{B}(\mathcal{H}_1)$, $C \in \mathcal{B}(\mathcal{H}_2, \mathcal{H}_1)$. We present some results from [**34**] that concern a necessary and sufficient condition for $M_{(X,Y)}$ to be right invertible for some $X \in \mathcal{B}(\mathcal{H}_1, \mathcal{H}_2)$ and $Y \in \mathcal{B}(\mathcal{H}_2)$. In addition, some results on left invertibility on $M_{(X,Y)}$ will be shown.

THEOREM 4.7 ([**34**]). *Let $A \in \mathcal{B}(\mathcal{H}_1)$ and $C \in \mathcal{B}(\mathcal{H}_2, \mathcal{H}_1)$ be given operators. Then $M_{(X,Y)}$ is a right invertible operator for some $X \in \mathcal{B}(\mathcal{H}_1, \mathcal{H}_2)$ and $Y \in \mathcal{B}(\mathcal{H}_2)$ if and only if $\mathcal{R}(A) + \mathcal{R}(C) = \mathcal{H}_1$ and $\dim(\mathcal{N}([\; A \quad C \;])) \geq \dim \mathcal{H}_2$.*

COROLLARY 4.8 ([**34**]). Let $A \in \mathcal{B}(\mathcal{H}_1)$ and $C \in \mathcal{B}(\mathcal{H}_2, \mathcal{H}_1)$ be given operators. Assume that $\dim \mathcal{H}_1 = \dim \mathcal{H}_2 = \infty$. Then $M_{(X,Y)}$ is a right invertible operator for some $X \in \mathcal{B}(\mathcal{H}_1, \mathcal{H}_2)$ and $Y \in \mathcal{B}(\mathcal{H}_2)$ if and only if $\mathcal{R}(A) + \mathcal{R}(C) = \mathcal{H}_1$ and one of the following conditions holds:

(i) $\mathcal{N}(A)$ is infinite dimensional.
(ii) $\mathcal{N}(C)$ is infinite dimensional.
(iii) $\mathcal{R}(A) \cap \mathcal{R}(C)$ is infinite dimensional.

COROLLARY 4.9 ([**34**]). Let $A \in \mathcal{B}(\mathcal{H}_1)$ and $C \in \mathcal{B}(\mathcal{H}_2, \mathcal{H}_1)$ be given operators. Assume that $\dim \mathcal{H}_1 < \infty$. Then $M_{(X,Y)}$ is a right invertible operator for some $X \in \mathcal{B}(\mathcal{H}_1, \mathcal{H}_2)$ and $Y \in \mathcal{B}(\mathcal{H}_2)$ if and only if $\mathcal{R}(A) + \mathcal{R}(C) = \mathcal{H}_1$.

COROLLARY 4.10 ([**34**]). Let $A \in \mathcal{B}(\mathcal{H}1)$ and $C \in \mathcal{B}(\mathcal{H}_2, \mathcal{H}_1)$ be given operators. Assume that $\dim \mathcal{H}_2 < \dim \mathcal{H}_1 = \infty$. Then $M_{(X,Y)}$ is a right invertible operator for some $X \in \mathcal{B}(\mathcal{H}_1, \mathcal{H}_2)$ and $Y \in \mathcal{B}(\mathcal{H}_2)$ if and only if $\mathcal{R}(A) + \mathcal{R}(C) = \mathcal{H}_1$ and $\dim(\mathcal{N}(A)) \geq \dim(\mathcal{R}(A)^{\perp})$.

COROLLARY 4.11 ([**34**]). Let $A \in \mathcal{B}(\mathcal{H}_1)$ and $C \in \mathcal{B}(\mathcal{H}_2, \mathcal{H}_1)$ be given operators.

(i) If $\dim \mathcal{H}_1 < \infty$ or $\dim(\mathcal{N}(C)) = \infty$, then

$$\bigcap_{X,Y} \sigma_r(M_{(X,Y)}) = \{\lambda \in \mathbb{C} \colon \mathcal{R}(A - \lambda I) + \mathcal{R}(C) \neq \mathcal{H}_1\}.$$

(ii) If $\dim \mathcal{H}_2 < \dim \mathcal{H}_1 = \infty$, then

$$\begin{aligned}
\bigcap_{X,Y} \sigma_r(M_{(X,Y)}) &= \{\lambda \in \mathbb{C} \colon \mathcal{R}(A - \lambda I) + \mathcal{R}(C) \neq \mathcal{H}_1\} \\
&\quad \cup \{\lambda \in \mathbb{C} \colon \dim(\mathcal{N}(A - \lambda I)) < \dim(\mathcal{R}(A - \lambda I)^{\perp}\}).
\end{aligned}$$

(iii) If $\dim(\mathcal{N}(C)) < \dim \mathcal{H}_1 = \dim \mathcal{H}_2 = \infty$, then

$$\begin{aligned}
\bigcap_{X,Y} \sigma_r(M_{(X,Y)}) &= \{\lambda \in \mathbb{C} \colon \mathcal{R}(A - \lambda I) + \mathcal{R}(C) \neq \mathcal{H}_1\} \\
&\quad \cup \{\lambda \in \mathbb{C} \colon \dim(\mathcal{N}(A - \lambda I)) < \infty,\ \dim(\mathcal{R}(A - \lambda I) \cap \mathcal{R}(C)) < \infty\}.
\end{aligned}$$

THEOREM 4.12 ([**34**]). *Let $A \in \mathcal{B}(\mathcal{H}_1)$ and $C \in \mathcal{B}(\mathcal{H}_2, \mathcal{H}_1)$ be given operators.*

(i) *If $\dim \mathcal{H}_2 = \infty$, then $M_{(X,Y)}$ is a left invertible operator for some $X \in \mathcal{B}(\mathcal{H}_1, \mathcal{H}_2)$ and $Y \in \mathcal{B}(\mathcal{H}_2)$.*

(ii) *If $\dim \mathcal{H}_2 < \infty$, then $M_{(X,Y)}$ is a left invertible operator for some $X \in \mathcal{B}(\mathcal{H}_1, \mathcal{H}_2)$ and $Y \in \mathcal{B}(\mathcal{H}_2)$ if and only if $\mathcal{R}(A)$ is a closed subspace and $\dim(\mathcal{N}([\begin{array}{cc} A & C \end{array}])) \leq \dim \mathcal{H}_2$.*

COROLLARY 4.13 ([**34**]). Let $A[\begin{array}{cc} A & C \end{array}] \in \mathcal{B}(\mathcal{H}_1)$ and $C \in \mathcal{B}(\mathcal{H}_2, \mathcal{H}_1)$ be given operators.

(i) If $\dim \mathcal{H}_2 = \infty$, then

$$\bigcap_{X,Y} \sigma_l(M_{(X,Y)}) = \emptyset.$$

(ii) If $\dim \mathcal{H}_2 < \infty$, then

$$\bigcap_{X,Y} \sigma_l(M_{(X,Y)}) = \{\lambda \in \mathbb{C} \colon \mathcal{R}(A - \lambda I) \text{ is not closed}\}$$
$$\cup \{\lambda \in \mathbb{C} \colon \dim(\mathcal{R}(A - \lambda I) \cap \mathcal{R}(C)) + \dim(\mathcal{N}(A - \lambda I)) > \dim(\mathcal{R}(C))\}.$$

4.2. Consistency of operator matrices $M_{(X,Y)}$ in invertibility

Let $\mathcal{H}$ be separable Hilbert space. Let $\mathcal{S}(\mathcal{H})$ denote a subset of $\mathcal{B}(\mathcal{H})$. An operator $A \in \mathcal{B}(\mathcal{H})$ is said to be $\mathcal{S}(\mathcal{H})$ consistent, or consistent in $\mathcal{S}(\mathcal{H})$, if

$$AB \in \mathcal{S}(\mathcal{H}) \Leftrightarrow BA \in \mathcal{S}(\mathcal{H}),$$

for every $B \in \mathcal{B}(\mathcal{H})$.

A characterization of operators consistent in invertibility (CI operators for short) and a characterization of CI operators invariant under compact perturbations were considered in [**86**]. Also, some work on consistency of operators and applications of it can be found in [**32**].

Motivation for the work presented in this section that is published in [**53**] was some papers in which various problems were considered of the completing of upper triangular operator matrices to different types of consistency. One of them was the paper of G. Hai et al. [**91**] in which for given operators $A \in \mathcal{B}(\mathcal{H})$ and $B \in \mathcal{B}(\mathcal{K})$ on separable Hilbert spaces, the existence of completions that are consistent in invertibility of the upper triangular operator matrix

$$\begin{bmatrix} A & ? \\ 0 & B \end{bmatrix} : \begin{bmatrix} \mathcal{H} \\ \mathcal{K} \end{bmatrix} \to \begin{bmatrix} \mathcal{H} \\ \mathcal{K} \end{bmatrix} \tag{4.1}$$

was considered.

In this section for given $A \in \mathcal{B}(\mathcal{H})$ and $C \in \mathcal{B}(\mathcal{K}, \mathcal{H})$, we consider the problem of existence of operators $X \in \mathcal{B}(\mathcal{H}, \mathcal{K})$ and $Y \in \mathcal{B}(\mathcal{K})$ such that

$$M_{(X,Y)} = \begin{bmatrix} A & C \\ X & Y \end{bmatrix} : \begin{bmatrix} \mathcal{H} \\ \mathcal{K} \end{bmatrix} \to \begin{bmatrix} \mathcal{H} \\ \mathcal{K} \end{bmatrix} \tag{4.2}$$

is consistent in invertibility.

Using the standard notations

$$\sigma_{CI}(A) = \{\lambda \in \mathbb{C}\colon\ A - \lambda \text{ is not consistent in invertibility}\}$$

for given $A \in \mathcal{B}(\mathcal{H})$ and $C \in \mathcal{B}(\mathcal{K}, \mathcal{H})$, we completely describe the following sets:

$$\bigcap_{X\in\mathcal{B}(\mathcal{H},\mathcal{K}),\ Y\in\mathcal{B}(\mathcal{K})} \sigma_{CI}(M_{(X,Y)}) \quad \text{and} \quad \bigcup_{X\in\mathcal{B}(\mathcal{H},\mathcal{K}),\ Y\in\mathcal{B}(\mathcal{K})} \sigma_{CI}(M_{(X,Y)}).$$

Before presenting the main results, we first give some auxiliary lemmas and theorems. We begin with some conditions that guarantee that an operator $X \in \mathcal{B}(\mathcal{H}, \mathcal{K})$ is consistent in invertibility.

THEOREM 4.14 ([**86**]). *An operator $X \in \mathcal{B}(\mathcal{H}, \mathcal{K})$ is a CI operator if and only if one of the following three mutually exclusive cases holds:*

(i) *X is invertible.*
(ii) *$\mathcal{R}(X)$ is not closed.*
(iii) *$\mathcal{N}(X) \neq \{0\}$ and $\mathcal{R}(X) = \overline{\mathcal{R}(X)} \neq \mathcal{K}$.*

We also need the following auxiliary result:

LEMMA 4.15. *Let $M_{(X,Y)} \in \mathcal{B}(\mathcal{H} \oplus \mathcal{K})$ be given by (4.2). If $\mathcal{R}(M_{(X,Y)})$ is not closed, then $\begin{bmatrix} A & C \end{bmatrix}$ is not left invertible.*

PROOF. Suppose that $\begin{bmatrix} A & C \end{bmatrix}$ is left invertible and take an arbitrary element $\begin{bmatrix} x & y \end{bmatrix} \in \overline{\mathcal{R}(M_{(X,Y)})}$. Then there exist sequences (x_n) from $\mathcal{H}$ and (y_n) from $\mathcal{K}$ such that

$$Ax_n + Cy_n \to x,\ n \to \infty,$$
$$Xx_n + Yy_n \to y,\ n \to \infty.$$

Since $\begin{bmatrix} A & C \end{bmatrix}$ is left invertible we get that sequence $(\begin{bmatrix} x_n & y_n \end{bmatrix})$ converges, so $\begin{bmatrix} x & y \end{bmatrix} \in \mathcal{R}(M_{(X,Y)})$. Hence, $\mathcal{R}(M_{(X,Y)})$ is closed. □

In the following theorem we consider the problem of the existence of operators $X \in \mathcal{B}(\mathcal{H}, \mathcal{K})$ and $Y \in \mathcal{B}(\mathcal{K})$ such that an operator matrix $M_{(X,Y)}$ given by (4.2) is a CI operator. We suppose that $\mathcal{H}, \mathcal{K} \neq \{0\}$ are separable Hilbert spaces.

THEOREM 4.16. *Let $A \in \mathcal{B}(\mathcal{H})$ and $C \in \mathcal{B}(\mathcal{K}, \mathcal{H})$. There exist operators $X \in \mathcal{B}(\mathcal{H}, \mathcal{K})$ and $Y \in \mathcal{B}(\mathcal{K})$ such that operator matrix $M_{(X,Y)}$ given by (4.2) is a CI operator if and only if $\begin{bmatrix} A & C \end{bmatrix}$ is not left invertible. Furthermore, if $M_{(X,Y)}$ is a CI operator for some $X \in \mathcal{B}(\mathcal{H}, \mathcal{K})$ and $Y \in \mathcal{B}(\mathcal{K})$, then $M_{(0,0)}$ is a CI operator.*

PROOF. Let $X \in \mathcal{B}(\mathcal{H}, \mathcal{K})$ and $Y \in \mathcal{B}(\mathcal{K})$ be such that $M_{(X,Y)}$ is a CI operator. According to Theorem 4.14 we have the following three possibilities:

(i) $M_{(X,Y)}$ is invertible: From Theorem 4.1 we have that $\dim(\mathcal{N}(\begin{bmatrix} A & C \end{bmatrix})) = \dim \mathcal{K}$. Since $\mathcal{K} \neq \{0\}$ it follows that $\begin{bmatrix} A & C \end{bmatrix}$ is not left invertible.

(ii) The range $\mathcal{R}(M_{(X,Y)})$ is not closed: By Lemma 4.15, we conclude that $\begin{bmatrix} A & C \end{bmatrix}$ is not left invertible.

(iii) $\mathcal{N}(M_{(X,Y)}) \neq \{0\}$ and $\mathcal{R}(M_{(X,Y)}) = \overline{\mathcal{R}(M_{(X,Y)})} \neq \mathcal{K}$: $\mathcal{N}(M_{(X,Y)}) \subseteq \mathcal{N}(\begin{bmatrix} A & C \end{bmatrix})$ implies that $\begin{bmatrix} A & C \end{bmatrix}$ is not injective and thus is not left invertible.

On the contrary, suppose that $\begin{bmatrix} A & C \end{bmatrix}$ is not left-invertible. If $\mathcal{R}(\begin{bmatrix} A & C \end{bmatrix})$ is not closed, then $\mathcal{R}(M_{(0,0)}) = \mathcal{R}(\begin{bmatrix} A & C \end{bmatrix})$ is not closed in $\mathcal{H}\oplus\mathcal{K}$ and by Theorem 4.14 we have that $M_{(0,0)}$ is CI. If $\mathcal{R}(\begin{bmatrix} A & C \end{bmatrix})$ is closed and $\mathcal{N}(\begin{bmatrix} A & C \end{bmatrix}) \neq \{0\}$, then we have that $\mathcal{R}(M_{(0,0)}) = \overline{\mathcal{R}(M_{(0,0)})} = \mathcal{R}(\begin{bmatrix} A & C \end{bmatrix}) \neq \mathcal{H} \oplus \mathcal{K}$ and that $\mathcal{N}(M_{(0,0)}) = \mathcal{N}(\begin{bmatrix} A & C \end{bmatrix}) \neq \{0\}$. Again, by Theorem 4.14 we have that $M_{(0,0)}$ is CI. □

In the following theorem we prove that in the case when $\mathcal{K}$ is an infinite-dimensional space, there always exist operators $X \in \mathcal{B}(\mathcal{H},\mathcal{K})$ and $Y \in \mathcal{B}(\mathcal{K})$ such that $M_{(X,Y)}$ is not CI.

THEOREM 4.17. *Let $A \in \mathcal{B}(\mathcal{H})$ and $C \in \mathcal{B}(\mathcal{K},\mathcal{H})$. If $\dim \mathcal{K} = \infty$, then there exist $X \in \mathcal{B}(\mathcal{H},\mathcal{K})$ and $Y \in \mathcal{B}(\mathcal{K})$ such that $M_{(X,Y)}$ given by (4.2) is not CI.*

PROOF. First, assume that there exist $X \in \mathcal{B}(\mathcal{H},\mathcal{K})$ and $Y \in \mathcal{B}(\mathcal{K})$ such that $M_{(X,Y)}$ is a right invertible operator. Then by Theorem 4.7, $\begin{bmatrix} A & C \end{bmatrix}$ is a right invertible operator and $\dim(\mathcal{N}(\begin{bmatrix} A & C \end{bmatrix})) = \infty$. Since $\dim \mathcal{K} = \infty$, there exist an infinite-dimensional closed subspace $\mathcal{S}$ of $\mathcal{K}$, $\mathcal{S} \neq \mathcal{K}$, and an operator $\begin{bmatrix} X & Y \end{bmatrix} : \begin{bmatrix} \mathcal{H} \\ \mathcal{K} \end{bmatrix} \to \mathcal{K}$ such that

$$\begin{bmatrix} X & Y \end{bmatrix}|_{\mathcal{N}(\begin{bmatrix} A & C \end{bmatrix})^{\perp}} = 0, \quad \mathcal{N}\left(\begin{bmatrix} X & Y \end{bmatrix}|_{\mathcal{N}(\begin{bmatrix} A & C \end{bmatrix})}\right) = \{0\},$$

and $\mathcal{R}(\begin{bmatrix} X & Y \end{bmatrix}) = \mathcal{S}$. Now, $\mathcal{N}(M_{(X,Y)}) = \{0\}$ and $\mathcal{R}(M_{(X,Y)}) = \mathcal{H} \oplus \mathcal{S} \neq \mathcal{H} \oplus \mathcal{K}$. By Theorem 4.14, we have that $M_{(X,Y)}$ is not a CI operator.

If on the other hand there are no $X \in \mathcal{B}(\mathcal{H},\mathcal{K})$ and $Y \in \mathcal{B}(\mathcal{K})$ such that $M_{(X,Y)}$ is a right invertible operator, we can find operators $X_0 \in \mathcal{B}(\mathcal{H},\mathcal{K})$ and $Y_0 \in \mathcal{B}(\mathcal{K})$ for which $M_{(X_0,Y_0)}$ is left invertible—the existence of these is guaranteed by Corollary 4.4. Since for $M_{(X_0,Y_0)}$ neither one of the cases (i)–(iii) of Theorem 4.14 holds, we can conclude that $M_{(X_0,Y_0)}$ is not a CI operator. □

If we compare the previous result with the analogous one concerning the case of upper triangular operator matrices, we can observe the following difference in them. Namely, if $\dim \mathcal{K} = \infty$, there is no pair of operators $A \in \mathcal{B}(\mathcal{H})$ and $C \in \mathcal{B}(\mathcal{K},\mathcal{H})$ such that $M_{(X,Y)}$ is a CI operator for all $X \in \mathcal{B}(\mathcal{H},\mathcal{K})$ and $Y \in \mathcal{B}(\mathcal{K})$ while in the case of upper traingular operator matrices the situation is different (see [**91**, Theorem 4]).

In the following result we consider the case when $\mathcal{K}$ is a finite-dimensional nontrivial Hilbert space and give necessary and sufficient conditions for $M_{(X,Y)}$ to be CI for any $X \in \mathcal{B}(\mathcal{H},\mathcal{K})$ and $Y \in \mathcal{B}(\mathcal{K})$.

THEOREM 4.18. *Let $A \in \mathcal{B}(\mathcal{H})$ and $C \in \mathcal{B}(\mathcal{K},\mathcal{H})$ be given operators and let $\dim(\mathcal{K}) < \infty$. The operator matrix $M_{(X,Y)}$ given by (4.2) is a CI operator for all $X \in \mathcal{B}(\mathcal{H},\mathcal{K})$ and $Y \in \mathcal{B}(\mathcal{K})$ if and only if one of the following conditions holds:*

(i) *$\mathcal{R}(A)$ is not closed.*
(ii) *$\mathcal{R}(A) + \mathcal{R}(C) = \mathcal{H}$, $\dim(\mathcal{N}(\begin{bmatrix} A & C \end{bmatrix})) = \dim \mathcal{K}$.*
(iii) *$\mathcal{R}(A) + \mathcal{R}(C) \neq \mathcal{H}$, $\dim(\mathcal{N}(\begin{bmatrix} A & C \end{bmatrix})) > \dim \mathcal{K}$.*

PROOF. We distinguish two cases: when $\mathcal{R}(A)$ is closed and when it is not. If $\mathcal{R}(A)$ is not closed, then for any $X \in \mathcal{B}(\mathcal{H},\mathcal{K})$ and $Y \in \mathcal{B}(\mathcal{K})$ we have that $\mathcal{R}(M_{(X,Y)})$ is not closed as the range of the sum of A and the finite range operator $\begin{bmatrix} 0 & C \\ X & Y \end{bmatrix}$. So by Theorem 4.14 we get that $M_{(X,Y)}$ is a CI operator for any

$X \in \mathcal{B}(\mathcal{H},\mathcal{K})$ and $Y \in \mathcal{B}(\mathcal{K})$. Now, suppose that $\mathcal{R}(A)$ is closed. For the above-mentioned reason, $\mathcal{R}(M_{(X,Y)})$ is closed. Now, we consider 5 cases:

Case 1: $\mathcal{R}(A) + \mathcal{R}(C) = \mathcal{H}$ and $\dim(\mathcal{N}(\begin{bmatrix} A & C \end{bmatrix})) > \dim \mathcal{K}$. There exist $X \in \mathcal{B}(\mathcal{H},\mathcal{K})$ and $Y \in \mathcal{B}(\mathcal{K})$, such that the restriction of the operator $\begin{bmatrix} X & Y \end{bmatrix}$ to the subspace $\mathcal{N}(\begin{bmatrix} A & C \end{bmatrix})$ is not injective but it is onto. This implies that $\mathcal{N}(M_{(X,Y)}) = \mathcal{N}(\begin{bmatrix} A & C \end{bmatrix}) \cap \mathcal{N}(\begin{bmatrix} X & Y \end{bmatrix}) \neq \{0\}$ and $\mathcal{R}(M_{(X,Y)}) = \mathcal{H} \oplus \mathcal{K}$. By Theorem 4.14 we get that for such choice of X and Y, $M_{(X,Y)}$ is not a CI operator.

Case 2: $\mathcal{R}(A)+\mathcal{R}(C) = \mathcal{H}$ and $\dim(\mathcal{N}(\begin{bmatrix} A & C \end{bmatrix})) < \dim \mathcal{K}$. Then there exist $X \in \mathcal{B}(\mathcal{H},\mathcal{K})$ and $Y \in \mathcal{B}(\mathcal{K})$, such that the restriction of the operator $\begin{bmatrix} X & Y \end{bmatrix}$ to the subspace $\mathcal{N}(\begin{bmatrix} A & C \end{bmatrix})$ is injective but it is not onto. This implies that $\mathcal{N}(M_{(X,Y)}) = \mathcal{N}(\begin{bmatrix} A & C \end{bmatrix}) \cap \mathcal{N}(\begin{bmatrix} X & Y \end{bmatrix}) = \{0\}$ and $\mathcal{R}(M_{(X,Y)}) \neq \mathcal{H} \oplus \mathcal{K}$. By Theorem 4.14 we get that for such choice of X and Y, $M_{(X,Y)}$ is not a CI operator.

Case 3: $\mathcal{R}(A) + \mathcal{R}(C) = \mathcal{H}$ and $\dim(\mathcal{N}(\begin{bmatrix} A & C \end{bmatrix})) = \dim \mathcal{K}$. If $X \in \mathcal{B}(\mathcal{H},\mathcal{K})$ and $Y \in \mathcal{B}(\mathcal{K})$ are such that the restriction of the operator $\begin{bmatrix} X & Y \end{bmatrix}$ to the subspace $\mathcal{N}(\begin{bmatrix} A & C \end{bmatrix})$ is bijective, then evidently $M_{(X,Y)}$ is invertible and by Theorem 4.14 we conclude that $M_{(X,Y)}$ is a CI operator. If $X \in \mathcal{B}(\mathcal{H},\mathcal{K})$ and $Y \in \mathcal{B}(\mathcal{K})$ are such that the restriction of the operator $\begin{bmatrix} X & Y \end{bmatrix}$ to the subspace $\mathcal{N}(\begin{bmatrix} A & C \end{bmatrix})$ is not bijective, then it is both: noninjective and nonsurjective. Hence, $\mathcal{N}(M_{(X,Y)}) = \mathcal{N}(\begin{bmatrix} A & C \end{bmatrix}) \cap \mathcal{N}(\begin{bmatrix} X & Y \end{bmatrix}) \neq \{0\}$. Also, from $\mathcal{R}\left(\begin{bmatrix} X & Y \end{bmatrix}|_{\mathcal{N}(\begin{bmatrix} A & C \end{bmatrix})}\right) \neq \mathcal{K}$, it follows that $\mathcal{R}(M_{(X,Y)}) \neq \mathcal{H} \oplus \mathcal{K}$. Now by Theorem 4.14, we conclude that $M_{(X,Y)}$ is a CI operator. Hence, $M_{(X,Y)}$ is a CI operator for any $X \in \mathcal{B}(\mathcal{H},\mathcal{K})$ and $Y \in \mathcal{B}(\mathcal{K})$.

Case 4: $\mathcal{R}(A) + \mathcal{R}(C) \neq \mathcal{H}$ and $\dim(\mathcal{N}(\begin{bmatrix} A & C \end{bmatrix})) \leq \dim \mathcal{K}$. Then we have that for any $X \in \mathcal{B}(\mathcal{H},\mathcal{K})$ and $Y \in \mathcal{B}(\mathcal{K})$, $\mathcal{R}(M_{(X,Y)}) \neq \mathcal{H} \oplus \mathcal{K}$. Also there exist $X \in \mathcal{B}(\mathcal{H},\mathcal{K})$ and $Y \in \mathcal{B}(\mathcal{K})$, such that the restriction of the operator $\begin{bmatrix} X & Y \end{bmatrix}$ to the subspace $\mathcal{N}(\begin{bmatrix} A & C \end{bmatrix})$ is injective. This implies that $\mathcal{N}(M_{(X,Y)}) = \mathcal{N}(\begin{bmatrix} A & C \end{bmatrix}) \cap \mathcal{N}(\begin{bmatrix} X & Y \end{bmatrix}) = \{0\}$ and by Theorem 4.14 we get that for such choice of X and Y, $M_{(X,Y)}$ is not a CI operator.

Case 5: $\mathcal{R}(A) + \mathcal{R}(C) \neq \mathcal{H}$ and $\dim(\mathcal{N}(\begin{bmatrix} A & C \end{bmatrix})) > \dim \mathcal{K}$. Then we have that for any $X \in \mathcal{B}(\mathcal{H},\mathcal{K})$ and $Y \in \mathcal{B}(\mathcal{K})$, $\mathcal{R}(M_{(X,Y)}) \neq \mathcal{H} \oplus \mathcal{K}$ and $\mathcal{N}(M_{(X,Y)}) = \mathcal{N}(\begin{bmatrix} A & C \end{bmatrix}) \cap \mathcal{N}(\begin{bmatrix} X & Y \end{bmatrix}) \neq \{0\}$. By Theorem 4.14, it follows that for any $X \in \mathcal{B}(\mathcal{H},\mathcal{K})$ and $Y \in \mathcal{B}(\mathcal{K})$, $M_{(X,Y)}$ is a CI operator. □

In the next corollary we consider the case when $\dim \mathcal{K} < \infty$ and $M_{(X,Y)}$ is not CI.

COROLLARY 4.19. Let $A \in \mathcal{B}(\mathcal{H})$ and $C \in \mathcal{B}(\mathcal{K},\mathcal{H})$. If $\dim \mathcal{K} < \infty$, then there exist operators $X \in \mathcal{B}(\mathcal{H},\mathcal{K})$ and $Y \in \mathcal{B}(\mathcal{K})$ such that $M_{(X,Y)}$ given by (4.2) is not CI if and only if one of the following mutually disjoint conditions holds:

(i) $\mathcal{R}(A) + \mathcal{R}(C) = \mathcal{H}$, $\dim(\mathcal{N}(\begin{bmatrix} A & C \end{bmatrix})) \neq \dim \mathcal{K}$.

(ii) $\mathcal{R}(A)$ is closed, $\mathcal{R}(A) + \mathcal{R}(C) \neq \mathcal{H}$, $\dim(\mathcal{N}(\begin{bmatrix} A & C \end{bmatrix})) \leq \dim \mathcal{K}$.

Now using Theorems 4.16–4.18, we can completely describe the sets

$$\bigcap_{X\in\mathcal{B}(\mathcal{H},\mathcal{K}),\ Y\in\mathcal{B}(\mathcal{K})} \sigma_{CI}(M_{(X,Y)}) \quad \text{and} \quad \bigcup_{X\in\mathcal{B}(\mathcal{H},\mathcal{K}),\ Y\in\mathcal{B}(\mathcal{K})} \sigma_{CI}(M_{(X,Y)}).$$

COROLLARY 4.20. Let $\mathcal{H},\mathcal{K}\neq\{0\}$ and let $A\in\mathcal{B}(\mathcal{H})$ and $C\in\mathcal{B}(\mathcal{K},\mathcal{H})$. Then

$$\bigcap_{X\in\mathcal{B}(\mathcal{H},\mathcal{K}),\ Y\in\mathcal{B}(\mathcal{K})}\sigma_{CI}(M_{(X,Y)})=\{\lambda\in\mathbb{C}\colon \begin{bmatrix} A-\lambda I & C\end{bmatrix} \text{ is not left invetible}\}.$$

COROLLARY 4.21. Let $A\in\mathcal{B}(\mathcal{H})$ and $C\in\mathcal{B}(\mathcal{K},\mathcal{H})$.

(i) If $\dim\mathcal{K}=\infty$, then $\displaystyle\bigcup_{X\in\mathcal{B}(\mathcal{H},\mathcal{K}),\ Y\in\mathcal{B}(\mathcal{K})}\sigma_{CI}(M_{(X,Y)})=\mathbb{C}.$

(ii) If $\dim\mathcal{K}<\infty$, then

$$\begin{aligned}\bigcup_{X\in\mathcal{B}(\mathcal{H},\mathcal{K}),\ Y\in\mathcal{B}(\mathcal{K})}\sigma_{CI}(M_{(X,Y)})=\{\lambda\in\mathbb{C}\colon &\mathcal{R}(A-\lambda)+\mathcal{R}(C)=\mathcal{H},\\ &\dim(\mathcal{N}(\begin{bmatrix} A-\lambda & C\end{bmatrix}))\neq\dim\mathcal{K}\}\\ \cup\,\{\lambda\in\mathbb{C}\colon \mathcal{R}(A-\lambda)=\overline{\mathcal{R}(A-\lambda)},\ &\mathcal{R}(A-\lambda)+\mathcal{R}(C)\neq\mathcal{H},\\ &\dim(\mathcal{N}(\begin{bmatrix} A-\lambda & C\end{bmatrix}))\leq\dim\mathcal{K}\}.\end{aligned}$$

In [**91**] it was proved that in the case of the upper triangular operator matrix given by

$$M_C=\begin{bmatrix} A & C\\ 0 & B\end{bmatrix}:\begin{bmatrix}\mathcal{H}\\ \mathcal{K}\end{bmatrix}\to\begin{bmatrix}\mathcal{H}\\ \mathcal{K}\end{bmatrix}, \tag{4.3}$$

we have that for any given operators $A\in\mathcal{B}(\mathcal{H})$ and $B\in\mathcal{B}(\mathcal{K})$,

$$\bigcap_{C\in\mathcal{B}(\mathcal{K},\mathcal{H})}\sigma_{CI}(M_C)=\sigma_{CI}(M_0).$$

As we see in the next example, this is not the case if we consider operator matrices defined by (4.2).

EXAMPLE 4.22. Let $\dim\mathcal{H}=\infty$ and let $\lambda\in\mathbb{C}\setminus\{0\}$. Assume that $A\in\mathcal{B}(\mathcal{H})$ is such that $A-\lambda$ is left invertible, but not right invertible, and let $C\in\mathcal{B}(\mathcal{K},\mathcal{H})$ be a noninjective operator. We prove that $\lambda\in\sigma_{CI}(M_{(0,0)})$, but on the other side $\lambda\notin\bigcap_{X\in\mathcal{B}(\mathcal{H},\mathcal{K}),\ Y\in\mathcal{B}(\mathcal{K})}\sigma_{CI}(M_{(X,Y)})$. Since C is noninjective, it follows that $\begin{bmatrix} A-\lambda & C\end{bmatrix}$ is not left invertible, which by Theorem 4.16 implies the existence of operators $X\in\mathcal{B}(\mathcal{H},\mathcal{K})$ and $Y\in\mathcal{B}(\mathcal{K})$ such that $M_{(X,Y)}-\lambda$ is a CI operator. Hence, $\lambda\notin\bigcap_{X\in\mathcal{B}(\mathcal{H},\mathcal{K}),\ Y\in\mathcal{B}(\mathcal{K})}\sigma_{CI}(M_{(X,Y)})$. To prove that $\lambda\in\sigma_{CI}(M_{(0,0)})$ using Theorem 4.14, it is sufficient to prove that $M_{(0,0)}-\lambda I$ is left invertible, but not right invertible. It is easy to see that

$$\begin{bmatrix} (A-\lambda)_l^{-1} & \frac{1}{\lambda}(A-\lambda)_l^{-1}C\\ 0 & -\frac{1}{\lambda}\end{bmatrix}$$

is a left inverse of $M_{(0,0)}-\lambda I$. In order to prove that $M_{(0,0)}-\lambda I$ is not right invertible, we assume to the contrary that there exists an operator matrix

$$\begin{bmatrix} X & Y\\ Z & W\end{bmatrix}:\begin{bmatrix}\mathcal{H}\\ \mathcal{K}\end{bmatrix}\to\begin{bmatrix}\mathcal{H}\\ \mathcal{K}\end{bmatrix},$$

such that

$$\begin{bmatrix} A-\lambda & C\\ 0 & \lambda\end{bmatrix}\begin{bmatrix} X & Y\\ Z & W\end{bmatrix}=\begin{bmatrix} I_{\mathcal{H}} & 0\\ 0 & I_{\mathcal{K}}\end{bmatrix}. \tag{4.4}$$

Evidently, by (4.4) it follows that $\lambda W = I$ and $\lambda Z = 0$, which in turn implies $(A-\lambda)X = I_{\mathcal{H}}$, which contradicts the assumption that $A-\lambda$ is not right invertible. Hence, $\lambda \in \sigma_{CI}(M_{(0,0)}) \setminus \bigcap_{X\in\mathcal{B}(\mathcal{H},\mathcal{K}),\ Y\in\mathcal{B}(\mathcal{K})} \sigma_{CI}(M_{(X,Y)})$.

4.3. Consistency of operator matrices $M_{(X,Y)}$ in Fredholmness

The problem of completing the upper triangular operator matrix given by (4.1) to a Fredholm consistent operator was considered in [**51**]. In this section for given $A \in \mathcal{B}(\mathcal{H})$ and $C \in \mathcal{B}(\mathcal{K},\mathcal{H})$, we will be interested in the problem of existence of completions of the operator matrix

$$\begin{bmatrix} A & C \\ ? & ? \end{bmatrix} : \begin{bmatrix} \mathcal{H} \\ \mathcal{K} \end{bmatrix} \to \begin{bmatrix} \mathcal{H} \\ \mathcal{K} \end{bmatrix}, \tag{4.5}$$

which are Fredholm consistent (for short FC) operators and we present the results that are mainly published in [**54**].

Using the standard notations

$$\sigma_{FC}(A) = \{\lambda \in \mathbb{C}\colon\ A-\lambda \text{ is not consistent in Fredholmness}\},$$

for given $A \in \mathcal{B}(\mathcal{H})$ and $C \in \mathcal{B}(\mathcal{K},\mathcal{H})$, we completely describe the sets

$$\bigcap_{X\in\mathcal{B}(\mathcal{H},\mathcal{K}),\ Y\in\mathcal{B}(\mathcal{K})} \sigma_{FC}(M_{(X,Y)}) \quad \text{and} \quad \bigcup_{X\in\mathcal{B}(\mathcal{H},\mathcal{K}),\ Y\in\mathcal{B}(\mathcal{K})} \sigma_{FC}(M_{(X,Y)}).$$

The necessary and sufficient conditions under which an operator $A \in \mathcal{B}(\mathcal{H})$ is a Fredholm consistent operator are given in the following theorem:

THEOREM 4.23. *Let $A \in \mathcal{B}(\mathcal{H})$. Then A is an FC operator if and only if one of the following conditions is satisfied:*

(i) $A \in \mathcal{F}(\mathcal{H})$.

(ii) *$\mathcal{R}(A)$ is closed, $\alpha(A) = d(A) = \infty$.*

(iii) *$\mathcal{R}(A)$ is not closed.*

In this section, for given $A \in \mathcal{B}(\mathcal{H})$ and $C \in \mathcal{B}(\mathcal{K},\mathcal{H})$, we consider the problem of completing the operator matrix (4.5) to a Fredholm consistent operator. Notice that this problem is strongly related to the one of completing the operator matrix (4.5) to different classes of Fredholm operators since evidently

$$A \in \mathcal{B}(\mathcal{H}) \text{ is not an FC operator } \Leftrightarrow A \in (\mathcal{F}_r(\mathcal{H}) \setminus \mathcal{F}_l(\mathcal{H})) \cup (\mathcal{F}_l(\mathcal{H}) \setminus \mathcal{F}_r(\mathcal{H})).$$

We assume that $\dim \mathcal{H} = \infty$ since otherwise the problems becomes trivial.

First, we state some auxiliary results which are valid in the more general setting of Banach spaces.

LEMMA 4.24. *Let $\mathcal{X},\mathcal{Y},\mathcal{Z}$ be Banach spaces and let $X \in \mathcal{B}(\mathcal{X},\mathcal{Y})$ and $Y \in \mathcal{B}(\mathcal{Y},\mathcal{Z})$. If $YX \in \mathcal{F}(\mathcal{X},\mathcal{Z})$, then the following holds: $X \in \mathcal{F}(\mathcal{X},\mathcal{Y})$ if and only if $Y \in \mathcal{F}(\mathcal{Y},\mathcal{Z})$.*

LEMMA 4.25. *Let $\mathcal{X},\mathcal{Y}$ be Banach spaces, let $X \in \mathcal{F}(\mathcal{X},\mathcal{Y})$, and let $F \in F(\mathcal{X},\mathcal{Y})$ be a finite rank operator. Then $X + F \in F(\mathcal{X},\mathcal{Y})$. The analogous holds for classes $\mathcal{F}_l$ and $\mathcal{F}_r$.*

LEMMA 4.26. *Let $\mathcal{X},\mathcal{Y}$ be Banach spaces, and let $X \in \mathcal{B}(\mathcal{X},\mathcal{Y})$. The operator X is left semi-Fredholm (right semi-Fredholm) if and only if there exists an operator $Y \in \mathcal{B}(\mathcal{Y},\mathcal{X})$ such that $YX = I_{\mathcal{X}} + F$ $(XY = I_{\mathcal{Y}} + F)$, where $F \in \mathcal{F}(\mathcal{X})(\mathcal{F}(\mathcal{Y}))$.*

The following theorems give conditions under which we can complete the operator matrix given by (4.5) to Fredholm, left semi-Fredholm, and right semi-Fredholm operators, respectively. It is interesting to note that the necessary and sufficient conditions for the existence of any of the three types of completions considered are the same.

THEOREM 4.27. *Let $A \in \mathcal{B}(\mathcal{H})$ and $C \in \mathcal{B}(\mathcal{K}, \mathcal{H})$ be given operators and let $\dim \mathcal{K} = \infty$. There exist $X \in \mathcal{B}(\mathcal{H}, \mathcal{K})$ and $Y \in \mathcal{B}(\mathcal{K})$ such that the operator matrix $M_{(X,Y)}$ given by (4.2) is a Fredholm operator if and only if $\begin{bmatrix} A & C \end{bmatrix} \in \mathcal{F}_r(\mathcal{H} \oplus \mathcal{K}) \setminus \mathcal{F}_l(\mathcal{H} \oplus \mathcal{K})$.*

PROOF. Suppose that $\begin{bmatrix} A & C \end{bmatrix} \in \mathcal{F}_r(\mathcal{H} \oplus \mathcal{K}) \setminus \mathcal{F}_l(\mathcal{H} \oplus \mathcal{K})$. Since $\alpha(\begin{bmatrix} A & C \end{bmatrix}) = \infty$, there exists a left invertible operator $Z \in \mathcal{B}(\mathcal{K}, \mathcal{H} \oplus \mathcal{K})$, such that $\mathcal{R}(Z) = \mathcal{N}(\begin{bmatrix} A & C \end{bmatrix})$. Evidently,

$$\begin{bmatrix} A & C \end{bmatrix} Z = 0. \tag{4.6}$$

Since $\mathcal{R}(Z)$ is closed, there exists the Moore-Penrose inverse $Z^\dagger \in \mathcal{B}(\mathcal{H} \oplus \mathcal{K}, \mathcal{K})$ of the operator Z. Similarly, $\mathcal{R}(\begin{bmatrix} A & C \end{bmatrix})$ is closed, so there exists the Moore-Penrose inverse of $\begin{bmatrix} A & C \end{bmatrix}$, which we denote by L. Obviously,

$$\begin{bmatrix} A & C \end{bmatrix} L = I_{\mathcal{H}} - P_{\mathcal{R}(\begin{bmatrix} A & C \end{bmatrix})^\perp}. \tag{4.7}$$

Since $\begin{bmatrix} A & C \end{bmatrix}$ is right semi-Fredholm, we have that $\dim \mathcal{R}(\begin{bmatrix} A & C \end{bmatrix})^\perp < \infty$ so $P_{\mathcal{R}(\begin{bmatrix} A & C \end{bmatrix})^\perp}$ is a finite rank operator. Furthermore, since

$$\mathcal{R}(L) = \mathcal{N}(\begin{bmatrix} A & C \end{bmatrix})^\perp \quad \text{and} \quad \mathcal{N}(Z^\dagger) = \mathcal{N}(\begin{bmatrix} A & C \end{bmatrix})^\perp,$$

we have that

$$Z^\dagger L = 0. \tag{4.8}$$

Now, suppose that Z, $Z^\dagger$, and L have the following matrix representations:

$$Z = \begin{bmatrix} E \\ G \end{bmatrix} : \mathcal{K} \to \begin{bmatrix} \mathcal{H} \\ \mathcal{K} \end{bmatrix},$$

$$Z^\dagger = \begin{bmatrix} X & Y \end{bmatrix} : \begin{bmatrix} \mathcal{H} \\ \mathcal{K} \end{bmatrix} \to \mathcal{K},$$

and

$$L = \begin{bmatrix} D \\ F \end{bmatrix} : \mathcal{H} \to \begin{bmatrix} \mathcal{H} \\ \mathcal{K} \end{bmatrix}.$$

From (4.6)–(4.8), we get that

$$\begin{bmatrix} A & C \\ X & Y \end{bmatrix} \begin{bmatrix} D & E \\ F & G \end{bmatrix} = \begin{bmatrix} I_{\mathcal{H}} & 0 \\ 0 & I_{\mathcal{K}} \end{bmatrix} - \begin{bmatrix} P_{\mathcal{R}(\begin{bmatrix} A & C \end{bmatrix})^\perp} & 0 \\ 0 & 0 \end{bmatrix}.$$

Since $P_{\mathcal{R}(\begin{bmatrix} A & C \end{bmatrix})^\perp}$ is a finite rank operator, by Lemma 4.25 we conclude that $M_{(X,Y)}$ is right semi-Fredholm. Since

$$\begin{aligned} \mathcal{N}(M_{(X,Y)}) &= \mathcal{N}(\begin{bmatrix} A & C \end{bmatrix}) \cap \mathcal{N}(\begin{bmatrix} X & Y \end{bmatrix}) \\ &= \mathcal{N}(\begin{bmatrix} A & C \end{bmatrix}) \cap \mathcal{N}(\begin{bmatrix} A & C \end{bmatrix})^\perp \\ &= \{0\}, \end{aligned}$$

we have that $M_{(X,Y)}$ is a Fredholm operator.

On the contrary, assume that there exist operators $X \in \mathcal{B}(\mathcal{H}, \mathcal{K})$ and $Y \in \mathcal{B}(\mathcal{K})$ such that the operator matrix $M_{(X,Y)}$ given by (4.2) is Fredholm. Then there exist $N \in \mathcal{F}(\mathcal{H} \oplus \mathcal{K})$ and $P \in \mathcal{F}(\mathcal{H} \oplus \mathcal{K})$ such that $M_{(X,Y)}N = I_{\mathcal{H}\oplus\mathcal{K}} + P$, which can be represented as

$$\begin{bmatrix} A & C \\ X & Y \end{bmatrix} \begin{bmatrix} D & E \\ F & G \end{bmatrix} = \begin{bmatrix} I_{\mathcal{H}} & 0 \\ 0 & I_{\mathcal{K}} \end{bmatrix} + \begin{bmatrix} P_{11} & P_{12} \\ P_{21} & P_{22} \end{bmatrix}, \tag{4.9}$$

where all the operators P_{ij} are finite rank operators. From (4.9), in particular we have that

$$\begin{bmatrix} A & C \end{bmatrix} \begin{bmatrix} D \\ F \end{bmatrix} = I_{\mathcal{H}} + P_{11},$$

which implies that $\begin{bmatrix} A & C \end{bmatrix}$ is right semi-Fredholm. Similarly, we have that $\begin{bmatrix} E \\ G \end{bmatrix}$ is left semi-Fredholm, so $\alpha\left(\begin{bmatrix} E \\ G \end{bmatrix}\right) < \infty$. Since P_{12} is a finite rank operator and $\mathcal{K}$ is infinite dimensional it follows that $\alpha(P_{12}) = \infty$. Now from

$$\begin{bmatrix} A & C \end{bmatrix} \begin{bmatrix} E \\ G \end{bmatrix} = P_{12} \tag{4.10}$$

and the fact that $\alpha\left(\begin{bmatrix} E \\ G \end{bmatrix}\right) < \infty$, we have that $\alpha(\begin{bmatrix} A & C \end{bmatrix}) = \infty$. Finally, we have that $\begin{bmatrix} A & C \end{bmatrix} \in \mathcal{F}_r(\mathcal{H} \oplus \mathcal{K}) \setminus \mathcal{F}_l(\mathcal{H} \oplus \mathcal{K})$. □

THEOREM 4.28. *Let $A \in \mathcal{B}(\mathcal{H})$ and $C \in \mathcal{B}(\mathcal{K}, \mathcal{H})$ be given operators and let $\dim \mathcal{K} = \infty$. There exist $X \in \mathcal{B}(\mathcal{H}, \mathcal{K})$ and $Y \in \mathcal{B}(\mathcal{K})$such that the operator matrix $M_{(X,Y)}$ given by (4.2) belongs to $\mathcal{F}_r(\mathcal{H} \oplus \mathcal{K})$ ($\mathcal{F}_r(\mathcal{H} \oplus \mathcal{K}) \setminus \mathcal{F}_l(\mathcal{H} \oplus \mathcal{K})$) if and only if $\begin{bmatrix} A & C \end{bmatrix} \in \mathcal{F}_r(\mathcal{H} \oplus \mathcal{K}) \setminus \mathcal{F}_l(\mathcal{H} \oplus \mathcal{K})$.*

THEOREM 4.29. *Let $A \in \mathcal{B}(\mathcal{H})$ and $C \in \mathcal{B}(\mathcal{K}, \mathcal{H})$ be given operators and let $\dim \mathcal{K} = \infty$. Then there exist operators $X \in \mathcal{B}(\mathcal{H}, \mathcal{K})$ and $Y \in \mathcal{B}(\mathcal{K})$ such that the operator matrix $M_{(X,Y)}$ given by (4.2) is left semi-Fredholm, but not right semi-Fredholm.*

PROOF. Since $\mathcal{H}$ and $\mathcal{K}$ are infinite-dimensional separable Hilbert spaces, we have that $\mathcal{K} \cong \mathcal{H} \oplus \mathcal{K}$, so the result follows by Theorem 3.2 from [**64**]. □

If $\mathcal{K}$ is finite dimensional, we have that $A \in \mathcal{B}(\mathcal{H})$ is Fredholm (left semi-Fredholm, right semi-Fredholm, has closed range, respectively) if and only if the operator matrix $\begin{bmatrix} A & 0 \\ 0 & 0 \end{bmatrix} \in \mathcal{B}(\mathcal{H} \oplus \mathcal{K})$ has the same corresponding property. Furthermore, in this case $M_{(X,Y)}$ given by (4.2) is a finite rank perturbation of $\begin{bmatrix} A & 0 \\ 0 & 0 \end{bmatrix}$ (the converse is true as well). From these facts we conclude that the following holds:

THEOREM 4.30. *Let $A \in \mathcal{B}(\mathcal{H})$ and $C \in \mathcal{B}(\mathcal{K}, \mathcal{H})$ be given operators and let $\dim \mathcal{K} < \infty$. The operator matrix $M_{(X,Y)}$ given by (4.2) is Fredholm (left semi-Fredholm, right semi-Fredholm, respectively) for all $X \in \mathcal{B}(\mathcal{H}, \mathcal{K})$ and $Y \in \mathcal{B}(\mathcal{K})$ if and only if A is Fredholm (left semi-Fredholm, right semi-Fredholm, respectively).*

In the case when $\dim \mathcal{K} = \infty$, the following theorem gives us certain necessary and sufficient conditions for $M_{(X,Y)}$ to be an FC operator.

THEOREM 4.31. *Let $A \in \mathcal{B}(\mathcal{H})$ and $C \in \mathcal{B}(\mathcal{K},\mathcal{H})$ be given operators. If $\dim \mathcal{K} = \infty$, the operator matrix $M_{(X,Y)}$ given by (4.2) is an FC operator for some $X \in \mathcal{B}(\mathcal{H},\mathcal{K})$ and $Y \in \mathcal{B}(\mathcal{K})$ if and only if $\begin{bmatrix} A & C \end{bmatrix}$ is not a left semi-Fredholm operator. Furthermore, if $M_{(X,Y)}$ is an FC operator for some $X \in \mathcal{B}(\mathcal{H},\mathcal{K})$ and $Y \in \mathcal{B}(\mathcal{K})$, then $M_{(0,0)}$ is an FC operator.*

PROOF. ($\Rightarrow$) We prove that if $\begin{bmatrix} A & C \end{bmatrix}$ is left semi-Fredholm, then $M_{(X,Y)}$ is not an FC operator for any choice of $X \in \mathcal{B}(\mathcal{H},\mathcal{K})$ and $Y \in \mathcal{B}(\mathcal{K})$. So suppose that $\begin{bmatrix} A & C \end{bmatrix}$ is left semi-Fredholm. We prove that for any choice $X \in \mathcal{B}(\mathcal{H},\mathcal{K})$ and $Y \in \mathcal{B}(\mathcal{K})$ none of the cases (i)–(iii) of Theorem 4.23 holds. Since $n\left(\begin{bmatrix} A & C \end{bmatrix}\right) < \infty$ and for any $X \in \mathcal{B}(\mathcal{H},\mathcal{K})$ and $Y \in \mathcal{B}(\mathcal{K})$, $\mathcal{N}(M_{(X,Y)}) \subseteq \mathcal{N}\left(\begin{bmatrix} A & C \end{bmatrix}\right)$ it follows that for any choice of $X \in \mathcal{B}(\mathcal{H},\mathcal{K})$ and $Y \in \mathcal{B}(\mathcal{K})$, $\alpha(M_{(X,Y)}) < \infty$. Hence, the case (ii) from Theorem 4.23 does not hold. That (i) of Theorem 4.23 is not satisfied for any choice of $X \in \mathcal{B}(\mathcal{H},\mathcal{K})$ and $Y \in \mathcal{B}(\mathcal{K})$ follows by Theorem 4.27. To prove that (iii) of Theorem 4.23 is not satisfied for any choice of $X \in \mathcal{B}(\mathcal{H},\mathcal{K})$ and $Y \in \mathcal{B}(\mathcal{K})$, it is sufficient to prove that for any $X \in \mathcal{B}(\mathcal{H},\mathcal{K})$ and $Y \in \mathcal{B}(\mathcal{K})$, the restriction of the operator $M_{(X,Y)}$ onto the subspace $\mathcal{N}\left(\begin{bmatrix} A & C \end{bmatrix}\right)^{\perp}$ has a closed range. But this directly follows by Lemma 4.15.

($\Leftarrow$) If $\begin{bmatrix} A & C \end{bmatrix}$ is not left semi-Fredholm, we consider two possible cases.

Case 1. $\mathcal{R}(\begin{bmatrix} A & C \end{bmatrix})$ is not closed. Obviously, $\mathcal{R}(M_{(0,0)})$ is not closed and thus by Theorem 4.23 we have that $M_{(0,0)}$ is an FC operator.

Case 2. $\mathcal{R}(\begin{bmatrix} A & C \end{bmatrix})$ is closed and $n\left(\begin{bmatrix} A & C \end{bmatrix}\right) = \infty$. Then $\alpha(M_{(0,0)}) = d(M_{(0,0)}) = \infty$, $\mathcal{R}(M_{(0,0)})$ is closed, and by Theorem 4.23 we have that $M_{(0,0)}$ is an FC operator. □

Through similar analysis as carried out for the consistency in invertibility, we show that when $\dim \mathcal{K} = \infty$, we can always find operators $X \in \mathcal{B}(\mathcal{H},\mathcal{K})$ and $Y \in \mathcal{B}(\mathcal{K})$ such that $M_{(X,Y)}$ is not an FC operator.

THEOREM 4.32. *Let $A \in \mathcal{B}(\mathcal{H})$ and $C \in \mathcal{B}(\mathcal{K},\mathcal{H})$. If $\dim \mathcal{K} = \infty$, then there exist operators $X \in \mathcal{B}(\mathcal{H},\mathcal{K})$ and $Y \in \mathcal{B}(\mathcal{K})$ such that $M_{(X,Y)}$ given by (4.2) is not FC.*

PROOF. From Theorem 4.29 we have that there exist operators $X \in \mathcal{B}(\mathcal{H},\mathcal{K})$ and $Y \in \mathcal{B}(\mathcal{K})$ such that $M_{(X,Y)}$ is left semi-Fredholm, but not right semi-Fredholm. According to Theorem 4.23 we have that $M_{(X,Y)}$ is not an FC operator. □

If $\mathcal{K}$ is finite dimensional, as in Theorem 4.30 we conclude that the following holds:

THEOREM 4.33. *Let $A \in \mathcal{B}(\mathcal{H})$ and $C \in \mathcal{B}(\mathcal{K},\mathcal{H})$. If $\dim \mathcal{K} < \infty$, then the operator matrix $M_{(X,Y)}$ given by (4.2) is an FC operator for all $X \in \mathcal{B}(\mathcal{H},\mathcal{K})$ and $Y \in \mathcal{B}(\mathcal{K})$ if and only if A is an FC operator.*

Using Theorems 4.31–4.33 we can completely describe the sets

$$\bigcap_{X\in\mathcal{B}(\mathcal{H},\mathcal{K}),\ Y\in\mathcal{B}(\mathcal{K})} \sigma_{FC}(M_{(X,Y)}) \quad \text{and} \quad \bigcup_{X\in\mathcal{B}(\mathcal{H},\mathcal{K}),\ Y\in\mathcal{B}(\mathcal{K})} \sigma_{FC}(M_{(X,Y)}).$$

COROLLARY 4.34. Let $A \in \mathcal{B}(\mathcal{H})$ and $C \in \mathcal{B}(\mathcal{K}, \mathcal{H})$.

(1) If $\dim \mathcal{K} = \infty$, then

$$\bigcap_{X\in\mathcal{B}(\mathcal{H},\mathcal{K}),\ Y\in\mathcal{B}(\mathcal{K})} \sigma_{FC}(M_{(X,Y)}) = \{\lambda \in \mathbb{C} \colon \begin{bmatrix} A-\lambda & C \end{bmatrix} \text{ is left semi-Fredholm}\}$$

and

$$\bigcup_{X\in\mathcal{B}(\mathcal{H},\mathcal{K}),\ Y\in\mathcal{B}(\mathcal{K})} \sigma_{FC}(M_{(X,Y)}) = \mathbb{C}.$$

(2) If $\dim \mathcal{K} < \infty$, then

$$\bigcap_{X\in\mathcal{B}(\mathcal{H},\mathcal{K}),\ Y\in\mathcal{B}(\mathcal{K})} \sigma_{FC}(M_{(X,Y)}) = \bigcup_{X\in\mathcal{B}(\mathcal{H},\mathcal{K}),\ Y\in\mathcal{B}(\mathcal{K})} \sigma_{FC}(M_{(X,Y)}) = \sigma_{FC}(A).$$

In [**51**] it is proved that for arbitrary operators $A \in \mathcal{B}(\mathcal{H})$ and $B \in \mathcal{B}(\mathcal{K})$ one has the equality

$$\bigcap_{C\in\mathcal{B}(\mathcal{K},\mathcal{H})} \sigma_{FC}(M_C) = \sigma_{FC}(M_0).$$

In the case of the operator matrix (4.2) an analogous result does not hold. In the next example we show that

$$\bigcap_{X\in\mathcal{B}(\mathcal{H},\mathcal{K}),\ Y\in\mathcal{B}(\mathcal{K})} \sigma_{FC}(M_{(X,Y)}) \neq \sigma_{FC}(M_{(0,0)}). \tag{4.11}$$

EXAMPLE 4.35. Suppose that $\dim \mathcal{K} = \dim \mathcal{H} = \infty$. Let $\lambda \in \mathbb{C} \setminus \{0\}$ be arbitrary. If we take an operator $A \in \mathcal{B}(\mathcal{H})$ such that $A - \lambda \in \mathcal{F}_l(\mathcal{H}) \setminus \mathcal{F}_r(\mathcal{H})$ and an arbitrary operator $C \in \mathcal{B}(\mathcal{K}, \mathcal{H})$ such that $\alpha(C) = \infty$, we have that $\begin{bmatrix} A-\lambda & C \end{bmatrix} \notin \mathcal{F}_l(\mathcal{H} \oplus \mathcal{K}, \mathcal{H})$. According to Theorem 4.31, there exist operators $X \in \mathcal{B}(\mathcal{H}, \mathcal{K})$ and $Y \in \mathcal{B}(\mathcal{K})$ such that $M_{(X,Y)} - \lambda$ is an FC operator, so

$$0 \notin \bigcap_{X\in\mathcal{B}(\mathcal{H},\mathcal{K}),\ Y\in\mathcal{B}(\mathcal{K})} \sigma_{FC}(M_{(X,Y)}). \tag{4.12}$$

Now, we prove that $M_{(0,0)} - \lambda I \in \mathcal{F}_l(\mathcal{H} \oplus \mathcal{K}) \setminus \mathcal{F}_r(\mathcal{H} \oplus \mathcal{K})$, meaning that $0 \in \sigma_{FC}(M_{(0,0)})$, which will in turn, together with (4.12), implies that (4.11) is satisfied. Since $A - \lambda$ is left semi-Fredholm, there exists $A_0 \in \mathcal{B}(\mathcal{H})$ such that $A_0(A - \lambda) = I_{\mathcal{H}} + F$, where $F \in \mathcal{F}(\mathcal{H})$. One has $M_{(0,0)} - \lambda I \in \mathcal{F}_l(\mathcal{H} \oplus \mathcal{K})$ since

$$\begin{bmatrix} A_0 & \frac{1}{\lambda}A_0 C \\ 0 & -\frac{1}{\lambda} \end{bmatrix} \begin{bmatrix} A-\lambda & C \\ 0 & -\lambda \end{bmatrix} = \begin{bmatrix} I_{\mathcal{H}}+F & 0 \\ 0 & I_{\mathcal{K}} \end{bmatrix}.$$

Let us prove that $M_{(0,0)} - \lambda I \notin \mathcal{F}_r(\mathcal{H} \oplus \mathcal{K})$. Assume, to the contrary, that $M_{(0,0)} - \lambda I \in \mathcal{F}_r(\mathcal{H} \oplus \mathcal{K})$. Then there exists an operator matrix

$$\begin{bmatrix} X & Y \\ Y & W \end{bmatrix} : \begin{bmatrix} \mathcal{H} \\ \mathcal{K} \end{bmatrix} \to \begin{bmatrix} \mathcal{H} \\ \mathcal{K} \end{bmatrix},$$

such that

$$\begin{bmatrix} A-\lambda & C \\ 0 & -\lambda \end{bmatrix} \begin{bmatrix} X & Y \\ Y & W \end{bmatrix} = \begin{bmatrix} I_{\mathcal{H}}+F_1 & F_2 \\ F_3 & I_{\mathcal{K}}+F_4 \end{bmatrix},$$

where the operators F_i are finite rank operators. Since $-\lambda Y = F_3$, it follows that Y is an operator of finite rank. Now, $(A-\lambda)X + CY = I_{\mathcal{H}} + F_1$ yields $A - \lambda \in \mathcal{F}_r(\mathcal{H})$, which is a contradiction. Hence, $M_{(0,0)} - \lambda I \notin \mathcal{F}_r(\mathcal{H} \oplus \mathcal{K})$.

CHAPTER 5

Applications of completion results

The primary purpose of this chapter is to obtain necessary and sufficient conditions for a linear combination $\alpha A+\beta B$ or simply for the sum $A+B$ of two operators $A, B \in \mathcal{B}(\mathcal{H})$ to belong to a certain class of operators, using the results concerning different properties of an upper triangular operator matrix of the form M_C. In the recently published literature we can find this kind of considerations but only for the special classes of operators such as idempotents or orthogonal projectors. Let us mention the papers [**9, 24, 67, 119, 120**], where mainly independence of invertibility of the linear combination $\alpha A+\beta B$ from the choice of complex nonzero scalars α and β was studied, in the cases when $A, B \in \mathcal{B}(\mathcal{H})$ are idempotents or orthogonal projectors. In this chapter we consider the general case, without the assumptions that $A, B \in \mathcal{B}(\mathcal{H})$ belong to any particular class of operators and also investigate other properties than invertibility. The basic idea lies in the fact that arbitrary $A, B \in \mathcal{B}(\mathcal{H})$ can be decomposed as

$$A=\begin{bmatrix} A_1 & A_2 \\ 0 & 0 \end{bmatrix}:\begin{bmatrix} \mathcal{N}(B) \\ \mathcal{P} \end{bmatrix}\to\begin{bmatrix} \overline{\mathcal{R}(A)} \\ \mathcal{Q} \end{bmatrix}, \tag{5.1}$$

$$B=\begin{bmatrix} 0 & B_1 \\ 0 & B_2 \end{bmatrix}:\begin{bmatrix} \mathcal{N}(B) \\ \mathcal{P} \end{bmatrix}\to\begin{bmatrix} \overline{\mathcal{R}(A)} \\ \mathcal{Q} \end{bmatrix}, \tag{5.2}$$

with respect to the decompositions $\mathcal{H}=\mathcal{N}(B)\oplus\mathcal{P}=\overline{\mathcal{R}(A)}\oplus\mathcal{Q}$.

For arbitrary $\alpha,\beta\in\mathbb{C}\setminus\{0\}$, using the above decompositions of $A, B \in \mathcal{B}(\mathcal{H})$, it follows that the linear combination $\alpha A+\beta B$ has a certain property if and only if the operator matrix

$$\begin{bmatrix} \alpha A_1 & \alpha A_2+\beta B_1 \\ 0 & \beta B_2 \end{bmatrix}:\begin{bmatrix} \mathcal{N}(B) \\ \mathcal{P} \end{bmatrix}\to\begin{bmatrix} \overline{\mathcal{R}(A)} \\ \mathcal{Q} \end{bmatrix} \tag{5.3}$$

has the same property. This fact allows us to exploit certain results concerning the upper triangular operator matrices in order to derive sought-after characterizations of the sum of operators. On the next pages, we present such results for the invertibility, injectivity, surjectivity, and Fredholmness.

5.1. Invertibility of linear combinations of operators

Invertibility of the linear combination of $\alpha A+\beta B$, in the cases when $A, B \in \mathcal{B}(\mathcal{H})$ are projectors or orthogonal projectors, were studied in [**9, 24, 67, 119, 120**]. In this section we consider the general case, without the assumptions that $A, B \in \mathcal{B}(\mathcal{H})$ belong to any particular classes of operators. As corollaries, we obtain results for certain special classes of operators. Specifically, we completely solve the problem of invertibility of the linear combination $\alpha A + \beta B$ in each of the following cases:

- if $A, B \in \mathcal{B}(\mathcal{H})$ are projectors or orthogonal projectors,
- if $\mathcal{R}(A) \cap \mathcal{R}(B) = \{0\}$,
- if $\overline{\mathcal{R}(AP_{\mathcal{N}(B)})} = \overline{\mathcal{R}(A)}$,
- if either one of $A, B \in \mathcal{B}(\mathcal{H})$ is injective.

The following well-known lemma is used throughout this section.

LEMMA 5.1. *Let $\mathcal{M}$ and $\mathcal{N}$ be subspaces of a Hilbert space $\mathcal{H}$. Then*

$$(\mathcal{M} + \mathcal{N})^{\perp} = \mathcal{M}^{\perp} \cap \mathcal{N}^{\perp}.$$

The basic idea which will be presented here is to use the results concerning the invertibility of an upper triangular operator matrix of the form M_C (see Section 1.2) in order to obtain the necessary and sufficient conditions for the invertibility of the linear combination $\alpha A + \beta B$. There are a few papers (see [**34, 47, 98**]) in which necessary and sufficient conditions are presented for the invertibility of M_C. In [**47**], for given $A \in \mathcal{B}(\mathcal{X})$ and $B \in \mathcal{B}(\mathcal{Y})$, additionally the set of all $C \in \mathcal{B}(\mathcal{Y}, \mathcal{X})$ such that M_C is invertible, denoted by $S(A, B)$, is completely described in the case when $\mathcal{X}$ and $\mathcal{Y}$ are Banach spaces (see Theorem 1.8). Here we give one modification of Theorem 1.8 in the case of separable Hilbert spaces.

THEOREM 5.2. *Let $\mathcal{H}$ and $\mathcal{K}$ be separable Hilbert spaces and let $A \in \mathcal{B}(\mathcal{H})$, $B \in \mathcal{B}(\mathcal{K})$, and $C \in \mathcal{B}(\mathcal{K}, \mathcal{H})$ be given operators. The operator matrix $\begin{bmatrix} A & C \\ 0 & B \end{bmatrix}$ is invertible if and only if A is left invertible, B is right invertible, and $P_{\mathcal{S},\mathcal{R}(A)}C|_{\mathcal{N}(B)}$ is an injective operator with (closed) range $\mathcal{S}$, where $\mathcal{H} = \mathcal{R}(A) \oplus \mathcal{S}$.*

Now, we reduce the problem of invertibility of the linear combination $\alpha A + \beta B$ to an equivalent one which concerns the invertibility of a certain upper triangular operator matrix with the idea of using the previously mentioned results.

THEOREM 5.3. *Let $A, B \in \mathcal{B}(\mathcal{H})$ be given operators and let $\alpha, \beta \in \mathbb{C} \setminus \{0\}$. Then $\alpha A + \beta B$ is invertible if and only if the conditions*

(i) $\mathcal{N}(A) \cap \mathcal{N}(B) = \{0\}$, $\mathcal{R}(A) + \mathcal{R}(B) = \mathcal{H}$,

(ii) $A|_{\mathcal{N}(B)}$ *has a closed range,*

(iii) $P_{\mathcal{S},\mathcal{R}(A|_{\mathcal{N}(B)})}(\alpha A + \beta B)|_{\mathcal{T}}$ *is an injective operator with range $\mathcal{S}$*

hold, where $\overline{\mathcal{R}(A)} = \mathcal{R}(A|_{\mathcal{N}(B)}) \oplus \mathcal{S}$, $\mathcal{T} = B^{-1}(\overline{\mathcal{R}(A)}) \cap \mathcal{P}$, and $\mathcal{H} = \mathcal{N}(B) \oplus \mathcal{P}$.

PROOF. Let $\mathcal{H} = \mathcal{N}(B) \oplus \mathcal{P} = \overline{\mathcal{R}(A)} \oplus \mathcal{Q}$ be decompositions of the space $\mathcal{H}$. With respect to these decompositions the given operators $A, B \in \mathcal{B}(\mathcal{H})$ have the following representations:

$$A = \begin{bmatrix} A_1 & A_2 \\ 0 & 0 \end{bmatrix} : \begin{bmatrix} \mathcal{N}(B) \\ \mathcal{P} \end{bmatrix} \to \begin{bmatrix} \overline{\mathcal{R}(A)} \\ \mathcal{Q} \end{bmatrix}, \tag{5.4}$$

$$B = \begin{bmatrix} 0 & B_1 \\ 0 & B_2 \end{bmatrix} : \begin{bmatrix} \mathcal{N}(B) \\ \mathcal{P} \end{bmatrix} \to \begin{bmatrix} \overline{\mathcal{R}(A)} \\ \mathcal{Q} \end{bmatrix}. \tag{5.5}$$

Take arbitrary $\alpha, \beta \in \mathbb{C} \setminus \{0\}$. Using the above decompositions of $A, B \in \mathcal{B}(\mathcal{H})$, it follows that the linear combination $\alpha A + \beta B$ is invertible if and only if the operator matrix

$$\begin{bmatrix} \alpha A_1 & \alpha A_2 + \beta B_1 \\ 0 & \beta B_2 \end{bmatrix} : \begin{bmatrix} \mathcal{N}(B) \\ \mathcal{P} \end{bmatrix} \to \begin{bmatrix} \overline{\mathcal{R}(A)} \\ \mathcal{Q} \end{bmatrix} \tag{5.6}$$

is invertible. Using Theorem 5.2, we have that this holds if and only if the following three conditions are satisfied:

(i) αA_1 is left invertible.
(ii) βB_2 is right invertible.
(iii) $P_{\mathcal{S},\mathcal{R}(A_1)}(\alpha A_2 + \beta B_1)|_{\mathcal{N}(B_2)}$ is an injective operator with range $\mathcal{S}$, where $\overline{\mathcal{R}(A)} = \mathcal{R}(A_1) \oplus \mathcal{S}$.

Evidently, (i) holds if and only if $\mathcal{N}(A) \cap \mathcal{N}(B) = \{0\}$ and $\mathcal{R}(A|_{\mathcal{N}(B)})$ is closed. Also, (ii) is satisfied if and only if $\mathcal{R}(P_{\mathcal{Q},\overline{\mathcal{R}(A)}}B) = \mathcal{Q}$. Since

$$\mathcal{R}(P_{\mathcal{Q},\overline{\mathcal{R}(A)}}B) = \mathcal{Q} \Leftrightarrow \mathcal{Q} \subseteq \overline{\mathcal{R}(A)} + \mathcal{R}(B) \Leftrightarrow \overline{\mathcal{R}(A)} + \mathcal{R}(B) = \mathcal{H},$$

we have that (ii) is equivalent with $\overline{\mathcal{R}(A)} + \mathcal{R}(B) = \mathcal{H}$.

To discuss the third condition notice that

$$\mathcal{N}(B_2) = \mathcal{N}(P_{\mathcal{Q},\overline{\mathcal{R}(A)}}B) \cap \mathcal{P} = B^{-1}(\overline{\mathcal{R}(A)}) \cap \mathcal{P}$$

and let $\mathcal{T} = B^{-1}(\overline{\mathcal{R}(A)}) \cap \mathcal{P}$. Evidently,

$$\mathcal{N}(P_{\mathcal{S},\mathcal{R}(A_1)}(\alpha A_2 + \beta B_1)|_{\mathcal{T}}) = \mathcal{N}(P_{\mathcal{S},\mathcal{R}(A_1)}(\alpha A + \beta B)|_{\mathcal{T}})$$

and

$$\mathcal{R}(P_{\mathcal{S},\mathcal{R}(A_1)}(\alpha A_2 + \beta B_1)|_{\mathcal{T}}) = \mathcal{R}(P_{\mathcal{S},\mathcal{R}(A_1)}(\alpha A + \beta B)|_{\mathcal{T}}).$$

Hence, we can conclude that $\alpha A + \beta B$ is invertible if and only if the following conditions hold:

(i) $\mathcal{N}(A) \cap \mathcal{N}(B) = \{0\}$, $\overline{\mathcal{R}(A)} + \mathcal{R}(B) = \mathcal{H}$.
(ii) $AP_{\mathcal{N}(B)}$ has closed range.
(iii) $P_{\mathcal{S},\mathcal{R}(A_1)}(\alpha A + \beta B)|_{\mathcal{T}}$ is an injective operator with range $\mathcal{S}$.

Notice that the second condition in (i), $\overline{\mathcal{R}(A)} + \mathcal{R}(B) = \mathcal{H}$, can be replaced by $\mathcal{R}(A) + \mathcal{R}(B) = \mathcal{H}$: Suppose that (i)–(iii) are satisfied. Since we have that $\mathcal{S} = \mathcal{R}(P_{\mathcal{S},\mathcal{R}(A_1)}(\alpha A + \beta B)|_{\mathcal{T}})$, it follows

$$\mathcal{S} \subseteq \mathcal{R}((\alpha A + \beta B)|_{\mathcal{T}}) + \mathcal{R}(A_1) \subseteq \mathcal{R}(A) + \mathcal{R}(B).$$

Now, $\overline{\mathcal{R}(A)} = \mathcal{R}(A_1) \oplus \mathcal{S}$ implies that $\overline{\mathcal{R}(A)} \subseteq \mathcal{R}(A) + \mathcal{R}(B)$. Hence, $\mathcal{R}(A) + \mathcal{R}(B) = \mathcal{H}$. Also, directly from the invertibility of $\alpha A + \beta B$ we can conclude that $\mathcal{R}(A) + \mathcal{R}(B) = \mathcal{H}$. □

In the special case when $\mathcal{S}$ is the orthogonal complement of $\mathcal{R}(A|_{\mathcal{N}(B)}) = \mathcal{R}(AP_{\mathcal{N}(B)})$ in $\overline{\mathcal{R}(A)}$ and $\mathcal{P} = \mathcal{N}(B)^{\perp}$, applying Theorem 5.3 we get the following result:

THEOREM 5.4. *Let $A, B \in \mathcal{B}(\mathcal{H})$ be given operators and let $\alpha, \beta \in \mathbb{C} \setminus \{0\}$. Then $\alpha A + \beta B$ is invertible if and only if the conditions*

(i) $\mathcal{N}(A) \cap \mathcal{N}(B) = \{0\}$, $\mathcal{R}(A) + \mathcal{R}(B) = \mathcal{H}$,
(ii) $A|_{\mathcal{N}(B)}$ *has a closed range,*
(iii) $P_{\mathcal{S}}(\alpha A + \beta B)|_{\mathcal{T}}$ *is an injectiv operator with range* $\mathcal{S}$

hold, where $\mathcal{S} = \mathcal{R}(AP_{\mathcal{N}(B)})^{\perp} \cap \overline{\mathcal{R}(A)}$ *and* $\mathcal{T} = \mathcal{N}(P_{\mathcal{R}(A)^{\perp}}B) \cap \mathcal{N}(B)^{\perp}$.

Evidently from Theorem 5.4, we can conclude that the invertibility of the linear combination $\alpha A + \beta B$ is possible for some constants $\alpha, \beta \in \mathbb{C} \setminus \{0\}$ only if

$$\dim(\mathcal{N}(P_{\mathcal{R}(A)^{\perp}}B) \cap \mathcal{N}(B)^{\perp}) = \dim(\mathcal{R}(AP_{\mathcal{N}(B)})^{\perp} \cap \overline{\mathcal{R}(A)}).$$

So we get the following result:

COROLLARY 5.5. Let $A, B \in \mathcal{B}(\mathcal{H})$ be given operators. If

$$\dim(\mathcal{N}(P_{\mathcal{R}(A)^\perp} B) \cap \mathcal{N}(B)^\perp) \neq \dim(\mathcal{R}(AP_{\mathcal{N}(B)})^\perp \cap \overline{\mathcal{R}(A)}),$$

then the linear combination $\alpha A + \beta B$ is not invertible, for any $\alpha, \beta \in \mathbb{C} \setminus \{0\}$.

Now we reconsider the condition (iii) from Theorem 5.3, which says that $\mathcal{R}(P_\mathcal{S}(\alpha A+\beta B)|_\mathcal{T}) = \mathcal{S}$ and $\mathcal{N}(P_\mathcal{S}(\alpha A+\beta B)|_\mathcal{T}) = \{0\}$. Suppose that $A, B \in \mathcal{B}(\mathcal{H})$ are given by (5.4) and (5.5), respectively, where $\mathcal{S}$ is the orthogonal complement of $\mathcal{R}(A|_{\mathcal{N}(B)}) = \mathcal{R}(AP_{\mathcal{N}(B)})$ in $\overline{\mathcal{R}(A)}$, $\mathcal{T} = \mathcal{N}(P_{\mathcal{R}(A)^\perp} B) \cap \mathcal{N}(B)^\perp$, and $\mathcal{P} = \mathcal{N}(B)^\perp$. The first condition is equivalent with

$$\overline{\mathcal{R}(A)} = \overline{\mathcal{R}(AP_{\mathcal{N}(B)})} + \overline{\mathcal{R}(A)} \cap \mathcal{R}((\alpha A + \beta B)P_{\mathcal{N}(B)^\perp}), \tag{5.7}$$

since $\mathcal{R}((\alpha A+\beta B)|_\mathcal{T}) = \overline{\mathcal{R}(A)} \cap \mathcal{R}((\alpha A+\beta B)P_{\mathcal{N}(B)^\perp})$. The second condition from (iii), $\mathcal{N}(P_\mathcal{S}(\alpha A + \beta B)|_\mathcal{T}) = \{0\}$, is equivalent with

$$\begin{aligned} &\mathcal{N}(\alpha A_2 + \beta B_1) \cap \mathcal{N}(B_2) = \{0\}, \\ &\mathcal{R}((\alpha A_2 + \beta B_1)|_\mathcal{T}) \cap \overline{\mathcal{R}(AP_{\mathcal{N}(B)})} = \{0\}. \end{aligned} \tag{5.8}$$

Evidently the first condition from (5.8) is equivalent with

$$\mathcal{N}(\alpha A + \beta B) \cap \mathcal{N}(B)^\perp = \{0\}$$

while the second one is equivalent with

$$\mathcal{R}((\alpha A + \beta B)P_{\mathcal{N}(B)^\perp}) \cap \overline{\mathcal{R}(AP_{\mathcal{N}(B)})} = \{0\}.$$

Now, in view of the previous two conditions and (5.7), we can conclude that the condition (iii) from Theorem 5.3 is equivalent with

$$\overline{\mathcal{R}(A)} = \overline{\mathcal{R}(AP_{\mathcal{N}(B)})} \oplus \overline{\mathcal{R}(A)} \cap \mathcal{R}((\alpha A + \beta B)P_{\mathcal{N}(B)^\perp})$$

and

$$\mathcal{N}(\alpha A + \beta B) \cap \mathcal{N}(B)^\perp = \{0\}.$$

Thus, we can formulate the following result:

THEOREM 5.6. *Let $A, B \in \mathcal{B}(\mathcal{H})$ be given operators and let $\alpha, \beta \in \mathbb{C}\setminus\{0\}$. Then the operator $\alpha A + \beta B$ is invertible if and only if the following conditions hold:*

(i) $\mathcal{N}(A) \cap \mathcal{N}(B) = \{0\}$, $\mathcal{R}(A) + \mathcal{R}(B) = \mathcal{H}$.
(ii) $AP_{\mathcal{N}(B)}$ *has closed range.*
(iii) $\overline{\mathcal{R}(A)} = \mathcal{R}(AP_{\mathcal{N}(B)}) \oplus \overline{\mathcal{R}(A)} \cap \mathcal{R}((\alpha A + \beta B)P_{\mathcal{N}(B)^\perp})$.
(iv) $\mathcal{N}(\alpha A + \beta B) \cap \mathcal{N}(B)^\perp = \{0\}$.

In Theorems 5.3 and 5.4, the problem of invertibility of a linear combination of two given operators is reduced to one in which yet another linear combination is required to be injective and to have a prescribed range, which at first glance might not strike the reader as much of an achievement. However, the conditions we have obtained (those given in Theorems 5.3 and 5.4) lend themselves for applications in further analysis of the initial problem for many special classes of operators where they will lead to its complete solution. This type of investigation will be the focus of our attention in the next section.

Since the condition that $\alpha A + \beta B$ be invertible is symmetrical in A and B, we can obtain new variants of the necessary and sufficient conditions in Theorems 5.3, 5.4, and 5.6 by interchanging the operators A and B in them.

Let us now consider invertibility of linear combinations for some special classes of operators using the results from the previous section.

(**1**) The problem of invertibility of $\alpha A+\beta B$, in the case when $A, B \in \mathcal{B}(\mathcal{H})$ are regular operators and $\alpha, \beta \in \mathbb{C}\setminus\{0\}$ was considered in [**96**].

THEOREM 5.7 ([**96**]). *Let $A, B \in \mathcal{B}(\mathcal{H})$ be given operators with closed ranges and let $\alpha, \beta \in \mathbb{C}\setminus\{0\}$. The operator $\alpha A+\beta B$ is invertible if and only if the conditions*

(i′) $\mathcal{N}(A)\cap\mathcal{N}(B)=\{0\}$, $\mathcal{R}(A)^{\perp}\cap\mathcal{R}(B)^{\perp}=\{0\}$,
(ii′) *both $A^{\dagger}A(I-B^{\dagger}B)$ and $(I-AA^{\dagger})BB^{\dagger}$ are closed range operators,*
(iii′) *$P_{\mathcal{L}}^{\mathtt{rst}}(\alpha AB^{\dagger}B+\beta AA^{\dagger}B)|_{\mathcal{M}}$ is invertible*

hold, where $\mathcal{L}=(A^*)^{\dagger}(R(A^*)\cap\mathcal{R}(B^*))$, $\mathcal{M}=B^{\dagger}(\mathcal{R}(A)\cap\mathcal{R}(B))$, and $P_{\mathcal{L}}^{\mathtt{rst}}\in\mathcal{B}(\mathcal{H},\mathcal{L})$ is defined by $P_{\mathcal{L}}^{\mathtt{rst}}x=P_{\mathcal{L}}x$, $x\in\mathcal{H}$.

As a corollary of Theorem 5.3 we get the main result from the paper [**96**, Theorem 5.7]. This cannot be seen directly and to explain why that is so, we first give the following lemma.

LEMMA 5.8. *Let $A, B \in \mathcal{B}(\mathcal{H})$ be given operators. If A and B have closed ranges, then:*

(i) $\mathcal{N}(P_{\mathcal{R}(A)^{\perp}}B)\cap\mathcal{N}(B)^{\perp}=B^{-1}(\mathcal{R}(A))\cap\mathcal{N}(B)^{\perp}=B^{\dagger}(\mathcal{R}(A)\cap\mathcal{R}(B))$.
(ii) $\mathcal{R}(A)\cap\mathcal{R}(AP_{\mathcal{N}(B)})^{\perp}=(A^*)^{\dagger}(R(A^*)\cap\mathcal{R}(B^*))$.

PROOF. (i) The first equality is evident. Let $x\in B^{-1}(\mathcal{R}(A))\cap\mathcal{N}(B)^{\perp}$. Then $Bx\in\mathcal{R}(A)$ and $x=B^{\dagger}Bx$. So, $x\in B^{\dagger}(\mathcal{R}(A)\cap\mathcal{R}(B))$. Now, suppose that $x\in B^{\dagger}(\mathcal{R}(A)\cap\mathcal{R}(B))$. Then for some $s,t\in\mathcal{H}$, we have that $x=B^{\dagger}Bt=B^{\dagger}As$ and $Bt=As$. Evidently, $x\in\mathcal{R}(B^*)=\mathcal{N}(B)^{\perp}$ and $Bx=Bt=As\in\mathcal{R}(A)$.

(ii) Let $y\in\mathcal{R}(A)\cap\mathcal{R}(AP_{\mathcal{N}(B)})^{\perp}$. Then $y=AA^{\dagger}y$ and $A^*y=B^{\dagger}BA^*y$. Hence, $A^*y=B^{\dagger}BA^*y\in\mathcal{R}(A^*)\cap\mathcal{R}(B^*)$. So,

$$y=(A^{\dagger})^*A^*y=(A^{\dagger})^*B^{\dagger}BA^*y.$$

Suppose that $x\in(A^*)^{\dagger}(\mathcal{R}(A^*)\cap\mathcal{R}(B^*))$. Then for some $s,t\in\mathcal{H}$ we have that $y=(A^{\dagger})^*A^*t=(A^{\dagger})^*B^*s$ and $A^*t=B^*s$. Evidently, $y\in\mathcal{R}(A)$ which implies $y=AA^{\dagger}y$. Now, we prove that $y\in\mathcal{R}(AP_{\mathcal{N}(B)})^{\perp}=\mathcal{N}(P_{\mathcal{N}(B)}A^*)$:

$$\begin{aligned}B^{\dagger}BA^*y&=B^{\dagger}BA^*(A^{\dagger})^*B^*s=B^{\dagger}BA^{\dagger}AB^*s\\&=B^{\dagger}BA^{\dagger}AA^*t=B^{\dagger}BB^*s=B^*s\\&=A^*t=A^*y.\end{aligned}$$

□

Now, we show that in the case when $A, B\in\mathcal{B}(\mathcal{H})$ are closed range operators, our Theorem 5.4 is equivalent with Theorem 2.1 from [**96**] which is not so evident at first sight:

THEOREM 5.9. *Let $A, B\in\mathcal{B}(\mathcal{H})$ be given operators with closed ranges and let $\alpha,\beta\in\mathbb{C}\setminus\{0\}$. The conditions* (i)–(iii) *from Theorem* 5.4 *are equivalent to the conditions* (i′)–(iii′) *from Theorem* 5.7.

PROOF. First, notice that by Lemma 5.8 we have that $\mathcal{M}=\mathcal{T}$ and $\mathcal{S}=\mathcal{L}$, for the subspaces $\mathcal{S},\mathcal{T}$ defined in Theorem 5.4 and subspaces $\mathcal{L},\mathcal{M}$ defined in Theorem 5.7. Also, $\mathcal{M}\subseteq\mathcal{R}(B^{\dagger})$ and $\mathcal{L}\subseteq\mathcal{R}((A^*)^{\dagger}A^*)=\mathcal{R}(A)$ which imply that

$AB^\dagger B|_{\mathcal{M}} = A|_{\mathcal{M}}$ and $P'_{\mathcal{L}} = AA^\dagger P'_{\mathcal{L}} = P'_{\mathcal{L}} AA^\dagger$. Hence, conditions (iii) and (iii′) are equivalent. Now, suppose that the conditions (i)–(iii) from Theorem 5.4 are satisfied. By Lemma 5.1 we have that $\mathcal{R}(A)^\perp \cap \mathcal{R}(B)^\perp = \{0\}$, so the condition (i′) holds. Since $\mathcal{H} = \mathcal{R}(A) + \mathcal{R}(B)$, it follows that $\mathcal{R}(A)^\perp \subseteq \mathcal{R}(A) + \mathcal{R}(B)$ which implies that $\mathcal{R}(P_{\mathcal{R}(A)^\perp} B) = \mathcal{R}(A)^\perp$. Hence, $P_{\mathcal{R}(A)^\perp} B$ has closed range which is equivalent with $(I - AA^\dagger)BB^\dagger$ having closed range. Evidently, $AP_{\mathcal{N}(B)}$ has closed range if and only if $A^\dagger A(I - B^\dagger B)$ is a closed range operator. So the condition (ii′) is satisfied. On the other side, suppose that the conditions (i′)–(iii′) from Theorem 5.7 hold. Then we only need to prove that $\mathcal{R}(A) + \mathcal{R}(B) = \mathcal{H}$. But this is satisfied since (i′)–(iii′) are equivalent with the invertibility of $\alpha A + \beta B$. □

(**2**) The problem of invertibility of projectors (idempotents) has been considered in several papers. Coming from that line of research we can single out the result that the invertibility of any linear combination $\alpha P + \beta Q$, where $\alpha, \beta \in \mathbb{C} \setminus \{0\}$, $\alpha + \beta \neq 0$, is in fact equivalent to the invertibility of $P + Q$, which means that it is independent of the choice of the scalars α and β. For the first time, this was realized by J. K. Baksalary et al. [**9**] for the finite-dimesional case, who proved that

$$\alpha P + \beta Q \text{ is invertible } \Leftrightarrow \mathcal{R}(P(I-Q)) \cap \mathcal{R}(Q(I-P)) = \mathcal{N}(P) \cap \mathcal{N}(Q) = \{0\},$$

and it was later generalized by H. K. Du et al. [**70**] to the case of idempotent operators on a Hilbert space and finally by J. J. Koliha et al. [**119**] to the Banach algebra case, without giving any necessary and sufficient conditions for the invertibility of $\alpha P + \beta Q$. The necessary and sufficient conditions for the invertibility of a linear combination of projectors P and Q on a Hilbert space are given later in another paper by J. J. Koliha et al. [**120**] (as well as for the elements of a unital ring):

THEOREM 5.10 ([**120**]). *Let $P, Q \in \mathcal{B}(\mathcal{H})$ be projectors on a Hilbert space $\mathcal{H}$. Then the following conditions are equivalent:*

(i) *$P + Q$ is invertible.*
(ii) *The range of $P + Q$ is closed and*

$$\mathcal{R}(P) \cap \mathcal{R}(Q(I-P)) = \mathcal{N}(P) \cap \mathcal{N}(Q) = \{0\},$$

$$\mathcal{R}(P^*) \cap \mathcal{R}(Q^*(I-P^*)) = \mathcal{N}(P^*) \cap \mathcal{N}(Q^*) = \{0\}.$$

In the case when $P, Q \in \mathcal{B}(\mathcal{H})$ are projectors, applying Theorem 5.3 to the decompositions $\mathcal{H} = \mathcal{N}(Q) \oplus \mathcal{R}(Q) = \mathcal{R}(P) \oplus \mathcal{N}(P)$, we get the main result from [**71**], which says that the invertibility of the linear combination $\alpha P + \beta Q$ is independent of the choice of the scalars $\alpha, \beta \in \mathbb{C}$, but additionally we also obtain necessary and sufficient conditions for the invertibility of the linear combination $\alpha P + \beta Q$ which are different from those given in Theorem 5.10.

THEOREM 5.11. *Let $P, Q \in \mathcal{B}(\mathcal{H})$ be given projectors and let $\alpha, \beta \in \mathbb{C} \setminus \{0\}$, $\alpha + \beta \neq 0$. Then $\alpha P + \beta Q$ is an invertible operator if and only if the following conditions hold:*

(i) *$\mathcal{N}(P) \cap \mathcal{N}(Q) = \{0\}$, $\mathcal{R}(P) + \mathcal{R}(Q) = \mathcal{H}$.*
(ii) *$\mathcal{R}(P) = \mathcal{R}(P) \cap \mathcal{R}(Q) \oplus \mathcal{R}(P|_{\mathcal{N}(Q)})$.*

PROOF. Indeed, in this case the subspace $\mathcal{T}$ that is defined in Theorem 5.3 by $\mathcal{T} = Q^{-1}(\mathcal{R}(P)) \cap \mathcal{R}(Q)$ is equal to $\mathcal{R}(P) \cap \mathcal{R}(Q)$. Hence, for any $x \in \mathcal{T}$, we have that $(\alpha P + \beta Q)x = (\alpha + \beta)x$ which implies that the injectivity of operator $P_{\mathcal{S},\mathcal{R}(P|_{\mathcal{N}(Q)})}(\alpha P + \beta Q)|_{\mathcal{T}}$ is equivalent with $\mathcal{R}(P|_{\mathcal{N}(Q)}) \cap \mathcal{T} = \{0\}$; i.e.,

$$\mathcal{R}(P|_{\mathcal{N}(Q)}) \cap \mathcal{R}(Q) = \{0\}. \tag{5.9}$$

Also, operator $P_{\mathcal{S},\mathcal{R}(P|_{\mathcal{N}(Q)})}(\alpha P + \beta Q)|_{\mathcal{T}}$ has range $\mathcal{S}$ if and only if $\mathcal{S} \subseteq \mathcal{T} + \mathcal{R}(P|_{\mathcal{N}(Q)})$, which is equivalent with $\mathcal{R}(P) = \mathcal{R}(P) \cap \mathcal{R}(Q) + \mathcal{R}(P|_{\mathcal{N}(Q)})$. Now, by (5.9), we have that

$$\mathcal{R}(P) = \mathcal{R}(P) \cap \mathcal{R}(Q) \oplus \mathcal{R}(P|_{\mathcal{N}(Q)}). \tag{5.10}$$

Using (5.10), the fact that the intersection of two operator ranges is an operator range, and Theorem 2.3 of [**77**], we conclude that $\mathcal{R}(P|_{\mathcal{N}(Q)})$ is closed. Now, the proof follows by Theorem 5.3. □

Obviously, from Theorem 5.11 we get the following corollary:

COROLLARY 5.12. Let $P, Q \in \mathcal{B}(\mathcal{H})$ be given projectors and let $\alpha, \beta \in \mathbb{C} \setminus \{0\}$, $\alpha + \beta \neq 0$. Then the invertibility of the linear combination $\alpha P + \beta Q$ is independent of the choice of the scalars $\alpha, \beta \in \mathbb{C} \setminus \{0\}$, $\alpha + \beta \neq 0$.

(**3**) The problem of invertibility of the linear combination $\alpha P + \beta Q$ when P and Q are orthogonal projectors has also received a lot of attention. In [**24**] Buckholtz considers the special case when $\alpha + \beta = 1$ and gives conditions under which the difference of projectors on a Hilbert space is invertible, as well as an explicit formula for its inverse. In the paper of J. J. Koliha et al. [**120**], the invertibility of the sum of two orthogonal projectors was considered which is, as we already know, equivalent with the invertibility of the linear combination $\alpha P + \beta Q$:

THEOREM 5.13 ([**120**]). *Let $P, Q \in \mathcal{B}(\mathcal{H})$ be orthogonal projectors on a Hilbert space $\mathcal{H}$. Then the following conditions are equivalent:*

(i) *$P + Q$ is invertible.*
(ii) *$\mathcal{R}(P + Q)$ is closed and*

$$\mathcal{R}(P) \cap \mathcal{R}(Q(I - P)) = \mathcal{N}(P) \cap \mathcal{N}(Q) = \{0\}.$$

First, we have to mention that the condition $\mathcal{R}(P) \cap \mathcal{R}(Q(I - P)) = \{0\}$ from Theorem 5.13 is superfluous, which will be proved in the following lemma (for more information on orthogonal projector please see the excellent paper of A. Böttcher et al. [**23**]).

LEMMA 5.14. *Let $P, Q \in \mathcal{B}(\mathcal{H})$ be orthogonal projectors on a Hilbert space $\mathcal{H}$. Then*

$$\mathcal{R}(P) \cap \mathcal{R}(Q(I - P)) = \{0\}.$$

PROOF. First, let us observe that

$$\mathcal{R}(P) \cap \mathcal{R}(Q(I - P)) = \mathcal{R}(P) \cap \mathcal{R}(Q(I - P)) \cap \mathcal{R}(Q).$$

So, it is sufficient to prove that $\mathcal{R}(P) \cap \mathcal{R}(Q(I - P)) \cap \mathcal{R}(Q) = \{0\}$. It can be easily checked that

$$\mathcal{R}(Q(I - P))^{\perp} \cap \mathcal{R}(Q) = \mathcal{N}((I - P)Q) \cap \mathcal{R}(Q) = \mathcal{R}(P) \cap \mathcal{R}(Q),$$

implying that $\mathcal{R}(P) \cap \mathcal{R}(Q) \subseteq \mathcal{R}(Q(I - P))^{\perp}$; i.e., $\mathcal{R}(P) \cap \mathcal{R}(Q) \cap \mathcal{R}(Q(I - P)) = \{0\}$. □

Hence, from Theorem 5.3 we obtain the following result:

THEOREM 5.15. *Let $P, Q \in \mathcal{B}(\mathcal{H})$ be given orthogonal projectors and let $\alpha, \beta \in \mathbb{C} \setminus \{0\}$, $\alpha + \beta \neq 0$. Then the following are equivalent:*

(i) *$\alpha P + \beta Q$ is invertible.*
(ii) *$\mathcal{R}(P) + \mathcal{R}(Q) = \mathcal{H}$.*
(iii) *$\mathcal{R}(P + Q)$ is closed and $\mathcal{N}(P) \cap \mathcal{N}(Q) = \{0\}$.*

PROOF. (i)⇔(ii): Notice that in the case when $P, Q \in \mathcal{B}(\mathcal{H})$ are orthogonal projectors, the subspaces $\mathcal{S}$ and $\mathcal{T}$ defined in Theorem 5.4 by $\mathcal{S} = \mathcal{R}(PP_{\mathcal{N}(Q)})^\perp \cap \mathcal{R}(P)$ and $\mathcal{T} = \mathcal{N}(P_{\mathcal{R}(P)^\perp} Q) \cap \mathcal{N}(Q)^\perp$ coincide and $\mathcal{S} = \mathcal{T} = \mathcal{R}(P) \cap \mathcal{R}(Q)$. Indeed, if $P, Q \in \mathcal{B}(\mathcal{H})$ are orthogonal projectors, then

$$\begin{aligned}\mathcal{S} &= \mathcal{R}(PP_{\mathcal{N}(Q)})^\perp \cap \mathcal{R}(P) = \mathcal{R}(P(I - Q))^\perp \cap \mathcal{R}(P)\\ &= \mathcal{N}((I - Q)P) \cap \mathcal{R}(P) = \mathcal{R}(P) \cap \mathcal{R}(Q)\end{aligned}$$

and

$$\begin{aligned}\mathcal{T} &= \mathcal{N}(P_{\mathcal{R}(P)^\perp} Q) \cap \mathcal{N}(Q)^\perp = \mathcal{N}((I - P)Q) \cap \mathcal{R}(Q)\\ &= \mathcal{R}(P) \cap \mathcal{R}(Q).\end{aligned}$$

Hence, for any $x \in \mathcal{T}$, we have that $(\alpha P + \beta Q)x = (\alpha + \beta)x$ and $P_{\mathcal{S}}(\alpha P + \beta Q)x = (\alpha + \beta)x$. So, the operator $P_{\mathcal{S}}(\alpha P + \beta Q)|_{\mathcal{T}}$ from item (iii) of Theorem 5.4 is an injective operator with range $\mathcal{S}$ if and only if $\alpha + \beta \neq 0$. Also, the condition $\mathcal{R}(P) + \mathcal{R}(Q) = \mathcal{H}$ implies $\mathcal{N}(P) \cap \mathcal{N}(Q) = \{0\}$. Now, from Theorem 5.4 we can conclude that in the case when $P, Q \in \mathcal{B}(\mathcal{H})$ are orthogonal projectors, $\alpha P + \beta Q$ is an invertible operator if and only if the following conditions hold:

(*) $\mathcal{R}(P) + \mathcal{R}(Q) = \mathcal{H}$.
(**) $P|_{\mathcal{N}(Q)}$ has closed range.

Notice that the condition (**) that $P|_{\mathcal{N}(Q)}$ has closed range can be replaced by the condition that $\mathcal{R}(P(I - Q))$ is closed. By Proposition 2.4 of [**108**], we have that $\mathcal{R}(P(I - Q))$ is closed if and only if $R(P + Q)$ is closed, which is by Corollary 3 of [**77**] equivalent with the fact that $\mathcal{R}(P) + \mathcal{R}(Q)$ is closed. Since the condition (*) guarantees closedness of $\mathcal{R}(P) + \mathcal{R}(Q)$, we conclude that condition (*) is necessary and sufficient for the invertibility of $\alpha P + \beta Q$.

(i)⇔(iii): It follows from Theorem 5.13 and Lemma 5.14. □

(**4**) Now we consider the invertibility of the linear combination $\alpha A + \beta B$ for given operators $A, B \in \mathcal{B}(\mathcal{H})$ in two special cases: when $\mathcal{R}(A) \cap \mathcal{R}(B) = \{0\}$ and when $\overline{\mathcal{R}(AP_{\mathcal{N}(B)})} = \overline{\mathcal{R}(A)}$. In both of these two cases, besides giving necessary and sufficient conditions for the invertibility of $\alpha A + \beta B$, we conclude that the invertibility of the linear combination $\alpha A + \beta B$ is independent of the choice of the scalars $\alpha, \beta \in \mathbb{C} \setminus \{0\}$.

In the special case when $A, B \in \mathcal{B}(\mathcal{H})$ are such that $\mathcal{R}(A) \cap \mathcal{R}(B) = \{0\}$, using Theorem 5.4 we get the following:

THEOREM 5.16. *Let $A, B \in \mathcal{B}(\mathcal{H})$ be given operators and let $\alpha, \beta \in \mathbb{C} \setminus \{0\}$. If $\mathcal{R}(A) \cap \mathcal{R}(B) = \{0\}$, then the operator $\alpha A + \beta B$ is invertible if and only if*

$$\mathcal{R}(A) \oplus \mathcal{R}(B) = \mathcal{H}, \ \mathcal{N}(A) \oplus \mathcal{N}(B) = \mathcal{H}. \tag{5.11}$$

PROOF. Suppose that $\alpha A + \beta B$ is invertible. By Theorem 5.4, we have that $\mathcal{R}(A) \oplus \mathcal{R}(B) = \mathcal{H}$ which by Theorem 2.3 of [**77**] gives that $\mathcal{R}(A)$ and $\mathcal{R}(B)$ are closed. Now $\mathcal{R}(A) \cap \mathcal{R}(B) = \{0\}$ together with the fact that $\mathcal{R}(A)$ is closed implies that $\mathcal{T} = \mathcal{N}(P_{\mathcal{R}(A)^\perp} B) \cap \mathcal{N}(B)^\perp = \{0\}$ which by the condition (iii) from Theorem 5.4 gives that $\mathcal{S} = \mathcal{R}(AP_{\mathcal{N}(B)})^\perp \cap \overline{\mathcal{R}(A)} = \{0\}$. Hence, $\mathcal{R}(A) = \mathcal{R}(AP_{\mathcal{N}(B)})$ which implies that $\mathcal{N}(A)^\perp \subseteq \mathcal{N}(B) + \mathcal{N}(A)$. So, $\mathcal{H} = \mathcal{N}(B) + \mathcal{N}(A)$. By the condition (i) of Theorem 5.4, we have that $\mathcal{N}(A) \cap \mathcal{N}(B) = \{0\}$, so $\mathcal{H} = \mathcal{N}(B) \oplus \mathcal{N}(A)$. On the other hand suppose that (5.11) holds. Evidently, $\mathcal{R}(A)$ and $\mathcal{R}(B)$ are closed and the first condition from Theorem 5.4 is satisfied. Also, $\mathcal{R}(A) = A(\mathcal{H}) = A(\mathcal{N}(A) \oplus \mathcal{N}(B)) = A(\mathcal{N}(B)) = \mathcal{R}(AP_{\mathcal{N}(B)})$, so (ii) of Theorem 5.4 is satisfied. To conclude that (iii) of Theorem 5.4 is true, simply notice that $\mathcal{T} = \mathcal{S} = \{0\}$. □

Similarly, we get the following:

THEOREM 5.17. *Let $A, B \in \mathcal{B}(\mathcal{H})$ be given operators and let $\alpha, \beta \in \mathbb{C} \setminus \{0\}$. If $\overline{\mathcal{R}(AP_{\mathcal{N}(B)})} = \overline{\mathcal{R}(A)}$, then the operator $\alpha A + \beta B$ is invertible if and only if the following conditions hold:*

(i) *$\mathcal{N}(A) \cap \mathcal{N}(B) = \{0\}$, $\mathcal{R}(A) \oplus \mathcal{R}(B) = \mathcal{H}$.*
(ii) *$AP_{\mathcal{N}(B)}$ has closed range.*

PROOF. If $\overline{\mathcal{R}(AP_{\mathcal{N}(B)})} = \overline{\mathcal{R}(A)}$, then for $\mathcal{S}$ defined in Theorem 5.3 we have that $\mathcal{S} = \{0\}$. So the condition (iii) from Theorem 5.3 is satisfied if and only if $\mathcal{T} = \{0\}$; i.e., $\overline{\mathcal{R}(A)} \cap \mathcal{R}(B) = \{0\}$. Now, the proof follows directly from Theorem 5.3. □

COROLLARY 5.18. Let $A, B \in \mathcal{B}(\mathcal{H})$ be given operators. If one of the conditions $\mathcal{R}(A) \cap \mathcal{R}(B) = \{0\}$ and $\overline{\mathcal{R}(AP_{\mathcal{N}(B)})} = \overline{\mathcal{R}(A)}$ holds, then the invertibility of the linear combination $\alpha A + \beta B$ is independent of the choice of the scalars $\alpha, \beta \in \mathbb{C} \setminus \{0\}$.

(**5**) Now we consider the case when either one of the operators $A, B \in \mathcal{B}(\mathcal{H})$ is injective. Since the condition that $\alpha A + \beta B$ is invertible is symmetrical in A and B, let us suppose that $B \in \mathcal{B}(\mathcal{H})$ is injective:

THEOREM 5.19. *Let $A, B \in \mathcal{B}(\mathcal{H})$ be given operators such that B is injective and let $\alpha, \beta \in \mathbb{C} \setminus \{0\}$. Then $\alpha A + \beta B$ is invertible if and only if the following conditions hold:*

(i) *$\mathcal{R}(A) + \mathcal{R}(B) = \mathcal{H}$.*
(ii) *$(\alpha A + \beta B)|_{B^{-1}(\overline{\mathcal{R}(A)})}$ is an injective operator with range $\overline{\mathcal{R}(A)}$.*

Considering some special classes of operators we have seen that the invertibility of the linear combination $\alpha A + \beta B$ is independent of the choice of the scalars $\alpha, \beta \in \mathbb{C} \setminus \{0\}$. Another instance of this phenomenon is provided by the following result.

THEOREM 5.20. *Let $A, B \in \mathcal{B}(\mathcal{H})$ be given operators and let $\alpha, \beta \in \mathbb{C} \setminus \{0\}$. If there exists a closed subspace $\mathcal{P}$ such that $\mathcal{H} = \mathcal{N}(B) \oplus \mathcal{P}$ and $A|_{\mathcal{P}} = 0$ or $P_{\overline{\mathcal{R}(A)}} B = 0$, then the invertibility of the linear combination $\alpha A + \beta B$ is independent of the choice of the scalars.*

PROOF. Using the representations (5.4) and (5.5) of the operators A and B and the representation (5.6) of $\alpha A + \beta B$, from Theorem 5.2 the desired conclusion is immediately reached. □

Now, we present one interesting example:

EXAMPLE 5.21. If the row operator $\begin{bmatrix} A & B \end{bmatrix} \in \mathcal{B}(\mathcal{H}\oplus\mathcal{H},\mathcal{H})$ is bijective, then for all $\alpha,\beta \in \mathbb{C}\setminus\{0\}$ the linear combination $\alpha A+\beta B$ is not invertible. This follows directly from the Proposition 3.1 of [**105**] and our Theorem 5.4.

It is worth mentioning that some interesting properties of the product of two projectors PQ as well as the invertibility of $PQ \pm QP$ were considered in the paper of O. M. Baksalary et al. [**11**]. Also, paper [**10**] provides characterizations of column spaces of $P \pm Q$ and $PQ \pm QP$, which supplement the considerations in [**11**].

5.2. Injectivity, right and left invertibility of the sum of two operators

With the help of Theorem 1.2 which treats injectivity of upper triangular operator matrices and Theorems 1.15 and 1.11 which treat left and right invertibility, using the same methods as in the previous sections, we give corresponding results for these properties of a sum of two operators. The results will be presented without proofs, which the reader can find in [**124**].

THEOREM 5.22. *Let $A, B \in \mathcal{B}(\mathcal{H})$ be given operators and let $\alpha,\beta \in \mathbb{C}\setminus\{0\}$. Then the operator $\alpha A+\beta B$ is injective if and only if the conditions*

(i) $\mathcal{N}(A)\cap\mathcal{N}(B)=\{0\}$,
(ii) $\mathcal{N}(\alpha A+\beta B)\cap\mathcal{N}(B)^{\perp}=\{0\}$,
(iii) $\mathcal{R}((\alpha A+\beta B)|_{\mathcal{T}})\cap\mathcal{R}(AP_{\mathcal{N}(B)})=\{0\}$

hold, where $\mathcal{T}=B^{-1}(\overline{\mathcal{R}(A)})\cap\mathcal{N}(B)^{\perp}$.

Now, we see in the next example that the injectivity of a linear combination of operators can depend on the choice of constants, which will serve as motivation to find special cases where injectivity of a linear combination of operators is independent of the choice of the constants.

EXAMPLE 5.23. Let $A, B \in \mathcal{B}(l_2)$ be operators defined by

$$Ae_n=\begin{cases}\frac{e_n+e_{n+1}}{2}, & n=3k-2,\\ \frac{e_n+e_{n-1}}{2}, & n=3k-1,\\ \frac{e_n}{2}, & n=3k,\end{cases} \qquad Be_n=\begin{cases}\frac{e_{n+1}-e_n}{2}, & n=3k-2,\\ \frac{e_{n-1}+e_n}{2}, & n=3k-1,\\ \frac{e_n}{2}, & n=3k,\end{cases}$$

where $k \in \mathbb{N}$ and e_n is the nth vector in the standard orthonormal basis of the space l_2. It is easy to see that $A+B$ is injective, whereas $A-B$ is not. If we analyze this situation using Theorem 5.22 (and its conditions) we see that

$$\begin{aligned}\mathcal{N}(A)&=\{(x_n)_{n\in\mathbb{N}} \mid x_{3k-1}=-x_{3k-2},\ x_{3k}=0,\ k\in\mathbb{N}\},\\ \mathcal{N}(B)&=\{(x_n)_{n\in\mathbb{N}} \mid x_{3k-1}=x_{3k-2},\ x_{3k}=0,\ k\in\mathbb{N}\},\\ \mathcal{N}(B)^{\perp}&=\{(x_n)_{n\in\mathbb{N}} \mid x_{3k-1}=-x_{3k-2},\ k\in\mathbb{N}\},\\ \overline{\mathcal{R}(A)}&=\mathcal{R}(A)=\{(x_n)_{n\in\mathbb{N}} \mid x_{3k-1}=x_{3k-2},\ k\in\mathbb{N}\},\\ \mathcal{T}&=\{0\}.\end{aligned}$$

Also, $\mathcal{N}(A)\cap\mathcal{N}(B)=\{0\}$, and since $\mathcal{T}=\{0\}$, condition (i) from Theorem 5.22 is satisfied, and condition (iii) trivially holds. It is easy to check that $A+B$ is injective on $\mathcal{N}(B)^{\perp}$, whereas $A-B$ is not, so by Theorem 5.22 we have that $A+B$ is injective and $A-B$ is not.

Some special cases in which the injectivity of the linear combination $\alpha A+\beta B$ is independent of the choice of scalars $\alpha, \beta \in \mathbb{C} \setminus \{0\}$ are investigated now:

THEOREM 5.24. *Let $A, B \in \mathcal{B}(\mathcal{H})$ be such that $\overline{\mathcal{R}(A)} \cap \overline{\mathcal{R}(B)} = \{0\}$, and let $\alpha, \beta \in \mathbb{C} \setminus \{0\}$. Then $\alpha A+\beta B$ is injective if and only if $\mathcal{N}(A) \cap \mathcal{N}(B) = \{0\}$.*

THEOREM 5.25. *Let $A, B \in \mathcal{B}(\mathcal{H})$ be such that $\mathcal{R}(A) = \mathcal{R}(AP_{\mathcal{N}(B)})$ and let $\alpha, \beta \in \mathbb{C} \setminus \{0\}$. Then $\alpha A+\beta B$ is injective if and only if $\mathcal{N}(A) \cap \mathcal{N}(B) = \mathcal{R}(A) \cap \mathcal{R}(B) = \{0\}$.*

From Theorems 5.24 and 5.25, we have the following corollary:

COROLLARY 5.26. Let $A, B \in \mathcal{B}(\mathcal{H})$ be given operators and let $\alpha, \beta \in \mathbb{C} \setminus \{0\}$. If one of the conditions $\overline{\mathcal{R}(A)} \cap \overline{\mathcal{R}(B)} = \{0\}$ and $\mathcal{R}(A) = \mathcal{R}(AP_{\mathcal{N}(B)})$ holds, then the injectivity of the linear combination $\alpha A+\beta B$ is independent of the choice of the scalars $\alpha, \beta \in \mathbb{C} \setminus \{0\}$.

Injectivity of linear combinations of projectors is investigated next (in a manner which will cover both the case of sums and that of differences):

THEOREM 5.27. *Let $P, Q \in \mathcal{B}(\mathcal{H})$ be given projectors and let $\alpha, \beta \in \mathbb{C} \setminus \{0\}$. Then the operator $\alpha P+\beta Q$ is injective if and only if*

$$\begin{cases} \mathcal{N}(P) \cap \mathcal{N}(Q) = \mathcal{R}(Q) \cap \mathcal{R}(P(I-Q)) = \{0\}, & \alpha+\beta \neq 0, \\ \mathcal{N}(P) \cap \mathcal{N}(Q) = \mathcal{R}(P) \cap \mathcal{R}(Q) = \{0\}, & \alpha+\beta = 0. \end{cases}$$

COROLLARY 5.28. Let $P, Q \in \mathcal{B}(\mathcal{H})$ be given projectors and let $\alpha, \beta \in \mathbb{C} \setminus \{0\}$, $\alpha \neq -\beta$. The linear combination $\alpha P+\beta Q$ is injective independently of the choice of constants α and β.

Conditions for injectivity in the case of orthogonal projectors are given as follows.

THEOREM 5.29. *Let $P, Q \in \mathcal{B}(\mathcal{H})$ be given orthogonal projectors and let $\alpha, \beta \in \mathbb{C} \setminus \{0\}$. Then the operator $\alpha P+\beta Q$ is injective if and only if*

$$\begin{cases} \mathcal{N}(P) \cap \mathcal{N}(Q) = \{0\}, & \alpha+\beta \neq 0, \\ \mathcal{N}(P) \cap \mathcal{N}(Q) = \mathcal{R}(P) \cap \mathcal{R}(Q) = \{0\}, & \alpha+\beta = 0. \end{cases}$$

Using Theorems 1.15 and 1.11, the necessary and sufficient conditions for left (right) invertibility of the linear combination $\alpha A+\beta B$, where $\alpha, \beta \in \mathbb{C} \setminus \{0\}$, can be obtained.

THEOREM 5.30. *Let $A, B \in \mathcal{B}(\mathcal{H})$ be given operators and let $\alpha, \beta \in \mathbb{C} \setminus \{0\}$. Then $\alpha A+\beta B$ is left invertible if and only if the conditions*

(i) $\mathcal{N}(A) \cap \mathcal{N}(B) = \{0\}$,
(ii) $\mathcal{R}(AP_{\mathcal{N}(B)})$ *is closed,*
(iii) $\mathcal{N}(B)^{\perp} = \mathcal{R}(B^* P_{\mathcal{R}(A)^{\perp}}) + \mathcal{R}((\alpha A+\beta B)^*|_{\mathcal{S}})$

hold, where $\mathcal{S} = \mathcal{R}(AP_{\mathcal{N}(B)})^{\perp} \cap \overline{\mathcal{R}(A)}$.

REMARK 5.31. Condition (iii) from Theorem 5.30 can be reformulated as

$$\begin{bmatrix} P_{\mathcal{S}}(\alpha A+\beta B)|_{\mathcal{N}(B)^\perp} \\ P_{\mathcal{R}(A)^\perp}B \end{bmatrix} : \mathcal{N}(B)^\perp \to \begin{bmatrix} \overline{\mathcal{R}(A)} \\ \mathcal{R}(A)^\perp \end{bmatrix}$$

is left invertible.

Similarly, we can give necessary and sufficient conditions for right invertibility:

THEOREM 5.32. *Let $A, B \in \mathcal{B}(\mathcal{H})$ be given operators and let $\alpha, \beta \in \mathbb{C}\setminus\{0\}$. Then $\alpha A+\beta B$ is right invertible if and only if the conditions*

(i) $\mathcal{R}(A)+\mathcal{R}(B)=\mathcal{H}$,
(ii) $\overline{\mathcal{R}(A)}=\mathcal{R}(AP_{\mathcal{N}(B)})+\mathcal{R}((\alpha A+\beta B)|_{\mathcal{T}})$

hold, where $\mathcal{T}=B^{-1}(\overline{\mathcal{R}(A)})\cap\mathcal{N}(B)^\perp$.

In conjunction with Theorems 5.30 and 5.32 we can examine some special cases where left (right) inveritiblity of the linear combination is independent of the choice of scalars $\alpha, \beta \in \mathbb{C}\setminus\{0\}$.

THEOREM 5.33. *Let $A, B \in \mathcal{B}(\mathcal{H})$ be such that $\mathcal{R}(A)=\mathcal{R}(AP_{\mathcal{N}(B)})$ and let $\alpha, \beta \in \mathbb{C}\setminus\{0\}$. Then $\alpha A+\beta B$ is left invertible if and only if the following conditions hold:*

(i) $\mathcal{N}(A)\cap\mathcal{N}(B)=\{0\}$, $\mathcal{R}(B)\cap\mathcal{R}(A)=\{0\}$.
(ii) $\mathcal{R}(A)$ *is closed.*
(iii) $\mathcal{R}(P_{\mathcal{R}(A)^\perp}B)$ *is closed.*

THEOREM 5.34. *Let $A, B \in \mathcal{B}(\mathcal{H})$ be given operators such that $\mathcal{R}(A)=\mathcal{R}(AP_{\mathcal{N}(B)})$ and let $\alpha, \beta \in \mathbb{C}\setminus\{0\}$. Then $\alpha A+\beta B$ is right invertible if and only if the conditions*

(i) $\mathcal{R}(A)+\mathcal{R}(B)=\mathcal{H}$,
(ii) $\overline{\mathcal{R}(A)}=\mathcal{R}(A)+\mathcal{R}(B|_{\mathcal{T}})$

hold, where $\mathcal{T}=\mathcal{N}(B_2)=B^{-1}(\overline{\mathcal{R}(A)})\cap\mathcal{N}(B)^\perp$.

THEOREM 5.35. *Let $A, B \in \mathcal{B}(\mathcal{H})$ be given operators such that $\overline{\mathcal{R}(A)}\cap\overline{\mathcal{R}(B)}=\{0\}$ and let $\alpha, \beta \in \mathbb{C}\setminus\{0\}$. Then $\alpha A+\beta B$ is right invertible if and only if the following conditions hold:*

(i) $\mathcal{R}(A)\oplus\mathcal{R}(B)=\mathcal{H}$.
(ii) $\mathcal{R}(AP_{\mathcal{N}(B)})=\overline{\mathcal{R}(A)}$.

In the next result we present some conditions for left (right) invertibility in the case of orthogonal projectors:

THEOREM 5.36. *Let $P, Q \in \mathcal{B}(\mathcal{H})$ be given orthogonal projectors and let $\alpha, \beta \in \mathbb{C}\setminus\{0\}$. Then $\alpha P+\beta Q$ is left invertible if and only if*

$$\begin{cases} \mathcal{R}(P)+\mathcal{R}(Q)=\mathcal{H}, & \alpha+\beta\neq 0, \\ \mathcal{R}(P)\oplus\mathcal{R}(Q)=\mathcal{H}, & \alpha+\beta=0. \end{cases}$$

Since from this theorem we see that the linear combinations $\alpha P + \beta Q$ and $\overline{\alpha}P + \overline{\beta}Q$ are simultaneously left invertible it is easily concluded that $\alpha P + \beta Q$ is invertible if and only if it is left invertible. This gives the following corollaries:

COROLLARY 5.37. Let $P, Q \in \mathcal{B}(\mathcal{H})$ be given orthogonal projectors and let $\alpha, \beta \in \mathbb{C} \setminus \{0\}$. Then $\alpha P + \beta Q$ is invertible if and only if

$$\begin{cases} \mathcal{R}(P) + \mathcal{R}(Q) = \mathcal{H}, & \alpha + \beta \neq 0, \\ \mathcal{R}(P) \oplus \mathcal{R}(Q) = \mathcal{H}, & \alpha + \beta = 0. \end{cases}$$

COROLLARY 5.38. Let $P, Q \in \mathcal{B}(\mathcal{H})$ be given orthogonal projectors and let $\alpha, \beta \in \mathbb{C} \setminus \{0\}$. Then the following statements are equivalent:

(i) $\alpha P + \beta Q$ is left (right) invertible.
(ii) $\alpha P + \beta Q$ is invertible.

COROLLARY 5.39. Let $P, Q \in \mathcal{B}(\mathcal{H})$ be given orthogonal projectors and let $\alpha, \beta \in \mathbb{C} \setminus \{0\}$ such that $\alpha \neq -\beta$. The linear combination $\alpha P + \beta Q$ is invertible, independent of the choice of constants α and β.

REMARK 5.40. In Corollary 4.3 from [**120**] it was proven that the sum of orthogonal projectors P and Q is invertible if and only if

$$\mathcal{N}(P) \cap \mathcal{N}(Q) = \mathcal{R}(P) \cap \mathcal{R}(Q(I - P)) = \{0\} \quad \text{and} \quad P + Q \text{ has closed range.}$$

This result is equivalent to Corollary 5.37 (for the case of sums of orthogonal projectors). Indeed, if $\mathcal{N}(P) \cap \mathcal{N}(Q) = \{0\}$, we have that $\overline{\mathcal{R}(P) + \mathcal{R}(Q)} = \mathcal{H}$. Furthermore, if $\mathcal{R}(P + Q)$ is closed, from Lemma 2.4 in [**119**], we have that $\mathcal{R}(P) + \mathcal{R}(Q)$ is closed as well and thus $\mathcal{R}(P) + \mathcal{R}(Q) = \mathcal{H}$. Conversely, if $\mathcal{R}(P) + \mathcal{R}(Q) = \mathcal{H}$, we have that $\mathcal{N}(P) \cap \mathcal{N}(Q) = \{0\}$ and, using Lemma 2.4 from [**119**] again, we have that $P + Q$ has closed range.

Some interesting results can be obtained for right invertiblity of linear combinations of oblique projectors as well:

THEOREM 5.41. *Let $P, Q \in \mathcal{B}(\mathcal{H})$ be given projectors and let $\alpha, \beta \in \mathbb{C} \setminus \{0\}$. Then $\alpha P + \beta Q$ is right invertible if and only if $\mathcal{R}(P) + \mathcal{R}(Q) = \mathcal{H}$ and*

$$\begin{cases} \mathcal{R}(P) = \mathcal{R}(P(I - Q)) + (\mathcal{R}(P) \cap \mathcal{R}(Q)), & \alpha \neq -\beta, \\ \mathcal{R}(P) = \mathcal{R}(P(I - Q)), & \alpha = \beta. \end{cases}$$

5.3. Fredholmness of the sum of two operators

In this section we address the question of the Fredholmness of the sum of two given operators $A, B \in \mathcal{B}(\mathcal{H}, \mathcal{K})$. Let us recall that if $\mathcal{M}$ is a closed subspace of $\mathcal{H}$, we use the symbol $P_{\mathcal{M}}^{\mathtt{rst}}$ to denote an operator from $\mathcal{B}(\mathcal{H}, \mathcal{M})$ defined by $P_{\mathcal{M}}^{\mathtt{rst}} x = P_{\mathcal{M}} x$, for all $x \in \mathcal{H}$, where $P_{\mathcal{M}}$ is the orthogonal projector onto $\mathcal{M}$.

THEOREM 5.42. *Let $A, B \in \mathcal{B}(\mathcal{H}, \mathcal{K})$ be given operators. Let $\mathcal{H} = \mathcal{N}(B) \oplus \mathcal{P}$ and $\mathcal{K} = \overline{\mathcal{R}(A)} \oplus \mathcal{Q}$. Then the operator $A + B$ is Fredholm if and only if the following conditions hold:*

(i) *$\dim(\mathcal{N}(A) \cap \mathcal{N}(B)) < \infty$ and $A|_{\mathcal{N}(B)}$ has closed range.*
(ii) *$\dim(\mathcal{N}(A') \cap \mathcal{N}(B')) < \infty$ and $P_{\mathcal{Q}, \overline{\mathcal{R}(A)}} B$ has closed range.*

(iii) $\dim(\mathcal{P}\cap\mathcal{N}(A+B))<\infty, \dim(\mathcal{R}(A|_{\mathcal{N}(B)})\cap\mathcal{R}((A+B)|_{\mathcal{P}}))<\infty$ *and* $\dim\overline{\mathcal{R}(A)}/(\mathcal{R}(A|_{\mathcal{N}(B)})+(\overline{\mathcal{R}(A)}\cap\mathcal{R}((A+B)|_{\mathcal{P}})))<\infty$.

Furthermore, if (i)–(iii) *hold, then*

$$\begin{aligned}\alpha(A+B) &= \dim(\mathcal{N}(A)\cap\mathcal{N}(B))+\dim(\mathcal{P}\cap\mathcal{N}(A+B))\\ &\quad+\dim(\mathcal{R}(A|_{\mathcal{N}(B)})\cap\mathcal{R}((A+B)|_{\mathcal{P}}))\\ \beta(A+B) &= \dim(\mathcal{N}(A')\cap\mathcal{N}(B'))\\ &\quad+\dim(\overline{\mathcal{R}(A)}/(\mathcal{R}(A|_{\mathcal{N}(B)})+(\overline{\mathcal{R}(A)}\cap\mathcal{R}((A+B)|_{\mathcal{P}})))).\end{aligned}$$

PROOF. With respect to the decompositions given below, $A, B\in\mathcal{B}(\mathcal{H},\mathcal{K})$ have the following representations:

$$A=\begin{bmatrix} A_1 & A_2\\ 0 & 0\end{bmatrix}:\begin{bmatrix}\mathcal{N}(B)\\ \mathcal{P}\end{bmatrix}\to\begin{bmatrix}\overline{\mathcal{R}(A)}\\ \mathcal{Q}\end{bmatrix}, \tag{5.12}$$

$$B=\begin{bmatrix} 0 & B_1\\ 0 & B_2\end{bmatrix}:\begin{bmatrix}\mathcal{N}(B)\\ \mathcal{P}\end{bmatrix}\to\begin{bmatrix}\overline{\mathcal{R}(A)}\\ \mathcal{Q}\end{bmatrix}. \tag{5.13}$$

So $A+B$ is a Fredholm operator if and only if the operator matrix given by

$$\begin{bmatrix} A_1 & B_1+A_2\\ 0 & B_2\end{bmatrix}:\begin{bmatrix}\mathcal{N}(B)\\ \mathcal{P}\end{bmatrix}\to\begin{bmatrix}\overline{\mathcal{R}(A)}\\ \mathcal{Q}\end{bmatrix}$$

is Fredholm.

Let $\overline{\mathcal{R}(A)}=\overline{\mathcal{R}(A_1)}\oplus\mathcal{S}$ and $\mathcal{P}=\mathcal{N}(B_2)\oplus\mathcal{W}$. By Theorem 1.55 we have that $A+B$ is a Fredholm operator if and only if the following three conditions are satisfied:

($*$) A_1 is left semi-Fredholm.
($**$) B_2 is right semi-Fredholm.
($***$) $P_{S,\overline{\mathcal{R}(A_1)}\oplus\mathcal{Q}}(A+B)|_{\mathcal{N}(B_2)}$ is Fredholm.

Evidently, ($*$) holds if and only if $\dim(\mathcal{N}(A)\cap\mathcal{N}(B))<\infty$ and $\mathcal{R}(A|_{\mathcal{N}(B)})$ is closed.

Also, ($**$) holds if and only if B_2' is left semi-Fredholm. Since $\mathcal{N}(B_2')=\mathcal{N}(B')\cap\mathcal{N}(A')$ it follows that ($**$) holds if and only if the range of $P_{\mathcal{Q},\overline{\mathcal{R}(A)}}B$ is closed and $\dim(\mathcal{N}(B')\cap\mathcal{N}(A'))<\infty$.

Now, we consider condition ($***$) taking into account that $\mathcal{R}(A_1)$ is closed in both directions. The condition ($***$) is satisfied if and only if

$$\dim(\mathcal{N}(P_{S,\mathcal{R}(A_1)\oplus\mathcal{Q}}(A+B)|_{\mathcal{N}(B_2)}))<\infty$$

and

$$\dim(\mathcal{S}/\mathcal{R}(P_{S,\mathcal{R}(A_1)\oplus\mathcal{Q}}(A+B)|_{\mathcal{N}(B_2)}))<\infty.$$

Since

$$\mathcal{N}(P_{S,\mathcal{R}(A_1)\oplus\mathcal{Q}}(A+B)|_{\mathcal{N}(B_2)})=\{x\in\mathcal{P}\mid(A+B)x\in\mathcal{R}(A_1)\},$$

we have that

$$\begin{aligned}\dim(\mathcal{N}(P_{S,\mathcal{R}(A_1)\oplus\mathcal{Q}}(A+B)|_{\mathcal{N}(B_2)})) &= \dim(\mathcal{R}(A|_{\mathcal{N}(B)})\cap\mathcal{R}((A+B)|_{\mathcal{P}}))\\ &\quad+\dim(\mathcal{P}\cap\mathcal{N}(A+B)).\end{aligned}$$

Also,

$$\dim\mathcal{S}/\mathcal{R}(P_{S,\mathcal{R}(A_1)\oplus\mathcal{Q}}(A+B)|_{\mathcal{N}(B_2)})<\infty,$$

if and only if there exists a finite-dimensional subspace $\mathcal{M}$ such that $\mathcal{S} = \mathcal{M} \oplus \mathcal{R}(P_{S,\mathcal{R}(A_1))\oplus\mathcal{Q}}(A+B)|_{\mathcal{N}(B_2)})$. The last assertion is equivalent with the existence of a finite dimensional subspace $\mathcal{N}$ such that

$$\overline{\mathcal{R}(A)} = \mathcal{N} \oplus (\mathcal{R}(A|_{\mathcal{N}(B)}) + (\overline{\mathcal{R}(A)} \cap \mathcal{R}((A+B)|_{\mathcal{P}})));$$

i.e.,

$$\dim(\overline{\mathcal{R}(A)}/(\mathcal{R}(A|_{\mathcal{N}(B)}) + (\overline{\mathcal{R}(A)} \cap \mathcal{R}((A+B)|_{\mathcal{P}}))) < \infty.$$

The rest of the proof follows by Remark 1.57. □

Notice that the condition $\dim(\mathcal{N}(A') \cap \mathcal{N}(B')) < \infty$ from item (ii) of the previous theorem can be replaced by $\dim(\mathcal{R}(A)^{\perp} \cap \mathcal{R}(B)^{\perp}) < \infty$.

In the special case when we take the orthogonal decompositions of spaces $\mathcal{H}, \mathcal{K}$ in the proof of Theorem 5.42, we get the following result:

THEOREM 5.43. *Let $A \in \mathcal{B}(\mathcal{H},\mathcal{K})$ and $B \in \mathcal{B}(\mathcal{H},\mathcal{K})$ be given operators. Then $A+B$ is Fredholm if and only if the conditions*

(i) $\dim(\mathcal{N}(A) \cap \mathcal{N}(B)) < \infty$, $\dim(\mathcal{R}(A)^{\perp} \cap \mathcal{R}(B)^{\perp}) < \infty$,
(ii) $AP_{\mathcal{N}(B)}$ *and* $P_{\mathcal{R}(A)^{\perp}}B$ *have closed ranges,*
(iii) $P^{\mathtt{rst}}_{\mathcal{S}}(A+B)|_{\mathcal{T}}$ *is Fredholm*

hold, where $\mathcal{S} = \mathcal{N}(P_{\mathcal{N}(B)}A^*) \cap \overline{\mathcal{R}(A)}$ *and* $\mathcal{T} = \mathcal{N}(P_{\mathcal{R}(A)^{\perp}}B) \cap \mathcal{N}(B)^{\perp}$.

Furthermore, if (i)–(iii) *hold, then*

$$\begin{aligned}\operatorname{ind}(A+B) = &\dim(\mathcal{N}(A) \cap \mathcal{N}(B)) - \dim(\mathcal{R}(A)^{\perp} \cap \mathcal{R}(B)^{\perp})\\ &+ \operatorname{ind}(P^{\mathtt{rst}}_{\mathcal{S}}(A+B)|_{\mathcal{T}}).\end{aligned}$$

PROOF. Let $\mathcal{H} = \mathcal{N}(B) \oplus \mathcal{N}(B)^{\perp}$ and $\mathcal{K} = \overline{\mathcal{R}(A)} \oplus \mathcal{R}(A)^{\perp}$. With respect to these decompositions $A, B \in \mathcal{B}(\mathcal{H},\mathcal{K})$ have the following representations:

$$A = \begin{bmatrix} A_1 & A_2 \\ 0 & 0 \end{bmatrix} : \begin{bmatrix} \mathcal{N}(B) \\ \mathcal{N}(B)^{\perp} \end{bmatrix} \to \begin{bmatrix} \overline{\mathcal{R}(A)} \\ \mathcal{R}(A)^{\perp} \end{bmatrix}, \tag{5.14}$$

$$B = \begin{bmatrix} 0 & B_1 \\ 0 & B_2 \end{bmatrix} : \begin{bmatrix} \mathcal{N}(B) \\ \mathcal{N}(B)^{\perp} \end{bmatrix} \to \begin{bmatrix} \overline{\mathcal{R}(A)} \\ \mathcal{R}(A)^{\perp} \end{bmatrix}. \tag{5.15}$$

Using the orthogonal decomposition $\overline{\mathcal{R}(A)} = \overline{\mathcal{R}(A_1)} \oplus (\mathcal{R}(A_1)^{\perp} \cap \overline{\mathcal{R}(A)})$ in the proof of Theorem 1.55 we get that the linear combination $A+B$ is a Fredholm operator if and only if the following three conditions are satisfied:

(i) A_1 is left semi-Fredholm.
(ii) B_2 is right semi-Fredholm.
(iii) $P^{\mathtt{rst}}_{\mathcal{R}(A_1)^{\perp}\cap\overline{\mathcal{R}(A)}}(A_2+B_1)|_{\mathcal{N}(B_2)}$ is Fredholm.

Evidently, (i) holds if and only if $\dim(\mathcal{N}(A) \cap \mathcal{N}(B)) < \infty$ and $AP_{\mathcal{N}(B)}$ has closed range. Also, (ii) holds if and only if B_2^* is upper semi-Fredholm. Since

$$\mathcal{N}(B_2^*) = \mathcal{N}(B^*) \cap \mathcal{R}(A)^{\perp} = \mathcal{R}(B)^{\perp} \cap \mathcal{R}(A)^{\perp},$$

it follows that (ii) holds if and only if the range of $P_{\mathcal{R}(A)^{\perp}}B$ is closed and $\dim(\mathcal{R}(A)^{\perp} \cap \mathcal{R}(B)^{\perp}) < \infty$. The third condition follows directly from the fact that $\mathcal{R}(A_1)^{\perp} \cap \overline{\mathcal{R}(A)} = \mathcal{N}(P_{\mathcal{N}(B)}A^*) \cap \overline{\mathcal{R}(A)}$ and $\mathcal{N}(B_2) = \mathcal{N}(P_{\mathcal{R}(A)^{\perp}}B) \cap \mathcal{N}(B)^{\perp}$. □

REMARK 5.44. If A_1 and B_2 from (5.1) and (5.2) are Fredholm, then $\mathcal{N}(B_2)$ and $\mathcal{S}$ defined in Theorem 5.42 are finite dimensional so the condition $(***)$ from the proof of Theorem 5.42 is satisfied. Hence, in this case taking αA instead of A and βB instead of B, we get that for all constants $\alpha, \beta \in \mathbb{C} \setminus \{0\}$, the linear combination $\alpha A + \beta B$ is a Fredholm operator.

Also if one of the operators A_1 and B_2 is Fredholm and the other one is not, at least one of the conditions $(*)$–$(***)$ from the proof of Theorem 5.42 is not satisfied. Indeed, if conditions $(*)$ and $(**)$ are satisfied, one of spaces $\mathcal{N}(B_2)$ and $\mathcal{S}$ is finite dimensional and the other one is infinite dimensional, so the condition $(***)$ is not satisfied and $A + B$ is not a Fredholm operator. Moreover, in this case the linear combination $\alpha A + \beta B$ is not a Fredholm operator for any constants $\alpha, \beta \in \mathbb{C} \setminus \{0\}$. Thus we have the following result:

THEOREM 5.45. *Let $A, B \in \mathcal{B}(\mathcal{H}, \mathcal{K})$ be given operators and let $\mathcal{H} = \mathcal{N}(B) \oplus \mathcal{P}$ and $\mathcal{K} = \overline{\mathcal{R}(A)} \oplus \mathcal{Q}$. Let $A_1 = P^{\mathtt{rst}}_{\overline{\mathcal{R}(A)},\mathcal{Q}} A|_{\mathcal{N}(B)}$ and $B_2 = P^{\mathtt{rst}}_{\mathcal{Q},\overline{\mathcal{R}(A)}} B|_{\mathcal{P}}$.*

(i) *If the operators A_1 and B_2 are Fredholm, then for all constants $\alpha, \beta \in \mathbb{C} \setminus \{0\}$ the linear combination $\alpha A + \beta B$ is a Fredholm operator.*
(ii) *If one of operators A_1 and B_2 is Fredholm and the other one is not, then for all constants $\alpha, \beta \in \mathbb{C} \setminus \{0\}$ the linear combination $\alpha A + \beta B$ is not a Fredholm operator.*

REMARK 5.46. If we take orthogonal decompositions of spaces $\mathcal{H}$ and $\mathcal{K}$, as in Theorem 5.43, we have that $\mathcal{S} = \mathcal{N}(A_1^*) = \{x \in \mathcal{N}(A^*)^\perp \mid A^*x \in \overline{\mathcal{R}(B^*)}\}$ and $\mathcal{T} = \mathcal{N}(B_2) = \{x \in \mathcal{N}(B)^\perp \mid Bx \in \overline{\mathcal{R}(A)}\}$. It can be checked that $\dim(\mathcal{N}(A_1^*)) = \dim(\mathcal{R}(A^*) \cap \overline{\mathcal{R}(B^*)})$ and $\dim(\mathcal{N}(B_2)) = \dim(\overline{\mathcal{R}(A)} \cap \mathcal{R}(B))$. So, we are ready to present some particular cases when the linear combination $\alpha A + \beta B$ is independent of the choice of scalars $\alpha, \beta \in \mathbb{C} \setminus \{0\}$:

THEOREM 5.47. *Let $A \in \mathcal{B}(\mathcal{H}, \mathcal{K})$ and $B \in \mathcal{B}(\mathcal{H}, \mathcal{K})$ and let $m = \dim(\mathcal{R}(A^*) \cap \overline{\mathcal{R}(B^*)})$ and $n = \dim(\overline{\mathcal{R}(A)} \cap \mathcal{R}(B))$. If one of the conditions*

(i) $\max\{m, n\} < \infty$,
(ii) $\min\{m, n\} < \max\{m, n\} = \infty$

holds, then Fredholmness of the linear combination $\alpha A + \beta B$ is independent of the choice of scalars $\alpha, \beta \in \mathbb{C} \setminus \{0\}$. Furthermore, we have in the case (i) *that $A + B$ is Fredholm if and only if* $\dim(\mathcal{N}(A) \cap \mathcal{N}(B)) < \infty$, $\dim(\mathcal{R}(A)^\perp \cap \mathcal{R}(B)^\perp) < \infty$, *and $AP_{\mathcal{N}(B)}$, $P_{\mathcal{R}(A)^\perp} B$ have closed ranges, while in the case* (ii) *we have that $A + B$ is not Fredholm.*

PROOF. The proof follows by Theorem 5.43 and Remark 5.46. □

From a previous theorem, we have that if m, n are both infinite or if one is finite and the other is infinite, then Fredholmness of the linear combination $\alpha A + \beta B$ is independent of the choice of scalars $\alpha, \beta \in \mathbb{C} \setminus \{0\}$. What happens in the case when both m, n are infinite? The next example gives an answer to this question:

EXAMPLE 5.48. Let $\mathcal{H}$ be infinite-dimensional Hilbert space and let $A = I_{\mathcal{H}}$ and $B = -2I_{\mathcal{H}}$. Evidently $\alpha A + \beta B$ is a Fredholm operator except in the case when $\alpha = 2\beta$ which means that Fredholmness of the linear combination $\alpha A + \beta B$ depends on the choice of $\alpha, \beta \in \mathbb{C} \setminus \{0\}$. On the other side if $A \in \mathcal{B}(H)$ is non-Fredholm and $B = A$, we get that $\alpha A + \beta B$, is a non-Fredholm operator for any $\alpha, \beta \in \mathbb{C} \setminus \{0\}$

which means that Fredholmness of the linear combination $\alpha A + \beta B$ is independent of the choice of scalars $\alpha, \beta \in \mathbb{C} \setminus \{0\}$.

From Theorem 5.43, we can deduce the following well-known result whose proof will be given presently since it is quite different from all others that can be found in the literature:

COROLLARY 5.49. Let $A \in \mathcal{B}(\mathcal{H}, \mathcal{K})$ be an operator of finite rank and let $B \in \mathcal{B}(\mathcal{H}, \mathcal{K})$. Then $A + B$ is Fredholm if and only if B is Fredholm.

PROOF. Since $\dim(\mathcal{R}(A)) < \infty$ from Theorem 5.47, we get that $A + B$ is Fredholm if and only if the following conditions hold:

(i) $\dim(\mathcal{N}(A) \cap \mathcal{N}(B)) < \infty$, $\dim(\mathcal{R}(A)^{\perp} \cap \mathcal{R}(B)^{\perp}) < \infty$.
(ii) $P_{\mathcal{R}(A)^{\perp}} B$ has closed range.

These conditions are satisfied if and only if B is Fredholm. Indeed, if B is Fredholm, the condition (i) is obviously satisfied. Also $P_{\mathcal{R}(A)^{\perp}} B$ is a left semi-Fredholm operator as the product of two such operators, so its range is closed. Conversely, if conditions (i) and (ii) are satisfied, then from the closedness of $\mathcal{R}(P_{\mathcal{R}(A)^{\perp}} B)$ we have that $\mathcal{R}(B^* P_{\mathcal{R}(A)^{\perp}})$ is closed. Now, from $\mathcal{R}(B^*) = \mathcal{R}(B^* P_{\mathcal{R}(A)^{\perp}}) + \mathcal{R}(B^* P_{\mathcal{R}(A)})$, we get that $\mathcal{R}(B)$ is closed. Using the fact that in any vector space $\mathcal{X}$ if $\mathcal{X}_1, \mathcal{X}_2$, and $\mathcal{M}$ are subspaces of $\mathcal{X}$ such that $\mathcal{X} = \mathcal{X}_1 \oplus \mathcal{X}_2$, $\dim \mathcal{X}_2 < \infty$, and $\dim(\mathcal{M} \cap \mathcal{X}_1) < \infty$, then $\dim \mathcal{M} < \infty$, from $\dim(\mathcal{N}(A) \cap \mathcal{N}(B)) < \infty$, $\mathcal{H} = \mathcal{N}(A) \oplus \mathcal{R}(A^*)$, and $\dim(\mathcal{R}(A^*)) < \infty$ it follows that $\dim(\mathcal{N}(B)) < \infty$. In the same manner from $\dim(\mathcal{R}(A)^{\perp} \cap \mathcal{R}(B)^{\perp}) < \infty, \mathcal{K} = \mathcal{R}(A)^{\perp} \oplus \mathcal{R}(A)$, and $\dim(\mathcal{R}(A)) < \infty$, we get that $\dim(\mathcal{R}(B)^{\perp}) < \infty$. So B is a Fredholm operator. $\square$

The Fredholmness of a difference, sum, and in general a linear combination of idempotents and orthogonal projectors has been considered in several papers (see [**67, 78, 79, 117**–**119, 167, 169**]). There are results in the literature pertaining to the independence of the linear combination $c_1 P_1 + c_2 P_2$ being left semi-Fredholm, right semi-Fredholm, and Fredholm from the constants c_1, c_2 in the case when P_1 and P_2 are orthogonal projectors or idempotents [**78, 119**]. Also, in the case when P_1 and P_2 are orthogonal projectors some necessary and sufficient conditions for $P_1 + P_2$ being Fredholm are known [**117**], which combined with the result given above furnish necessary and sufficient conditions for $c_1 P_1 + c_2 P_2$ being Fredholm as well as those under which the difference of two Hilbert space orthogonal projectors is a Fredholm operator. Here we present the results from [**50**] that give necessary and sufficient conditions for $c_1 P_1 + c_2 P_2$ to be Fredholm in the case when P_1 and P_2 are idempotents. We have the following two results whose proofs are given below as one:

THEOREM 5.50. *Let $P, Q \in \mathcal{B}(\mathcal{H})$ be idempotents and let $\alpha, \beta \in \mathbb{C} \setminus \{0\}$, $\alpha + \beta \neq 0$. Then $\alpha P + \beta Q$ is Fredholm if and only if the following conditions hold:*

(i) $\dim(\mathcal{N}(P) \cap \mathcal{N}(Q)) < \infty$ *and* $\dim(\mathcal{R}(P)^{\perp} \cap \mathcal{R}(Q)^{\perp}) < \infty$.
(ii) $P(I - Q)$ *and* $(I - P)Q$ *have closed ranges.*
(iii) $\dim(\mathcal{R}(P(I - Q)) \cap \mathcal{R}(Q)) < \infty$ *and* $\dim(\mathcal{R}(P)/(\mathcal{R}(P(I - Q)) + (\mathcal{R}(P) \cap \mathcal{R}(Q)))) < \infty$.

Furthermore, if (i)–(iii) *hold, then*

$$\begin{aligned}\alpha(\alpha P+\beta Q) &= \dim(\mathcal{N}(P)\cap\mathcal{N}(Q)) + \dim(\mathcal{R}(P(I-Q))\cap\mathcal{R}(Q)),\\ \beta(\alpha P+\beta Q) &= \dim(\mathcal{R}(P)^{\perp}\cap\mathcal{R}(Q)^{\perp})\\ &\qquad + \dim(\mathcal{R}(P)/(\mathcal{R}(P(I-Q)) + (\mathcal{R}(P)\cap\mathcal{R}(Q)))).\end{aligned}$$

THEOREM 5.51. *Let $P, Q\in\mathcal{B}(\mathcal{H})$ be idempotents. Then $P-Q$ is Fredholm if and only if the following conditions hold:*

(i) $\dim(\mathcal{N}(P)\cap\mathcal{N}(Q))<\infty$ *and* $\dim(\mathcal{R}(P)^{\perp}\cap\mathcal{R}(Q)^{\perp})<\infty$.
(ii) *$(I-P)Q$ has closed range.*
(iii) $\dim(\mathcal{R}(P)\cap\mathcal{R}(Q))<\infty$ *and* $\dim(\mathcal{R}(P)/\mathcal{R}(P(I-Q)))<\infty$.

Furthermore, if (i)–(iii) *hold, then*

$$\begin{aligned}\alpha(P-Q) &= \dim(\mathcal{N}(P)\cap\mathcal{N}(Q)) + \dim(\mathcal{R}(P)\cap\mathcal{R}(Q)),\\ \beta(P-Q) &= \dim(\mathcal{R}(P)^{\perp}\cap\mathcal{R}(Q)^{\perp}) + \dim(\mathcal{R}(P)/\mathcal{R}(P(I-Q))).\end{aligned}$$

PROOF. Let $\alpha,\beta\in\mathbb{C}\setminus\{0\}$. By Theorem 5.42, we have that $\alpha P+\beta Q$ is a Fredholm operator if and only if:

(i) $\dim(\mathcal{N}(P)\cap\mathcal{N}(Q))<\infty$ and $\dim(\mathcal{N}(P')\cap\mathcal{N}(Q'))<\infty$.
(ii) $P(I-Q)$ and $(I-P)Q$ have closed ranges.
(iii) $\dim(\mathcal{R}(Q)\cap\mathcal{N}(\alpha P+\beta Q))<\infty$, $\dim(\mathcal{R}(P|_{\mathcal{N}(Q)})\cap\mathcal{R}((\alpha P+\beta Q)|_{\mathcal{R}(Q)}))<\infty$, and $\dim(\mathcal{R}(P)\setminus(\mathcal{R}(P|_{\mathcal{N}(Q)})+(\mathcal{R}(P)\cap\mathcal{R}((\alpha P+\beta Q)|_{\mathcal{R}(Q)}))))<\infty$.

Notice that

$$\mathcal{N}(P')\cap\mathcal{N}(Q') = \mathcal{R}(P)^{\circ}\cap\mathcal{R}(Q)^{\circ} = (\mathcal{R}(P)+\mathcal{R}(Q))^{\circ}$$

which implies that

$$\dim(\mathcal{N}(P')\cap\mathcal{N}(Q')) = \dim(\mathcal{R}(P)+\mathcal{R}(Q))^{\circ} = \dim(\mathcal{H}/\overline{\mathcal{R}(P)+\mathcal{R}(Q)}).$$

Hence, we have that $\dim(\mathcal{N}(P')\cap\mathcal{N}(Q'))<\infty$ if and only if $\dim(\mathcal{R}(P)^{\perp}\cap\mathcal{R}(Q)^{\perp})<\infty$.

Now the theorems follow from the following equalities:

$$\mathcal{R}(Q)\cap\mathcal{N}(\alpha P+\beta Q) = \begin{cases}\{0\} & \text{if } \alpha+\beta\neq 0,\\ \mathcal{R}(P)\cap\mathcal{R}(Q) & \text{if } \alpha+\beta=0,\end{cases}$$

$$\mathcal{R}(P|_{\mathcal{N}(Q)})\cap\mathcal{R}((\alpha P+\beta Q)|_{\mathcal{R}(Q)}) = \begin{cases}\mathcal{R}(P(I-Q))\cap\mathcal{R}(Q) & \text{if } \alpha+\beta\neq 0,\\ \{0\} & \text{if } \alpha+\beta=0,\end{cases}$$

$$\mathcal{R}(P)\cap\mathcal{R}((\alpha P+\beta Q)|_{\mathcal{R}(Q)}) = \begin{cases}\mathcal{R}(P)\cap\mathcal{R}(Q) & \text{if } \alpha+\beta\neq 0,\\ \{0\} & \text{if } \alpha+\beta=0.\end{cases}$$

In the case $\alpha+\beta=0$, the condition

$$\dim(\mathcal{R}(P)/(\mathcal{R}(P|_{\mathcal{N}(Q)})+(\mathcal{R}(P)\cap\mathcal{R}((\alpha P+\beta Q)|_{\mathcal{R}(Q)}))))<\infty$$

is equivalent to the condition $\dim(\mathcal{R}(P)/\mathcal{R}(P(I-Q)))<\infty$, which implies that $\mathcal{R}(P(I-Q))$ is closed. □

In the case when $P, Q \in \mathcal{B}(\mathcal{H})$ are orthogonal projectors, the well-known result below follows from our Theorems 5.50 and 5.51:

THEOREM 5.52. *Let $P, Q \in \mathcal{B}(\mathcal{H})$ be orthogonal projectors on a Hilbert space $\mathcal{H}$.*

(1) *If $\alpha, \beta \in \mathbb{C} \setminus \{0\}$, $\alpha + \beta \neq 0$, then the following are equivalent:*
   (i) *$\alpha P + \beta Q$ is Fredholm.*
   (ii) *$\mathcal{R}(P) + \mathcal{R}(Q)$ is closed and $\dim(\mathcal{N}(P) \cap \mathcal{N}(Q)) < \infty$.*

(2) *$P - Q$ is Fredholm if and only if $P + Q$ is Fredholm and $\dim(\mathcal{R}(P) \cap \mathcal{R}(Q)) < \infty$.*

Furthermore, if $\alpha P + \beta Q$ is Fredholm, then $\operatorname{ind}(\alpha P + \beta Q) = 0$.

PROOF. It is evident that condition (i) from Theorem 5.50 is equivalent with $\dim(\mathcal{N}(P) \cap \mathcal{N}(Q)) < \infty$ while using Lemma 2.4 of [**117**] we have that the condition (ii) from Theorem 5.50 is equivalent with the fact that $\mathcal{R}(P) + \mathcal{R}(Q)$ is closed. Now, the proof follows having in mind that

$$\mathcal{R}(P(I-Q)) \cap \mathcal{R}(Q) = \{0\}$$

and from the fact that $\mathcal{R}(P) + \mathcal{R}(Q)$ is closed implies

$$\mathcal{R}(P)/(\mathcal{R}(P(I-Q)) + (\mathcal{R}(P) \cap \mathcal{R}(Q))) = \{0\}. \qquad \square$$

Let us mention that using the result from [**119**] which says that for any two idempotents P, Q and $\alpha, \beta \in \mathbb{C} \setminus \{0\}$ such that $\alpha + \beta \neq 0$,

$$\alpha(\alpha P + \beta Q) = \dim(\mathcal{N}((I-P)Q) \cap \mathcal{N}(P)),$$

and our Theorem 5.50, we get the following result:

THEOREM 5.53. *Let $P, Q \in \mathcal{B}(\mathcal{H})$ be idempotents and let $\mathcal{R}(P(I-Q))$ be closed. Then*

$$\dim(\mathcal{N}(P) \cap \mathcal{N}(Q)) + \dim(\mathcal{R}(P(I-Q)) \cap \mathcal{R}(Q)) = \dim(\mathcal{N}((I-P)Q) \cap \mathcal{N}(P)).$$

Notice that if we remove the condition that $\mathcal{R}(P(I-Q))$ is closed, the equality from the previous theorem is still valid.

From Theorem 5.50 we easily obtain some known results:

THEOREM 5.54 ([**78**]). *Let P and Q be idempotents on a Hilbert space $\mathcal{H}$, and let $\alpha, \beta \in \mathbb{C} \setminus \{0\}$ such that $\alpha + \beta \neq 0$. Then $\alpha P + \beta Q$ is Fredholm if and only if $P + Q$ is Fredholm and $\operatorname{ind}(\alpha P + \beta Q) = \operatorname{ind}(P + Q)$.*

COROLLARY 5.55 ([**78**]). Let P and Q be idempotents on $\mathcal{H}$. If $P - Q$ is Fredholm, then so is $P + Q$. Also, if $\dim(\mathcal{R}(P) \cap \mathcal{R}(Q)) < \infty$, then $P - Q$ is Fredholm if and only if $P + Q$ is.

The next example shows that the formula given in Corollary 4 from [**78**] stating that $\operatorname{ind}(P - Q) = \operatorname{ind}(P + Q) + \dim(\mathcal{R}(P) \cap \mathcal{R}(Q))$, is not true in the case when $P, Q \in \mathcal{B}(\mathcal{H})$ are orthogonal projectors:

EXAMPLE 5.56. Let $P, Q \in \mathcal{B}(l_2)$ be defined by

$$Px = (x_1, 0, x_3, 0, x_5, \ldots),$$
$$Qx = (x_1, x_2, 0, x_4, 0, x_6, \ldots),$$

for any $x = (x_n)_{n=1}^{\infty} \in l_2$. It is easy to see that P, Q are orthogonal projectors and that $\operatorname{ind}(P + Q) = \operatorname{ind}(P - Q) = 0$ and $\dim(\mathcal{R}(P) \cap \mathcal{R}(Q)) = 1$. Hence, $\operatorname{ind}(P - Q) \neq \operatorname{ind}(P + Q) + \dim(\mathcal{R}(P) \cap \mathcal{R}(Q))$.

Unlike the case of orthogonal projectors and idempotents, for some other classes of operators Fredholmness of linear combinations of operators in general depends on the choice of scalars $\alpha, \beta \in \mathbb{C} \setminus \{0\}$ such that $\alpha + \beta \neq 0$. The following examples illustrate this fact for the classes of k-potent, nilpotent, and partial isometry operators.

EXAMPLE 5.57 (*k-potent operators*). Let $A, B \in \mathcal{B}(l_2)$ be defined by

$$Ax = \left(\frac{1}{2}x_1 - \frac{\sqrt{3}}{2}x_2, \frac{\sqrt{3}}{2}x_1 + \frac{1}{2}x_2, \dots, \frac{1}{2}x_k - \frac{\sqrt{3}}{2}x_{k+1}, \frac{\sqrt{3}}{2}x_k + \frac{1}{2}x_{k+1}, \dots\right),$$

$$Bx = \left(\frac{1}{2}x_1 + \frac{\sqrt{3}}{2}x_2, \frac{\sqrt{3}}{2}x_1 - \frac{1}{2}x_2, \dots, \frac{1}{2}x_k + \frac{\sqrt{3}}{2}x_{k+1}, \frac{\sqrt{3}}{2}x_k - \frac{1}{2}x_{k+1}, \dots\right),$$

for any $x = (x_n)_{n=1}^{\infty} \in l_2$. It is easy to see that $A^7 = A$ and $B^3 = B$, so A and B are k-potent operators. Also, for $\alpha, \beta \in \mathbb{C} \setminus \{0\}$, if $\alpha \neq \pm\beta$, we have that $\alpha A + \beta B$ is invertible, so it is Fredholm. On the other hand, $A+B$ and $A-B$ are not Fredholm since $\alpha(A+B) = \alpha(A-B) = \infty$.

EXAMPLE 5.58 (*Nilpotent operators*). Let $A, B \in \mathcal{B}(l_2)$ be defined by

$$\begin{aligned} Ax &= (x_2 + x_3, x_3, 0, x_5 + x_6, x_6, 0, \dots, x_{3k-1} + x_{3k}, x_{3k}, 0, \dots), \\ Bx &= (0, 2x_3, x_1, 0, 2x_6, x_4, \dots, 0, 2x_{3k}, x_{3k-2}, \dots), \end{aligned}$$

for any $x = (x_n)_{n=1}^{\infty} \in l_2$. Then $A^3 = B^3 = 0$. It is easy to see that $A + B$ is invertible, so it is Fredholm, and that $2A - B$ is not Fredholm since $\alpha(2A - B) = \beta(2A - B) = \infty$.

EXAMPLE 5.59 (*Partial isometry*). Let $A, B \in \mathcal{B}(l_2)$ be defined by

$$Ax = \left(\frac{1}{2}x_1 - \frac{\sqrt{3}}{2}x_2, -\frac{\sqrt{3}}{2}x_1 - \frac{1}{2}x_2, \dots, \frac{1}{2}x_k - \frac{\sqrt{3}}{2}x_{k+1}, -\frac{\sqrt{3}}{2}x_k - \frac{1}{2}x_{k+1}, \dots\right),$$

$$Bx = \left(\frac{1}{2}x_1 - \frac{\sqrt{3}}{2}x_2, \frac{\sqrt{3}}{2}x_1 + \frac{1}{2}x_2, \dots, \frac{1}{2}x_k - \frac{\sqrt{3}}{2}x_{k+1}, \frac{\sqrt{3}}{2}x_k + \frac{1}{2}x_{k+1}, \dots\right),$$

for any $x = (x_n)_{n=1}^{\infty} \in l_2$. It is easy to check that $A+B$ is not a Fredholm operator while $A + 2B$ is Fredholm.

Bibliography

[1] P. Aiena, *Fredholm and local spectral theory, with applications to multipliers*, Kluwer Academic Publishers, Dordrecht, 2004. MR2070395

[2] P. Aiena, *Fredholm and local spectral theory II, with application to Weyl-type theorems*, Lecture Notes in Mathematics, vol. 2235, Springer, Cham, 2018, DOI 10.1007/978-3-030-02266-2. MR3839610

[3] M. Amouch, M. Benharrat, and B. Messirdi, *Spectral mapping theorem for generalized Kato spectrum*, J. Math. Anal. Appl. **423** (2015), no. 1, 1–9, DOI 10.1016/j.jmaa.2014.09.043. MR3273162

[4] C. Apostol, *The reduced minimum modulus*, Michigan Math. J. **32** (1985), no. 3, 279–294, DOI 10.1307/mmj/1029003239. MR803833

[5] B. Aupetit, *A primer on spectral theory*, Universitext, Springer-Verlag, New York, 1991, DOI 10.1007/978-1-4612-3048-9. MR1083349

[6] M. Bakonyi and H. J. Woerdeman, *The central method for positive semi-definite, contractive and strong Parrott type completion problems*, Operator theory and complex analysis (Sapporo, 1991), Oper. Theory Adv. Appl., vol. 59, Birkhäuser, Basel, 1992, pp. 78–95. MR1246810

[7] M. Bakonyi and H. J. Woerdeman, *On the strong Parrott completion problem*, Proc. Amer. Math. Soc. **117** (1993), no. 2, 429–433, DOI 10.2307/2159179. MR1145412

[8] M. Bakonyi and C. R. Johnson, *The Euclidean distance matrix completion problem*, SIAM J. Matrix Anal. Appl. **16** (1995), no. 2, 646–654, DOI 10.1137/S0895479893249757. MR1321802

[9] J. K. Baksalary and O. M. Baksalary, *Nonsingularity of linear combinations of idempotent matrices*, Linear Algebra Appl. **388** (2004), 25–29, DOI 10.1016/j.laa.2004.02.025. MR2077846

[10] O. M. Baksalary and G. Trenkler, *Column space equalities for orthogonal projectors*, Appl. Math. Comput. **212** (2009), no. 2, 519–529, DOI 10.1016/j.amc.2009.02.042. MR2531978

[11] O. M. Baksalary and G. Trenkler, *Revisitation of the product of two orthogonal projectors*, Linear Algebra Appl. **430** (2009), no. 10, 2813–2833, DOI 10.1016/j.laa.2009.01.020. MR2509860

[12] B. A. Barnes, *Riesz points of upper triangular operator matrices*, Proc. Amer. Math. Soc. **133** (2005), no. 5, 1343–1347, DOI 10.1090/S0002-9939-04-07811-6. MR2111940

[13] M. Barraa and M. Boumazgour, *On the perturbations of spectra of upper triangular operator matrices*, J. Math. Anal. Appl. **347** (2008), no. 1, 315–322, DOI 10.1016/j.jmaa.2008.05.073. MR2433850

[14] M. Barraa and M. Boumazgour, *A note on the spectrum of an upper triangular operator matrix*, Proc. Amer. Math. Soc. **131** (2003), no. 10, 3083–3088, DOI 10.1090/S0002-9939-03-06862-X. MR1993217

[15] W. Barrett, C. R. Johnson, and P. Tarazaga, *The real positive definite completion problem for a simple cycle*, Computational linear algebra in algebraic and related problems (Essen, 1992), Linear Algebra Appl. **192** (1993), 3–31, DOI 10.1016/0024-3795(93)90234-F. MR1236734

[16] W. W. Barrett, C. R. Johnson, and R. Loewy, *The real positive definite completion problem: cycle completability*, Mem. Amer. Math. Soc. **122** (1996), no. 584, viii+69, DOI 10.1090/memo/0584. MR1342017

[17] C. Benhida, E. H. Zerouali, and H. Zguitti, *Spectra of upper triangular operator matrices*, Proc. Amer. Math. Soc. **133** (2005), no. 10, 3013–3020, DOI 10.1090/S0002-9939-05-07812-3. MR2159780

[18] S. K. Berberian, *An extension of Weyl's theorem to a class of not necessarily normal operators*, Michigan Math. J. **16** (1969), 273–279. MR250094

[19] S. K. Berberian, *The Weyl spectrum of an operator*, Indiana Univ. Math. J. **20** (1970/71), 529–544, DOI 10.1512/iumj.1970.20.20044. MR279623

[20] M. Berkani, *Index of B-Fredholm operators and generalization of a Weyl theorem*, Proc. Amer. Math. Soc. **130** (2002), no. 6, 1717–1723, DOI 10.1090/S0002-9939-01-06291-8. MR1887019

[21] M. Berkani and A. Arroud, *Generalized Weyl's theorem and hyponormal operators*, J. Aust. Math. Soc. **76** (2004), no. 2, 291–302, DOI 10.1017/S144678870000896X. MR2041251

[22] M. Boumazgour, *Drazin invertibility of upper triangular operator matrices*, Linear Multilinear Algebra **61** (2013), no. 5, 627–634, DOI 10.1080/03081087.2012.698618. MR3005643

[23] A. Böttcher and I. M. Spitkovsky, *A gentle guide to the basics of two projections theory*, Linear Algebra Appl. **432** (2010), no. 6, 1412–1459, DOI 10.1016/j.laa.2009.11.002. MR2580440

[24] D. Buckholtz, *Inverting the difference of Hilbert space projections*, Amer. Math. Monthly **104** (1997), no. 1, 60–61, DOI 10.2307/2974825. MR1426419

[25] E. J. Candès and B. Recht, *Exact matrix completion via convex optimization*, Found. Comput. Math. **9** (2009), no. 6, 717–772, DOI 10.1007/s10208-009-9045-5. MR2565240

[26] E. J. Candès and T. Tao, *The power of convex relaxation: near-optimal matrix completion*, IEEE Trans. Inform. Theory **56** (2010), no. 5, 2053–2080, DOI 10.1109/TIT.2010.2044061. MR2723472

[27] E. J. Candès and Y. Plan, *Matrix completion with noise*, Proc. IEEE **98**(6) (2010), 925–936.

[28] X. Cao and B. Meng, *Essential approximate point spectra and Weyl's theorem for operator matrices*, J. Math. Anal. Appl. **304** (2005), no. 2, 759–771, DOI 10.1016/j.jmaa.2004.09.053. MR2127605

[29] X. Cao, M. Guo, and B. Meng, *Weyl's theorem for upper triangular operator matrices*, Linear Algebra Appl. **402** (2005), 61–73, DOI 10.1016/j.laa.2004.12.005. MR2141074

[30] X. H. Cao, M. Z. Guo, and B. Meng, *Semi-Fredholm spectrum and Weyl's theorem for operator matrices*, Acta Math. Sin. (Engl. Ser.) **22** (2006), no. 1, 169–178, DOI 10.1007/s10114-004-0505-1. MR2200773

[31] X. Cao, *Browder spectra for upper triangular operator matrices*, J. Math. Anal. Appl. **342** (2008), no. 1, 477–484, DOI 10.1016/j.jmaa.2007.11.059. MR2440813

[32] X. Cao, H. Zhang, and Y. Zhang, *Consistent invertibility and Weyl's theorem*, J. Math. Anal. Appl. **369** (2010), no. 1, 258–264, DOI 10.1016/j.jmaa.2010.03.023. MR2643864

[33] Y. Cao and Y. Xie, *Poisson matrix recovery and completion*, IEEE Trans. Signal Process. **64** (2016), no. 6, 1609–1620, DOI 10.1109/TSP.2015.2500192. MR3548877

[34] A. Chen and G. Hai, *Perturbations of the right and left spectra for operator matrices*, J. Operator Theory **67** (2012), no. 1, 207–214. MR2881539

[35] A. Chen, G.-L. Hou, and G.-J. Hai, *Perturbation of spectra for a class of* 2×2 *operator matrices*, Acta Math. Appl. Sin. Engl. Ser. **28** (2012), no. 4, 711–720, DOI 10.1007/s10255-012-0195-x. MR3023879

[36] X. Chen, S. Zhang, and H. Zhong, *On the filling in holes problem for operator matrices*, Linear Algebra Appl. **430** (2009), no. 1, 558–563, DOI 10.1016/j.laa.2008.08.022. MR2460538

[37] Y. Chen, Y. Chi, J. Fan, C. Ma, and Y. Yan, *Noisy matrix completion: understanding statistical guarantees for convex relaxation via nonconvex optimization*, SIAM J. Optim. **30** (2020), no. 4, 3098–3121, DOI 10.1137/19M1290000. MR4167625

[38] E. C. Chi and T. Li, *Matrix completion from a computational statistics perspective*, Wiley Interdiscip. Rev. Comput. Stat. **11** (2019), no. 5, e1469, 25, DOI 10.1002/wics.1469. MR3999530

[39] M.-D. Choi, J. Hou, and P. Rosenthal, *Completion of operator partial matrices to square-zero contractions*, Linear Algebra Appl. **256** (1997), 1–30, DOI 10.1016/S0024-3795(96)00774-4. MR1438223

[40] L. A. Coburn, *Weyl's theorem for nonnormal operators*, Michigan Math. J. **13** (1966), 285–288. MR201969

[41] R. W. Cross, *On the continuous linear image of a Banach space*, J. Austral. Math. Soc. Ser. A **29** (1980), no. 2, 219–234. MR566287

[42] A. S. Cvetković and G. V. Milovanović, *On Drazin inverse of operator matrices*, J. Math. Anal. Appl. **375** (2011), no. 1, 331–335, DOI 10.1016/j.jmaa.2010.08.080. MR2735718

[43] D. S. Cvetković Ilić and Y. Wei, *Algebraic properties of generalized inverses*, Developments in Mathematics, vol. 52, Springer, Singapore, 2017, DOI 10.1007/978-981-10-6349-7. MR3645470

[44] D. S. Cvetković Ilić, *A note on the representation for the Drazin inverse of* 2×2 *block matrices*, Linear Algebra Appl. **429** (2008), no. 1, 242–248, DOI 10.1016/j.laa.2008.02.019. MR2419153

[45] D. S. Cvetković Ilić and Y. Wei, *Representations for the Drazin inverse of bounded operators on Banach space*, Electron. J. Linear Algebra **18** (2009), 613–627, DOI 10.13001/1081-3810.1334. MR2565876

[46] D. S. Cvetković Ilić, *The point, residual and continuous spectrum of an upper triangular operator matrix*, Linear Algebra Appl. **459** (2014), 357–367, DOI 10.1016/j.laa.2014.07.010. MR3247230

[47] D. S. Cvetković Ilić, *Invertible and regular completions of operator matrices*, Electron. J. Linear Algebra **30** (2015), 530–549, DOI 10.13001/1081-3810.3126. MR3483432

[48] D. S. Cvetković Ilić, G. Hai, and A. Chen, *Some results on Fredholmness and boundedness below of an upper triangular operator matrix*, J. Math. Anal. Appl. **425** (2015), no. 2, 1071–1082, DOI 10.1016/j.jmaa.2015.01.024. MR3303908

[49] D. S. Cvetković Ilić, *Completions of upper-triangular matrices to left-Fredholm operators with non-positive index*, J. Operator Theory **76** (2016), no. 1, 93–106, DOI 10.7900/jot.2015sep07.2078. MR3527982

[50] D. Cvetković Ilić and J. Milošević, *Fredholmness of a linear combination of operators*, J. Math. Anal. Appl. **458** (2018), no. 1, 555–565, DOI 10.1016/j.jmaa.2017.09.027. MR3711920

[51] D. S. Cvetković Ilić, *Fredholm consistence of upper-triangular operator matrices*, J. Spectr. Theory **7** (2017), no. 4, 1023–1038, DOI 10.4171/JST/184. MR3737887

[52] D. S. Cvetković Ilić and V. Pavlović, *Drazin invertibility of upper triangular operator matrices*, Linear Multilinear Algebra **66** (2018), no. 2, 260–267, DOI 10.1080/03081087.2017.1295914. MR3750588

[53] D. S. Cvetković Ilić and M. Kostadinov, *Consistency in invertibility and Fredholmness of operator matrices*, J. Spectr. Theory **9** (2019), no. 2, 635–649, DOI 10.4171/JST/258. MR3950664

[54] D. S. Cvetković Ilić and M. Kostadinov, *Invertibility of linear combinations in* $B(\mathcal{H})$, Linear Multilinear Algebra **66** (2018), no. 11, 2139–2150, DOI 10.1080/03081087.2017.1388354. MR3851182

[55] D. S. Cvetković Ilić and V. Pavlović, *An analogue to a result of Takahashi*, J. Math. Anal. Appl. **446** (2017), no. 1, 264–275, DOI 10.1016/j.jmaa.2016.08.060. MR3554726

[56] D. S. Cvetković Ilić, *An analogue to a result of Takahashi II*, J. Math. Anal. Appl. **479** (2019), no. 1, 1266–1280, DOI 10.1016/j.jmaa.2019.06.078. MR3987084

[57] D. S. Cvetković Ilić, Q. W. Wang, and Y. Wei, *Different completions of* $A+CX$, J. Spectr. Theory **11** (2021), no. 3, 905–914, DOI 10.4171/jst/356. MR4322025

[58] D. S. Cvetković Ilić, G. Hai, and A. Chen, *Some results on Fredholmness and boundedness below of an upper triangular operator matrix*, J. Math. Anal. Appl. **425** (2015), no. 2, 1071–1082, DOI 10.1016/j.jmaa.2015.01.024. MR3303908

[59] D. S. Cvetković Ilić and V. Pavlović, *On the point and residual spectrum of the operator matrix* M_X, Linear Multilinear Algebra **65** (2017), no. 1, 35–43, DOI 10.1080/03081087.2016.1165171. MR3575876

[60] C. Deng, *On the invertibility of Hilbert space idempotents*, Acta Math. Sci. Ser. B (Engl. Ed.) **34** (2014), no. 2, 523–536, DOI 10.1016/S0252-9602(14)60025-4. MR3174098

[61] F. Deutsch, *The angle between subspaces of a Hilbert space*, Approximation theory, wavelets and applications (Maratea, 1994), NATO Adv. Sci. Inst. Ser. C: Math. Phys. Sci., vol. 454, Kluwer Acad. Publ., Dordrecht, 1995, pp. 107–130. MR1340886

[62] D. S. Djordjević, *Perturbations of spectra of operator matrices*, J. Operator Theory **48** (2002), no. 3, suppl., 467–486. MR1962466

[63] D. S. Djordjević, B. P. Duggal, and R. E. Harte, *Consistent in invertibility operators and SVEP*, Filomat **27** (2013), no. 5, 767–776, DOI 10.2298/FIL1305767D. MR3186097

[64] D. S. Djordjević and M. Z. Kolundžija, *Right and left Fredholm operator matrices*, Bull. Korean Math. Soc. **50** (2013), no. 3, 1021–1027, DOI 10.4134/BKMS.2013.50.3.1021. MR3066245

[65] S. V. Djordjević and H. Zguitti, *Essential point spectra of operator matrices through local spectral theory*, J. Math. Anal. Appl. **338** (2008), no. 1, 285–291, DOI 10.1016/j.jmaa.2007.05.031. MR2386415

[66] S. V. Djordjević and Y. M. Han, *Browder's theorems and spectral continuity*, Glasg. Math. J. **42** (2000), no. 3, 479–486, DOI 10.1017/S0017089500030147. MR1793814

[67] Y.-N. Dou, G.-C. Du, C.-F. Shao, and H.-K. Du, *Closedness of ranges of upper-triangular operators*, J. Math. Anal. Appl. **356** (2009), no. 1, 13–20, DOI 10.1016/j.jmaa.2009.02.014. MR2524211

[68] R. G. Douglas, *On majorization, factorization, and range inclusion of operators on Hilbert space*, Proc. Amer. Math. Soc. **17** (1966), 413–415, DOI 10.2307/2035178. MR203464

[69] R. G. Douglas, *Banach algebra techniques in operator theory*, 2nd ed., Graduate Texts in Mathematics, vol. 179, Springer-Verlag, New York, 1998, DOI 10.1007/978-1-4612-1656-8. MR1634900

[70] H.-K. Du, C.-Y. Deng, M. Mbekhta, and V. Müller, *On spectral properties of linear combinations of idempotents*, Studia Math. **180** (2007), no. 3, 211–217, DOI 10.4064/sm180-3-2. MR2314077

[71] H. Du, X. Yao, and C. Deng, *Invertibility of linear combinations of two idempotents*, Proc. Amer. Math. Soc. **134** (2006), no. 5, 1451–1457, DOI 10.1090/S0002-9939-05-08091-3. MR2199192

[72] H. K. Du and P. Jin, *Perturbation of spectrums of* 2×2 *operator matrices*, Proc. Amer. Math. Soc. **121** (1994), no. 3, 761–766, DOI 10.2307/2160273. MR1185266

[73] B. P. Duggal, S. V. Djordjević, and C. Kubrusly, *Kato type operators and Weyl's theorem*, J. Math. Anal. Appl. **309** (2005), no. 2, 433–441, DOI 10.1016/j.jmaa.2004.08.018. MR2154126

[74] B. P. Duggal, *Upper triangular operator matrices, SVEP and Browder, Weyl theorems*, Integral Equations Operator Theory **63** (2009), no. 1, 17–28, DOI 10.1007/s00020-008-1648-8. MR2480636

[75] B. P. Duggal, I. H. Jeon, and I. H. Kim, *Continuity of the spectrum on a class of upper triangular operator matrices*, J. Math. Anal. Appl. **370** (2010), no. 2, 584–587, DOI 10.1016/j.jmaa.2010.04.068. MR2651679

[76] H. Elbjaoui and E. H. Zerouali, *Local spectral theory for* 2×2 *operator matrices*, Int. J. Math. Math. Sci. **42** (2003), 2667–2672, DOI 10.1155/S0161171203012043. MR2005905

[77] P. A. Fillmore and J. P. Williams, *On operator ranges*, Advances in Math. **7** (1971), 254–281, DOI 10.1016/S0001-8708(71)80006-3. MR293441

[78] H.-L. Gau and P. Y. Wu, *Fredholmness of linear combinations of two idempotents*, Integral Equations Operator Theory **59** (2007), no. 4, 579–583, DOI 10.1007/s00020-007-1531-z. MR2370049

[79] H.-L. Gau, C.-J. Wang, and N.-C. Wong, *Invertibility and Fredholmness of linear combinations of quadratic, k-potent and nilpotent operators*, Oper. Matrices **2** (2008), no. 2, 193–199, DOI 10.7153/oam-02-13. MR2420723

[80] C. Genes, I. Esnaola, S. M. Perlaza, L. F. Ochoa, and D. Coca, *Recovering missing data via matrix completion in electricity distribution systems*, 2016 IEEE 17th International Workshop on Signal Processing Advances in Wireless Communications (SPAWC), 2016, 1–6.

[81] C. Genes, *Novel matrix completion methods for missing data recovery in urban systems*, PhD thesis, The University of Sheffield, 2019.

[82] I. Gohberg, *New aspects in interpolation and completion theories*, Operator Theory: Advances and Applications **64**, Birkhäuser Verlag, Basel, 1993.

[83] I. Gohberg, M. A. Kaashoek, and F. van Schagen, *Partially specified matrices and operators: classification, completion, applications*, Operator Theory: Advances and Applications, vol. 79, Birkhäuser Verlag, Basel, 1995, DOI 10.1007/978-3-0348-9100-4. MR1409814

[84] S. Goldberg, *Unbounded linear operators: Theory and applications*, McGraw-Hill Book Co., New York-Toronto, Ont.-London, 1966. MR0200692

[85] A. Goldberg, B. Recht, J. Xu, R. Nowak, and X. Zhu, *Transduction with matrix completion: Three birds with one stone*, Advances in Neural Information Processing Systems **23** (2010), 757–765.

[86] W. B. Gong and D. G. Han, *Spectrum of the products of operators and compact perturbations*, Proc. Amer. Math. Soc. **120** (1994), no. 3, 755–760, DOI 10.2307/2160467. MR1197538

[87] S. Grabiner and J. Zemánek, *Ascent, descent, and ergodic properties of linear operators*, J. Operator Theory **48** (2002), no. 1, 69–81. MR1926044

[88] J. Groß and G. Trenkler, *Nonsingularity of the difference of two oblique projectors*, SIAM J. Matrix Anal. Appl. **21** (1999), no. 2, 390–395, DOI 10.1137/S0895479897320277. MR1718336

[89] G. Hai and A. Chen, *The residual spectrum and the continuous spectrum of upper triangular operator matrices*, Filomat **28** (2014), no. 1, 65–71, DOI 10.2298/FIL1401065H. MR3359982

[90] G. Hai and A. Chen, *On the right (left) invertible completions for operator matrices*, Integral Equations Operator Theory **67** (2010), no. 1, 79–93, DOI 10.1007/s00020-010-1771-1. MR2629978

[91] G. Hai and A. Chen, *Consistent invertibility of upper triangular operator matrices*, Linear Algebra Appl. **455** (2014), 22–31, DOI 10.1016/j.laa.2014.04.027. MR3217398

[92] G. Hai and A. Chen, *The semi-Fredholmness of* 2×2 *operator matrices*, J. Math. Anal. Appl. **352** (2009), no. 2, 733–738, DOI 10.1016/j.jmaa.2008.11.035. MR2501918

[93] G. J. Hai and A. Chen, *Invertible completions for a class of operator partial matrices* (Chinese, with English and Chinese summaries), Acta Math. Sinica (Chinese Ser.) **52** (2009), no. 6, 1219–1224. MR2640951

[94] G. Hai and A. Chen, *On the invertibility of upper triangular operator matrices*, Linear Multilinear Algebra **62** (2014), no. 4, 538–547, DOI 10.1080/03081087.2013.786718. MR3177047

[95] G. Hai and A. Chen, *Invertibility of operator matrices on a Banach space*, Complex Anal. Oper. Theory **7** (2013), no. 6, 1807–1818, DOI 10.1007/s11785-012-0250-x. MR3129893

[96] G. Hai, C. Bao, and A. Chen, *The invertibility for linear combinations of bounded linear operators with closed range*, Oper. Matrices **11** (2017), no. 3, 715–723, DOI 10.7153/oam-11-48. MR3655678

[97] G. Hai and D. S. Cvetković Ilić, *Generalized left and right Weyl spectra of upper triangular operation matrices*, Electron. J. Linear Algebra **32** (2017), 41–50, DOI 10.13001/1081-3810.3373. MR3627005

[98] J. K. Han, H. Y. Lee, and W. Y. Lee, *Invertible completions of* 2×2 *upper triangular operator matrices*, Proc. Amer. Math. Soc. **128** (2000), no. 1, 119–123, DOI 10.1090/S0002-9939-99-04965-5. MR1618686

[99] R. Harte, *Invertibility and singularity for bounded linear operators*, Monographs and Textbooks in Pure and Applied Mathematics, vol. 109, Marcel Dekker, Inc., New York, 1988. MR920812

[100] R. Harte, *Taylor exactness and Kato's jump*, Proc. Amer. Math. Soc. **119** (1993), no. 3, 793–801, DOI 10.2307/2160514. MR1152982

[101] R. Harte, *On Kato non-singularity*, Studia Math. **117** (1996), no. 2, 107–114, DOI 10.4064/sm-117-2-107-114. MR1367438

[102] H. G. Heuser, *Functional analysis*, translated from the German by John Horváth, A Wiley-Interscience Publication, John Wiley & Sons, Ltd., Chichester, 1982. MR640429

[103] J. Hou, *Completion of operator partial matrices to projections*, Linear Algebra Appl. **246** (1996), 71–82, DOI 10.1016/0024-3795(94)00336-X. MR1407659

[104] J. Hou, H. Radjavi, and P. Rosenthal, *Idempotent completions of operator partial matrices*, Acta Math. Sin. (Engl. Ser.) **15** (1999), no. 3, 333–346, DOI 10.1007/BF02650728. MR1736683

[105] J. Huang, J. Sun, A. Chen, and C. Trunk, *Invertibity of* 2×2 *operator matrices*, Math. Nachr. **292** (2019), no. 11, 2411–2426, DOI 10.1002/mana.201800351. MR4033014

[106] J. Huang, Y. Huang, and H. Wang, *Closed range and Fredholm properties of upper-triangular operator matrices*, Ann. Funct. Anal. **6** (2015), no. 3, 42–52, DOI 10.15352/afa/06-3-4. MR3336903

[107] I. S. Hwang and W. Y. Lee, *The boundedness below of* 2×2 *upper triangular operator matrices*, Integral Equations Operator Theory **39** (2001), no. 3, 267–276, DOI 10.1007/BF01332656. MR1818061

[108] S. Izumino, *The product of operators with closed range and an extension of the reverse order law*, Tohoku Math. J. (2) **34** (1982), no. 1, 43–52, DOI 10.2748/tmj/1178229307. MR651705

[109] D. Jannach, P. Resnick, A. Tuzhilin, and M. Zanker, *Recommender systems: beyond matrix completion*, Communications of the ACM **59** (2016), no. 11, 94–102.

[110] H. Ji, E. O'Saben, R. Lambi, and Y. H. Li, *Matrix completion based model v2.0: Predicting the winning probabilities of march madness matches*, Proc. Modeling, Simulation, and Visualization Student Capstone Conf., Suffolk, VA, USA, 2016.

[111] Y. Q. Ji, *Quasitriangular + small compact = strongly irreducible*, Trans. Amer. Math. Soc. **351** (1999), no. 11, 4657–4673, DOI 10.1090/S0002-9947-99-02307-7. MR1603910

[112] A. Kapur, K. Marwah, and G. Alterovitz, *Gene expression prediction using low-rank matrix completion*, Springer, BMC Bioinformatics **17** (2016), Article number: 243.

[113] T. Kato, *Perturbation theory for nullity, deficiency and other quantities of linear operators*, J. Analyse Math. **6** (1958), 261–322, DOI 10.1007/BF02790238. MR107819

[114] T. Kato, *Perturbation theory for linear operators*, 2nd ed., Grundlehren der Mathematischen Wissenschaften, Band 132, Springer-Verlag, Berlin-New York, 1976. MR0407617

[115] J. J. Koliha, *A generalized Drazin inverse*, Glasgow Math. J. **38** (1996), no. 3, 367–381, DOI 10.1017/S0017089500031803. MR1417366

[116] J. J. Koliha, *The Drazin and Moore-Penrose inverse in C^*-algebras*, Math. Proc. R. Ir. Acad. **99A** (1999), no. 1, 17–27. MR1883060

[117] J. J. Koliha and V. Rakočević, *Fredholm properties of the difference of orthogonal projections in a Hilbert space*, Integral Equations Operator Theory **52** (2005), no. 1, 125–134, DOI 10.1007/s00020-003-1274-4. MR2138700

[118] J. J. Koliha and V. Rakočević, *The nullity and rank of linear combinations of idempotent matrices*, Linear Algebra Appl. **418** (2006), no. 1, 11–14, DOI 10.1016/j.laa.2006.01.011. MR2257572

[119] J. J. Koliha and V. Rakočević, *Stability theorems for linear combinations of idempotents*, Integral Equations Operator Theory **58** (2007), no. 4, 597–601, DOI 10.1007/s00020-007-1495-z. MR2329138

[120] J. J. Koliha and V. Rakočević, *Invertibility of the sum of idempotents*, Linear Multilinear Algebra **50** (2002), no. 4, 285–292, DOI 10.1080/03081080290004960. MR1936951

[121] J. J. Koliha and V. Rakočević, *On the norm of idempotents in C^*-algebras*, Rocky Mountain J. Math. **34** (2004), no. 2, 685–697, DOI 10.1216/rmjm/1181069874. MR2072801

[122] J. J. Koliha, V. Rakočević, and I. Straškraba, *The difference and sum of projectors*, Linear Algebra Appl. **388** (2004), 279–288, DOI 10.1016/j.laa.2004.03.008. MR2077865

[123] J. J. Koliha and V. Rakočević, *Invertibility of the difference of idempotents*, Linear Multilinear Algebra **51** (2003), no. 1, 97–110, DOI 10.1080/030810802100023499. MR1950415

[124] M. Kostadinov, *Injectivity of linear combinations in $\mathcal{B}(\mathcal{H})$*, Electron. J. Linear Algebra **37** (2021), 359–369. MR4284783

[125] C. S. Kubrusly, *Spectral theory of operators on Hilbert spaces*, Birkhäuser/Springer, New York, 2012, DOI 10.1007/978-0-8176-8328-3. MR2951608

[126] M. Laurent, *A tour d'horizon on positive semidefinite and Euclidean distance matrix completion problems*, Topics in semidefinite and interior-point methods (Toronto, ON, 1996), Fields Inst. Commun., vol. 18, Amer. Math. Soc., Providence, RI, 1998, pp. 51–76. MR1607310

[127] D. C. Lay, *Spectral analysis using ascent, descent, nullity and defect*, Math. Ann. **184** (1969/70), 197–214, DOI 10.1007/BF01351564. MR259644

[128] W. Y. Lee, *Weyl spectra of operator matrices*, Proc. Amer. Math. Soc. **129** (2001), no. 1, 131–138, DOI 10.1090/S0002-9939-00-05846-9. MR1784020

[129] S. G. Lee and H. G. Seol, *A survey on the matrix completion problem*, Trends in Mathematics **4** (2001), no. 1, 38–43.

[130] G. Li, G. Hai, and A. Chen, *Generalized Weyl spectrum of upper triangular operator matrices*, Mediterr. J. Math. **12** (2015), no. 3, 1059–1067, DOI 10.1007/s00009-014-0426-x. MR3376831

[131] T. Li, *A note on the statistical view of matrix completion*, arXiv:1605.03040.

[132] T. Li and C. Deng, *On the invertibility and range closedness of the linear combinations of a pair of projections*, Linear Multilinear Algebra **65** (2017), no. 3, 613–622, DOI 10.1080/03081087.2016.1198302. MR3589624

[133] Y. Li and H. K. Du, *On essential spectra of 2×2 operator matrices*, J. Math. Res. Exposition **28** (2008), no. 2, 359–365. MR2419749

[134] Y. Li and H. Du, *The intersection of essential approximate point spectra of operator matrices*, J. Math. Anal. Appl. **323** (2006), no. 2, 1171–1183, DOI 10.1016/j.jmaa.2005.11.032. MR2260171

[135] Y. Li, X.-H. Sun, and H.-K. Du, *Intersections of the left and right essential spectra of* 2×2 *upper triangular operator matrices*, Bull. London Math. Soc. **36** (2004), no. 6, 811–819, DOI 10.1112/S0024609304003492. MR2083757

[136] Y. Li, X.-H. Sun, and H.-K. Du, *The intersection of left (right) spectra of* 2×2 *upper triangular operator matrices*, Linear Algebra Appl. **418** (2006), no. 1, 112–121, DOI 10.1016/j.laa.2006.01.020. MR2257582

[137] Y. Li, X. H. Sun, and H. K. Du, *A note on the left essential spectra of operator matrices*, Acta Math. Sin. (Engl. Ser.) **23** (2007), no. 12, 2235–2240, DOI 10.1007/s10114-007-0966-0. MR2357457

[138] J. Liu, P. Musialski, P. Wonka, and J. P. Ye, *Tensor completion for estimating missing values in visual data*, IEEE Trans. Pattern Anal. Mach. Intell. **35** (2013), no. 1, 208–220.

[139] C. Q. Lu, M. Y. Yang, F. Luo, F. X. Wu, M. Li, Y. Pan, Y. H. Li, and J. X. Wang, *Prediction of IncRNA-disease associations based on inductive matrix completion*, Bioinformatics **34** (2018), no. 19, 3357–3364.

[140] M. Mbekhta, *Décomposition de Kato généralisée* (French, with English summary), C. R. Acad. Sci. Paris Sér. I Math. **303** (1986), no. 20, 979–982, DOI 10.1017/S0017089500006807. MR877185

[141] M. Mbekhta, *Généralisation de la décomposition de Kato aux opérateurs paranormaux et spectraux* (French), Glasgow Math. J. **29** (1987), no. 2, 159–175, DOI 10.1017/S0017089500006807. MR901662

[142] M. Mbekhta, *Résolvant généralisé et théorie spectrale* (French), J. Operator Theory **21** (1989), no. 1, 69–105. MR1002122

[143] M. Mbekhta, *Spectre généralisé, algèbre de Calkin et problèmes d'Apostol*, PUB. IRMA, LILE **19** (1989), no. 1.

[144] M. Mbekhta, *Habilation a diriger des recherches en sciences option Mathematique*, Universite de Nice, 1989.

[145] M. Mbekhta, *Local spectrum and generalized spectrum*, Proc. Amer. Math. Soc. **112** (1991), no. 2, 457–463, DOI 10.2307/2048740. MR1045142

[146] M. Mbekhta and A. Ouahab, *Opérateur s-régulier dans un espace de Banach et théorie spectrale* (French, with English summary), Acta Sci. Math. (Szeged) **59** (1994), no. 3-4, 525–543. MR1317171

[147] V. Müler, *On the regular spectrum*, J. Operator Theory **31** (1994), no. 2, 363–380. MR1331783

[148] V. Müller, *Spectral theory of linear operators and spectral systems in Banach algebras*, 2nd ed., Operator Theory: Advances and Applications, vol. 139, Birkhäuser Verlag, Basel, 2007. MR2355630

[149] K. K. Oberai, *On the Weyl spectrum*, Illinois J. Math. **18** (1974), 208–212. MR333762

[150] V. Pavlović and D. S. Cvetković Ilić, *Applications of completions of operator matrices to reverse order law for {1}-inverses of operators on Hilbert spaces*, Linear Algebra Appl. **484** (2015), 219–236, DOI 10.1016/j.laa.2015.06.015. MR3385061

[151] V. Pavlović and D. S. Cvetković Ilić, *Completions of upper-triangular matrices to Kato nonsingularity*, J. Math. Anal. Appl. **433** (2016), no. 2, 1230–1242, DOI 10.1016/j.jmaa.2015.08.041. MR3398761

[152] D. S. Cvetković Ilić and V. Pavlović, *On surjectivity and denseness of range of the operator* $A + CX$, J. Spectr. Theory **8** (2018), no. 3, 871–881, DOI 10.4171/JST/215. MR3831149

[153] A. Ramlatchan, M. Yang, Q. Liu, M. Li, J. Wang, and Y. Li, *A survey of matrix completion methods for recommendation systems*, Big Data Mining and Analytics **1** (2018), no. 4, 308–323.

[154] B. Recht and C. Ré, *Parallel stochastic gradient algorithms for large-scale matrix completion*, Math. Program. Comput. **5** (2013), no. 2, 201–226, DOI 10.1007/s12532-013-0053-8. MR3069879

[155] F.-G. Ren, H.-K. Du, and H.-X. Cao, *The intersection of the spectra of operator completions*, Linear Algebra Appl. **371** (2003), 103–109, DOI 10.1016/S0024-3795(03)00422-1. MR1997365

[156] H. Schneider, *Theorems on M-splittings of a singular M-matrix which depend on graph structure*, Linear Algebra Appl. **58** (1984), 407–424, DOI 10.1016/0024-3795(84)90222-2. MR739297

[157] D. Shin, S. Cetintas, K. C. Lee, and I. S. Dhillon, *Tumblr blog recommendation with boosted inductive matrix completion*, Proc. 24th ACM Int. Conf. on Information and Knowledge Management, Melbourne, Australia, 2015, 203–212.

[158] B. Sturmfels and C. Uhler, *Multivariate Gaussian, semidefinite matrix completion, and convex algebraic geometry*, Ann. Inst. Statist. Math. **62** (2010), no. 4, 603–638, DOI 10.1007/s10463-010-0295-4. MR2652308

[159] T.-Y. Tam and W. Yan, *Unitary completions of complex symmetric and skew symmetric matrices*, Appl. Math. E-Notes **7** (2007), 84–92. MR2295691

[160] K. Takahashi, *Invertible completions of operator matrices*, Integral Equations Operator Theory **21** (1995), no. 3, 355–361, DOI 10.1007/BF01299969. MR1316548

[161] A. E. Taylor and D. C. Lay, *Introduction to functional analysis*, 2nd ed., John Wiley & Sons, New York-Chichester-Brisbane, 1980. MR564653

[162] Y. Tian, *Completing block Hermitian matrices with maximal and minimal ranks and inertias*, Electron. J. Linear Algebra **21** (2010), 124–141, DOI 10.13001/1081-3810.1419. MR2735877

[163] Y. Tian and Y. Takane, *The inverse of any two-by-two nonsingular partitioned matrix and three matrix inverse completion problems*, Comput. Math. Appl. **57** (2009), no. 8, 1294–1304, DOI 10.1016/j.camwa.2009.01.025. MR2512228

[164] C. Tretter, *Spectral theory of block operator matrices and applications*, Imperial College Press, London, 2008, DOI 10.1142/9781848161122. MR2463978

[165] H. Weyl, *Über beschränkte quadratische Formen, deren Differenz vollstetig ist*, Rend. Circ. Mat. Palermo **27** (1909), 373–392.

[166] D. Williams and L. Carin, *Analytical kernel matrix completion with incomplete multi-view data*, Proceedings of the ICML Workshop on Learning with Multiple Views, 2005.

[167] T. Xie and K. Zuo, *Fredholmness of combinations of two idempotents*, Eur. J. Pure Appl. Math. **3** (2010), no. 4, 678–685, DOI 10.3934/krm.2010.3.685. MR2680520

[168] Z. Xiong and Z. Liu, *Applications of completions of operator matrices to some properties of operator products on Hilbert spaces*, Complex Anal. Oper. Theory **12** (2018), no. 1, 123–140, DOI 10.1007/s11785-016-0600-1. MR3741672

[169] Z. Yang, X. Feng, M. Chen, C. Deng, and J. J. Koliha, *Fredholm stability results for linear combinations of m-potent operators*, Oper. Matrices **6** (2012), no. 1, 193–199, DOI 10.7153/oam-06-14. MR2952445

[170] X. Yao, *A spectral completion of* 2×2 *operator matrices*, Int. Math. Forum **2** (2007), no. 53-56, 2737–2745. MR2380116

[171] E. H. Zerouali and H. Zguitti, *Perturbation of spectra of operator matrices and local spectral theory*, J. Math. Anal. Appl. **324** (2006), no. 2, 992–1005, DOI 10.1016/j.jmaa.2005.12.065. MR2265096

[172] H. Zguitti, *On the Drazin inverse for upper triangular operator matrices*, Bull. Math. Anal. Appl. **2** (2010), no. 2, 27–33. MR2658125

[173] H.-Y. Zhang and H.-K. Du, *Browder spectra of upper-triangular operator matrices*, J. Math. Anal. Appl. **323** (2006), no. 1, 700–707, DOI 10.1016/j.jmaa.2005.10.073. MR2262238

[174] H. Y. Zhang, *Spectra of* 2×2 *upper-triangular operator matrices*, Applied Mathematics **4** (2013), 22–25.

[175] S. Zhang and Z. Wu, *Characterizations of perturbations of spectra of* 2×2 *upper triangular operator matrices*, J. Math. Anal. Appl. **392** (2012), no. 2, 103–110, DOI 10.1016/j.jmaa.2012.03.031. MR2917290

[176] S. Zhang and H. Zhong, *A note on Browder spectrum of operator matrices*, J. Math. Anal. Appl. **344** (2008), no. 2, 927–931, DOI 10.1016/j.jmaa.2008.03.048. MR2426321

[177] S. Zhang, H. Zhong, and Q. Jiang, *Drazin spectrum of operator matrices on the Banach space*, Linear Algebra Appl. **429** (2008), no. 8-9, 2067–2075, DOI 10.1016/j.laa.2008.06.002. MR2446641

[178] S.-F. Zhang, H.-J. Zhong, and L.-Q. Lin, *Generalized Drazin spectrum of operator matrices*, Appl. Math. J. Chinese Univ. Ser. B **29** (2014), no. 2, 162–170, DOI 10.1007/s11766-014-3142-1. MR3229241

[179] S. Zhang, Z. Wu, and H. Zhong, *Continuous spectrum, point spectrum and residual spectrum of operator matrices*, Linear Algebra Appl. **433** (2010), no. 3, 653–661, DOI 10.1016/j.laa.2010.03.036. MR2653829

[180] K. Zuo, O. M. Baksalary, and D. Cvetković Ilić, *Further characterizations of the co-EP matrices*, Linear Algebra Appl. **616** (2021), 66–83, DOI 10.1016/j.laa.2020.12.029. MR4203589

Index

Selected Published Titles in This Series

267 **Dragana S. Cvetković Ilić,** Completion Problems on Operator Matrices, 2022
265 **Nabil H. Mustafa,** Sampling in Combinatorial and Geometric Set Systems, 2022
264 **J. Scott Carter and Seiichi Kamada,** Diagrammatic Algebra, 2021
263 **Vugar E. Ismailov,** Ridge Functions and Applications in Neural Networks, 2021
262 **Ragnar-Olaf Buchweitz,** Maximal Cohen–Macaulay Modules and Tate Cohomology, 2021
261 **Shiri Artstein-Avidan, Apostolos Giannopoulos, and Vitali D. Milman,** Asymptotic Geometric Analysis, Part II, 2021
260 **Lindsay N. Childs, Cornelius Greither, Kevin P. Keating, Alan Koch, Timothy Kohl, Paul J. Truman, and Robert G. Underwood,** Hopf Algebras and Galois Module Theory, 2021
259 **William Heinzer, Christel Rotthaus, and Sylvia Wiegand,** Integral Domains Inside Noetherian Power Series Rings, 2021
258 **Pramod N. Achar,** Perverse Sheaves and Applications to Representation Theory, 2021
257 **Juha Kinnunen, Juha Lehrbäck, and Antti Vähäkangas,** Maximal Function Methods for Sobolev Spaces, 2021
256 **Michio Jimbo, Tetsuji Miwa, and Fedor Smirnov,** Local Operators in Integrable Models I, 2021
255 **Alexandre Boritchev and Sergei Kuksin,** One-Dimensional Turbulence and the Stochastic Burgers Equation, 2021
254 **Karim Belabas and Henri Cohen,** Numerical Algorithms for Number Theory, 2021
253 **Robert R. Bruner and John Rognes,** The Adams Spectral Sequence for Topological Modular Forms, 2021
252 **Julie Déserti,** The Cremona Group and Its Subgroups, 2021
251 **David Hoff,** Linear and Quasilinear Parabolic Systems, 2020
250 **Bachir Bekka and Pierre de la Harpe,** Unitary Representations of Groups, Duals, and Characters, 2020
249 **Nikolai M. Adrianov, Fedor Pakovich, and Alexander K. Zvonkin,** Davenport–Zannier Polynomials and Dessins d'Enfants, 2020
248 **Paul B. Larson and Jindrich Zapletal,** Geometric Set Theory, 2020
247 **István Heckenberger and Hans-Jürgen Schneider,** Hopf Algebras and Root Systems, 2020
246 **Matheus C. Bortolan, Alexandre N. Carvalho, and José A. Langa,** Attractors Under Autonomous and Non-autonomous Perturbations, 2020
245 **Aiping Wang and Anton Zettl,** Ordinary Differential Operators, 2019
244 **Nabile Boussaïd and Andrew Comech,** Nonlinear Dirac Equation, 2019
243 **José M. Isidro,** Jordan Triple Systems in Complex and Functional Analysis, 2019
242 **Bhargav Bhatt, Ana Caraiani, Kiran S. Kedlaya, Peter Scholze, and Jared Weinstein,** Perfectoid Spaces, 2019
241 **Dana P. Williams,** A Tool Kit for Groupoid C^*-Algebras, 2019
240 **Antonio Fernández López,** Jordan Structures in Lie Algebras, 2019
239 **Nicola Arcozzi, Richard Rochberg, Eric T. Sawyer, and Brett D. Wick,** The Dirichlet Space and Related Function Spaces, 2019
238 **Michael Tsfasman, Serge Vlăduţ, and Dmitry Nogin,** Algebraic Geometry Codes: Advanced Chapters, 2019
237 **Dusa McDuff, Mohammad Tehrani, Kenji Fukaya, and Dominic Joyce,** Virtual Fundamental Cycles in Symplectic Topology, 2019

For a complete list of titles in this series, visit the
AMS Bookstore at **www.ams.org/bookstore/survseries/**.